AF777535

Certified Wireless Network Administrator (CWNA®)

Official Study Guide

(CWNA-108)

Certified Wireless Network Professionals

Errata, when available, for this study guide can be found at: www.cwnp.com/errata/

First printing: September 2020, version 1.0

ISBN: 978-0-9993048-1-5

CWNP would like to thank the many CWNE and CWNP-certified contributors who participated in the development of the CWNA-108 materials. This participation included Job Task Analysis (JTA) participation, objectives review, courseware authoring and editing, study guide authoring and editing, exam item review, and practice exam review. Individuals who participated in one or more of these processes include Robert Bartz, Tom Carpenter, Peter Mackenzie, Mitch Dickey, Jonathan Davis, Jennifer Minella, Manon Lessard, Phil Morgan, Landon Foster, Rowell Dionicio, Djamel Ramoul and Sean Stallings. Thank you all for your exceptional contributions.

Table of Contents

Introduction

The Certified Wireless Network Administrator (CWNA) is an individual who can install, administer, secure, and troubleshoot an enterprise wireless local area network (WLAN). The CWNA-108 exam tests the candidate's knowledge to verify his or her ability to perform the duties of a CWNA.

A CWNA will be able to explain, configure, optimize, troubleshoot, upgrade and monitor an enterprise WLAN. The skills acquired in preparation for the CWNA exam will assist the CWNA in performing these actions, regardless of the vendor equipment used. In the suite of CWNP certifications, the CWNA certification is deeper than either CWS or CWT and broader than CWSP, CWAP, or CWDP. In fact, the CWNA should know everything a CWS or CWT-certified professional would know, and much more. While the depth of knowledge required of the professional level certifications (CWSP, CWAP, and CWDP) is not required of the CWNA, the CWNA should know the basics of WLAN security, analysis (for the purpose of troubleshooting) and design.

The CWNA-108 exam consists of 60 multiple-choice, single-correct-answer questions and is delivered through Pearson VUE. The candidate can register for the exam at the Pearson VUE website (https://home.pearsonvue.com/). The candidate will have 90 minutes to take the exam and must achieve a score of 70% or greater to earn the CWNA certification. If the candidate desires to become a Certified Wireless Network Trainer (CWNT), the passing score must be 80% or greater. A CWNT is authorized to teach official CWNP courses for certifications in which they hold the CWNT credential.

Book Features

The "CWNA Official Study Guide" includes the following features:

- End-of-chapter "Points to Remember." These lists of important facts help you retain the information learned in the chapter.
- End-of-chapter review quizzes. These quizzes help you test the knowledge you've acquired from the chapter. Each chapter contains 10 quiz items.
- Notes with special indicators. The notes throughout the book fall into one of three categories as outlined in Table i.1.
- CWNP official glossary. A glossary of terms provided at the end of the book that helps you as a reference while reading.
- Complete coverage of the CWNA-108 objectives.
- Periodic "Beyond the Exam" sections that go deeper into important areas or areas of interest to the reader.

Icon	Description
	**Note:** A general note related to the current topic.
	**Defined Note:** A note providing a concise definition of a term or concept.
	**Exam Note:** A note providing tips for exam preparation.

Table i.1: Book Note Icons

About the Authors

Tom Carpenter is the CTO of CWNP and has more than 20 years of experience in the Information Technology industry. He has written 18 previous books and developed more than 50 eLearning programs in the past 15 years. He is a CWNE and holds several other industry certifications, as well. As the CTO of CWNP, Tom is responsible for setting the direction for certifications and managing product development projects through their lifecycles. He can be reached at tom@cwnp.com and is heard on a monthly webinar series presented by CWNP and archived at YouTube on the CWNPTV channel. Tom lives in Ohio with his wife Julie and loves books and all things tech.

Mitch Dickey, Defender of Wrongly Accused, is a Wi-Fi and Senior Network Engineer at one of the fastest growing school districts in the United States, serving over 80,000 students in Loudoun County Public Schools. He holds several Cisco routing, switching, and wireless certifications, as well as all Certified Wireless Network Professional certifications; CWNA, CWAP, CWDP, CWSP, and hopefully CWNE in the very near future. He is also an Ekahau Certified Survey Engineer. Mitch is a husband, daddy, hunter, and fisherman. He also enjoys Amateur Radio and holds a General Class Amateur Radio license (W4LAN). He provided a significant section on Single Channel Architecture (SCA) deployments, which is an area in which he has extensive experience.

CWNA-108 Objectives

The CWNA-108 exam tests your knowledge against seven knowledge domains, as documented in Table i.2. The CWNA candidate should understand these domains before taking the exam. The CWNA-108 objectives follow.

Knowledge Domain	Percentage
Radio Frequency (RF) Technologies	15%
WLAN Regulations and Standards	20%
WLAN Protocols and Devices	20%
WLAN Network Architecture and Design Concepts	15%
WLAN Network Security	10%
RF Validation	10%
WLAN Troubleshooting	10%

Table i.2: CWNA-108 Exam Knowledge Domains with Percentage of Questions in Each Domain

1.0 Radio Frequency (RF) Technologies – 15%

1.1 Define and explain the basic characteristics of RF and RF behavior

- Wavelength, frequency, amplitude, phase, sine waves
- RF propagation and coverage
- Reflection, refraction, diffraction and scattering
- Multipath and RF interference
- Gain and loss
- Amplification
- Attenuation
- Absorption
- Voltage Standing Wave Ratio (VSWR)
- Return Loss
- Free Space Path Loss (FSPL)

1.2 Apply the basic concepts of RF mathematics and measurement

- Watt and milliwatt
- Decibel (dB)
- dBm and dBi
- Noise floor
- SNR
- RSSI
- dBm to mW conversion rules of 10 and 3
- Equivalent Isotropically Radiated Power (EIRP)

1.3 Identify RF signal characteristics as they relate to antennas.

- RF and physical line of sight and Fresnel zone clearance
- Beamwidths
- Passive gain
- Polarization
- Antenna diversity types
- Radio chains
- Spatial multiplexing (SM)
- Transmit Beamforming (TxBF)
- Maximal Ratio Combining (MRC)
- MIMO

1.4 Explain and apply the functionality of RF antennas, antenna systems, and accessories available

- Omni-directional antennas
- Semi-directional antennas
- Highly directional antennas
- Reading Azimuth and Elevation charts for different antenna types
- Antenna orientation
- RF cables and connectors
- Lightning arrestors and grounding rods/wires

2.0 WLAN Regulations and Standards – 20%

2.1 Explain the roles of WLAN and networking industry organizations

- IEEE
- Wi-Fi Alliance
- IETF
- Regulatory domains and agencies

2.2 Explain and apply the various Physical Layer (PHY) solutions of the IEEE 802.11-2016 standard as amended including supported channel widths, spatial streams, data rates.

- DSSS – 802.11
- HR-DSSS – 802.11b
- OFDM – 802.11a
- ERP – 802.11g
- Wi-Fi 4 - HT – 802.11n
- Wi-Fi 5 - VHT – 802.11ac
- Wi-Fi 6 - HE - 802.11ax

2.3 Understand spread spectrum technologies, Modulation and Coding Schemes (MCS)

- DSSS
- OFDM
- OFDMA and Resource Units
- BPSK
- QPSK
- QAM (16, 64, 256,1024)

2.4 Identify and apply 802.11 WLAN functional concepts

- Primary channels
- Adjacent overlapping and non-overlapping channels
- Throughput vs. data rate
- Bandwidth
- Guard Interval

2.5 Describe the OSI model layers affected by the 802.11-2016 standard and amendments

2.6 Identify and comply with regulatory domain requirements and constraints (specifically in 2.4 GHz and 5 GHz)

- Frequency bands used by the 802.11 PHYs
- Available channels
- Regulatory power constraints
- Dynamic Frequency Selection (DFS)
- Transmit Power Control (TPC)

2.7 Explain basic use case scenarios for 802.11 wireless networks

- Wireless LAN (WLAN) – BSS and ESS
- Wireless bridging
- Wireless Ad-Hoc (IBSS)
- Wireless Mesh

3.0 WLAN Protocols and Devices – 20%

3.1 Describe the components and functions that make up an 802.11 wireless service set

- Stations (STAs)
- Basic Service Set (BSS) (Infrastructure mode)
- SSID
- BSSID
- Extended Service Set (ESS)
- IBSS (Ad-Hoc)
- Distribution System (DS)
- Distribution System Media (DSM)

3.2 Define terminology related to the 802.11 MAC and PHY

- MSDU, MPDU, PSDU, and PPDU
- A-MSDU and A-MPDU
- PHY preamble and header

3.3 Identify and explain the MAC frame format

- MAC frame format
- MAC addressing

3.4 Identify and explain the purpose of the three main 802.11 frame types

- Management
- Control
- Data

3.5 Explain the process used to locate and connect to a WLAN

- Scanning (active and passive)

- Authentication
- Association
- Open System Authentication and Shared Key authentication
- Connecting to 802.1X/EAP and Pre-Shared Key authentication networks
- BSS selection
- Connecting to hidden SSIDs

3.6 Explain 802.11 channel access methods

- DCF
- EDCA
- RTS/CTS
- CTS-to-Self
- NAV
- Interframe spaces (SIFS, DIFS, EIFS, AIFS)
- Physical carrier sense and virtual carrier sense
- Hidden node

3.7 Explain 802.11 MAC operations

- Roaming
- Power save modes and frame buffering
- Protection mechanisms

3.8 Describe features of, select, and install WLAN devices, control, and management systems

- Access Points (APs)
- WLAN controllers
- Wireless network management systems
- Wireless bridge and mesh APs
- Client devices

4.0 WLAN Network Architecture and Design Concepts– 15%

4.1 Describe and implement Power over Ethernet (PoE) 802.3af, 802.3at, 802.3bt

- Power Source Equipment
- Powered Device
- Midspan and endpoint PSEs
- Power classes to include power differences between PSE and PD
- Power budgets and powered port density

4.2 Define and describe differences, advantages and constraints of the different wireless LAN architectures

- Centralized data forwarding
- Distributed data forwarding
- Control, Management and Data planes
- Scalability and availability solutions
- Tunneling, QoS and VLANs

4.3 Describe design considerations for common deployment scenarios in wireless such as coverage requirements, roaming considerations, throughput, capacity and security

- Design considerations for data
- Design considerations for voice
- Design considerations for video
- Design considerations for location services including Real-Time Location Services (RTLS)
- Design considerations for highly mobile devices (e.g. tablets and smartphones)
- Capacity planning for high and very high-density environments
- Design considerations for guest access/BYOD
- Design considerations for supporting legacy 802.11 devices

4.4 Demonstrate awareness of common proprietary features in wireless networks.

- AirTime Fairness
- Band steering
- Dynamic power and channel management features

4.5 Determine and configure required network services supporting the wireless network

- DHCP for client addressing, AP addressing and/or controller discovery
- DNS for address resolution for clients and APs
- Time synchronization protocols (e.g. NTP, SNTP)

- VLANs for segmentation
- Authentication services (e.g. RADIUS, LDAP)
- Access Control Lists for segmentation
- Wired network capacity requirements

5.0 WLAN Network Security – 10%

5.1 Identify weak security options that should not be used in enterprise WLANs

- WEP
- Shared Key authentication
- SSID hiding as a security mechanism
- MAC filtering
- Use of deprecated security methods (e.g. WPA and/or WPA2 with TKIP)
- Wi-Fi Protected Setup (WPS)

5.2 Identify and configure effective security mechanisms for enterprise WLANs

- Application of AES with CCMP for encryption and integrity
- WPA2-Personal including limitations and best practices for pre-shared (PSK) use
- WPA2-Enterprise -configuring wireless networks to use 802.1X including connecting to RADIUS servers and appropriate EAP methods

5.3 Understand basic concepts of WPA3 and Opportunistic Wireless Encryption (OWE) and enhancements compared to WPA2

- Understand basic security enhancements in WPA3 vs. WPA2
- Understand basic security enhancements of encryption and integrity in WPA3 (e.g. CCMP, GCMP, AES)
- Simultaneous Authentication of Equals (SAE) in WPA3 as an enhancement for legacy pre-shared key technology
- Understand the purpose of Opportunistic Wireless Encryption (OWE) for public and guest networks

5.4 Describe common security options and tools used in wireless networks

- Access control solutions (e.g. captive portals, NAC, BYOD)
- Protected management frames
- Fast Secure Roaming methods
- Wireless Intrusion Prevention System (WIPS) and/or rogue AP detection
- Protocol and spectrum analyzers
- Best practices in secure management protocols (e.g. encrypted management HTTPS, SNMPv3, SSH2, VPN and password management)

6.0 RF Validation – 10%

6.1 Verify and document that design requirements are met including coverage, throughput, roaming, and connectivity with a post-implementation validation survey

6.2 Locate and identify sources of RF interference

- Identify RF disruption from 802.11 wireless devices including contention vs. interference and causes/sources of both including co-channel contention (CCC), overlapping channels, and 802.11 wireless device proximity

- Identify sources of RF interference from non-802.11 wireless devices based on the investigation of airtime and frequency utilization
- Understand interference mitigation options including removal of interference source or change of wireless channel usage

6.3 Perform application testing to validate WLAN performance

- Network and service availability
- VoIP testing
- Real-time application testing
- Throughput testing

6.4 Understand and use the basic features of validation tools

- Use of throughput testers for validation tasks
- Use of wireless validation software (specifically survey software and wireless scanners)
- Use of protocol analyzers for validation tasks
- Use of spectrum analyzers for validation tasks

7.0 WLAN Troubleshooting – 10%

7.1 Describe and apply common troubleshooting tools used in WLANs

- Use of protocol analyzers for troubleshooting tasks
- Use of spectrum analyzers for identifying sources of interference
- Use of management, monitoring and logging systems for troubleshooting tasks

- Use of wireless LAN scanners for troubleshooting tasks

7.2 Identify and troubleshoot common wireless issues

- Identify causes of insufficient throughput in the wireless distribution systems including LAN port speed/duplex misconfigurations, insufficient PoE budget, and insufficient Internet or WAN bandwidth
- Identify and solve RF interference using spectrum analyzers
- Identify wireless performance issues using SNR, retransmissions, and airtime utilization statistics
- Identify causes of wireless issues related to network services including DHCP, DNS, and time protocols including using native interface and IP configuration tools (e.g. pings, DNS lookups, interface configuration)
- Identify wireless issues related to security configuration mismatches

Extended Table of Contents

Chapter 1 — WLAN and Networking Industry Organizations

When I first started working with 802.11-based wireless networking, it was an interesting technology used by a few companies. I loved the deep knowledge required to work with it and became a student of all things Wi-Fi. Today, however, things have changed drastically. Few companies do not have wireless networks, and even those companies often have them and don't realize it, thanks to determined users. For the common user today, Wi-Fi or wireless has become synonymous with the Internet. They will say, "Do you have Wi-Fi?" and what they mean is, "Do you have wireless Internet?" While those of us who specialize in wireless networking may laugh at the point of confusion, it is understandable, given that wireless cellphones have made Internet access ubiquitous, and users may wonder why you would even have a wireless network if it was not also providing Internet access. In organizations, however, it is not uncommon to use wireless networking for internal access only.

Given that both internal network access and Internet access are in high demand across our wireless networks, it is essential that the wireless network administrator or CWNA, understand how the 802.11 standards come about and who creates these standards and manages the usage of the radio frequency (RF) spectrum. This chapter introduces you to both of these important concepts. First, you will explore a bit of history related to wireless communications. Next, you will examine the industry organizations who shape all things wireless. Then, you will examine the IEEE standard creation process. Finally, you will learn about common wireless network types. Let's get started and create a solid foundation for the rest of this book and your learning experience.

A Brief History of Wireless Communications

Over the air, electromagnetic wave-based communications (simply called wireless) have been utilized for many decades. In fact, radio and television both depend on these electromagnetic waves. Additionally, electromagnetic waves — or radio waves — have been used for purposes such as wireless voice conversations (today, we call these cellphones) and data communications. Militaries used wireless communications for many decades before it was commonly used in the private sector for personal and business purposes. The first available wireless equipment was expensive proprietary equipment and was not interoperable with equipment from other vendors as it used unique and non-standard modulation methods.

Long ago, in 1880, the first wireless phone conversation took place. The Photophone was used to communicate between the rooftops of two buildings in Washington D.C. The Photophone was invented by Alexander Graham Bell, and light was used as the communications medium rather than radio frequency waves.

Wireless communications over great distances all started with the letter "S." It was this letter that Guglielmo Marconi transmitted, received, and printed with the Morse inker across the Atlantic in the first decade of the 20th century. Figure 1.1 shows Marconi with wireless equipment similar to that used to transmit his wireless message. Though Marconi was not the first to communicate over distance without wires, this event stirred interest throughout the government and business communities and eventually resulted in the many uses of wireless technology we see today.

Radio waves were in heavy use for telecommunications by the 1920s. The first transatlantic telephone service became available in 1927 from New York to London. Twenty-one years earlier, in 1906, Reginald Fessenden successfully communicated from land to sea over a distance of 11 miles using radio waves to carry voice communications. Bell Laboratories had created a mobile two-way voice-carrying radio wave device by 1924, but mobile voice technology was not perfected and used widely until the 1940s.

If you were involved in the early days of computing and used bulletin board systems (BBS), or the early internet, you have used modems. If that's the case, you are aware that computer data can be transferred over land-based telephone lines using these devices. A modem modulates the binary data into analog signals and demodulates the analog signals into binary data. This allows two computers to talk to each other across such landlines. As you can imagine, the leap to communicating digital data across wireless connections is not a large leap. From the early use of radio technology for broadcasting (radio and television) and voice communications to today's massive data transfer over wireless links, radio wave communication has evolved rapidly.

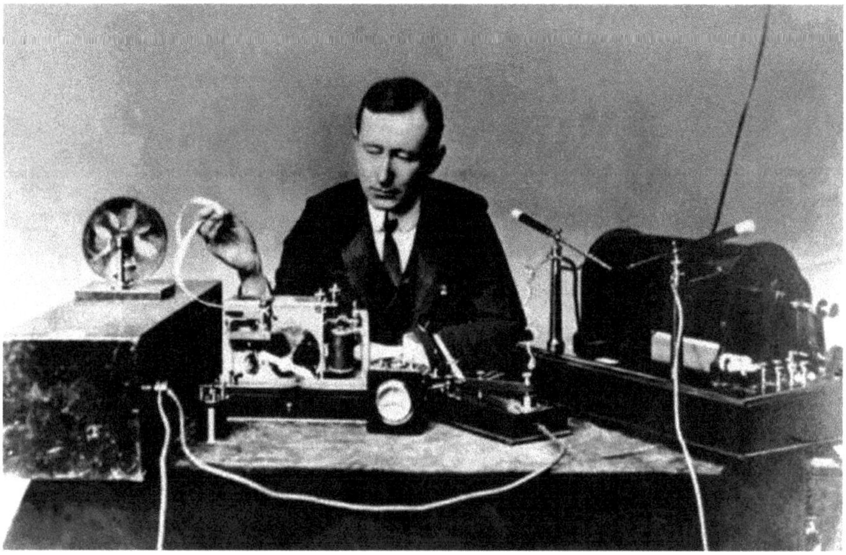

Figure 1.1: Guglielmo Marconi with his wireless equipment

One of the greatest problems with these early technologies was the proprietary nature of the devices. Like humans, if two devices are to communicate with each other, they must possess a shared language. Without standards, each company created devices that communicated in ways they either thought were best, or simply in the only ways their engineers knew how to implement. This scenario resulted in incompatibilities among the different devices. Many standards have been developed by organizations like the IEEE, IETF and ANSI that have helped overcome this hurdle. When it comes to wireless data communications in Local Area Networks (LANs), the IEEE 802.11 standard is the one that is most widely used.

The Institute of Electrical and Electronics Engineers (IEEE) ratified the first 802.11 standard in 1997. Today, we fondly call this 802.11-Prime. It was capable of 1 or 2 Mbps data rates and, even for that time, this was a rather slow transmission rate.

However, in 1999, the IEEE ratified the 802.11a and 802.11b amendments, which provided maximum data rates of 54 Mbps and 11 Mbps respectively. This increased speed also saw increased adoption of the technology.

The term ***Mbps*** stands for megabits per second and should not be confused with MBps, which stands for megabytes per second. 1 Mbps is equal to one million bits transmitted in one second.

Industry Organizations

Three types of organizations guide the wireless industry. They are regulation, standardization and compatibility/certification. The Federal Communications Commission (FCC) and the European Telecommunications Standards Institute (ETSI) are examples of regulatory bodies that provide regulations in North America and Europe respectively. The Institute of Electrical and Electronics Engineers (IEEE) and Internet Engineering Task Force (IETF) are examples of standards development organizations. The Wi-Fi Alliance is a compatibility testing and certification group. Of course, CWNP is a knowledge certification company.

It is important to understand what these organizations do, and it is also important to understand how they work together. For example, consider the interdependency between the FCC and the IEEE, or the relationship between the Wi-Fi Alliance and the IEEE. The FCC sets the boundaries within which the IEEE standards may function. The Wi-Fi Alliance tests equipment based on portions of IEEE standards and certifies it as being interoperable. These three organizations provide regulation, standardization and compatibility services for wireless Local Area Network (WLAN) technologies within the North America.

The benefits of these organizations to the consumer are clear and are depicted in Figure 1.1. When regulations are in place, such as power output limits, it is possible to implement local wireless networks with less interference from nearby networks. When standards are in place, like the IEEE 802.11 standard, it is possible to purchase devices that are compatible even though they come from different vendors. When certifications are in place to validate interoperability, consumers may buy products with confidence that those devices sharing the same certifications should be interoperable and fewer man hours will be required for compatibility testing.

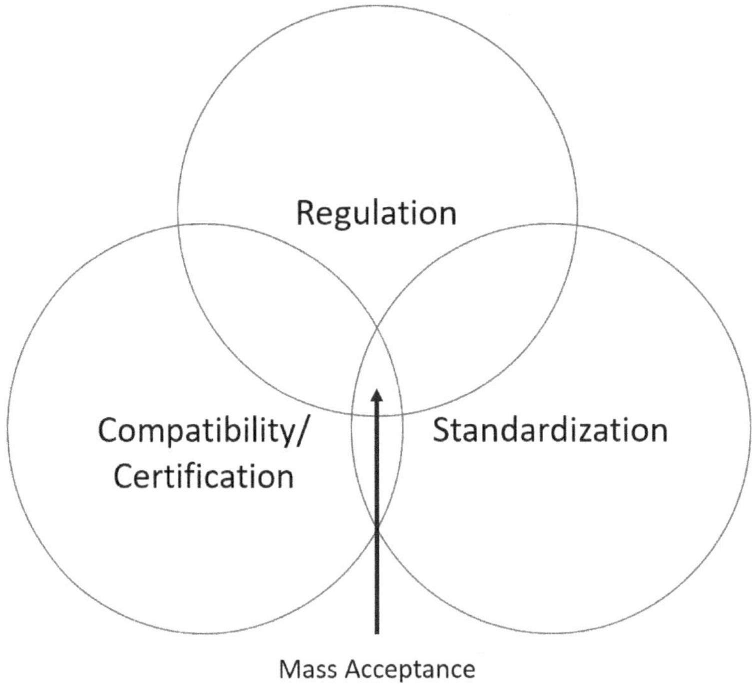

Figure 1.2: Requirements of Mass Acceptance

In an ideal world, we would get all these benefits with exact perfection. In the real world, however, this is not the case. In the real world; interference is reduced, but not eliminated, and a trained professional must use their skills to remedy the issues that can arise in this field. A common example could be hardware that is interoperable with a system, but not necessarily fully compatible. In this situation; a skilled professional can carry out the required careful implementation, reduce testing time, and properly configure systems in the given environment. If you are installing a wireless network in an office that shares space with other offices, you may still encounter interference — even with the lower output power constraints of regulation. If you are working with devices from different vendors, you may encounter specific compatibility issues outside the standards upon which the devices are based, and careful selection of firmware and driver versions may be needed. If you are implementing hardware that has been certified by the Wi-Fi Alliance, you should still test it with your hardware to ensure that no compatibility issues exist. Particularly, compatibility issues that arise from proprietary non-standard features in the equipment (such as the implementation of 802.11ac modulation in 2.4 GHz). Even considering these realities, the benefits that the regulatory standards and compatibility organizations bring to the wireless industry are valuable.

Regulatory Domain Governing Agencies

A *regulatory domain* is defined as a geographically bounded area that is controlled by a set of laws or policies. Currently, governing bodies exist at the city, county, state and country level within the United States forming a hierarchical regulatory domain system. In other countries, governments exist with similar hierarchies or with a single-level of authority at the top level of the country or group of countries. In many cases, these governments have assigned the responsibility of managing communications to a specific organization that is responsible to the government. In the United States, this managing organization is the Federal Communications Commission. In the UK, it is the Office of Communications. In Australia, it is the Australian Communications and Media Authority. The following sections outline four such governing bodies and the roles they play in the wireless networking industry of their respective regulatory domains.

> For the CWNA exam, you will not be required to know the specific regulatory constraints of any agency in any domain. Too many organizations exist throughout the world to place this demand on the CWNA candidate. However, you should understand the role played by a regulator agency, such as the FCC or ETSI.

FCC

The Federal Communications Commission (FCC) was born out of the Communications Act of 1934. Charged with the regulation of interstate and international communications by radio, television, cable, satellite and wire, the FCC has a large body of responsibility. The regulatory domain covered by the FCC includes all 50 of the United States as well as the District of Columbia and other U.S. possessions, like the Virgin Islands and Guam.

Because WLAN devices use radio wave communications, they fall under the regulatory control of the FCC. The factors regulated by the FCC include:

- Radio frequencies available
- Output power levels
- Indoor and outdoor usage

A radio frequency is specified in hertz. The hertz is the measurement of wave cycles per second; therefore, a radio frequency of 2.412 gigahertz (GHz) cycles 2,412,000,000 times per second. The FCC regulates which frequencies may be used within the regulatory domain it manages. For example, the FCC provides two types of license-free bands for radio communications: The Industrial Scientific Medical (ISM) bands and the Unlicensed National Information Infrastructure (U-NII — usually pronounced you-knee) bands. Currently, 11 ISM bands are used in various frequencies throughout the RF spectrum, but only the one starting at 2.4 GHz is used by the IEEE 802.11 standard. This ISM frequency band is most familiar to WLAN administrators. The four U-NII bands exist in the 5 GHz frequency range and are all used by 802.11. Table 1.1 provides a summary of the ISM and U-NII license-free bands used by 802.11. 802.11 networks do not use all of the frequency space in these bands for primary communications, but Table 1.1 specifies the entire range of the bands in which 802.11 networks operate.

Frequency Band	Total Band Width	License-Free Band
2,400-2,500 MHz	100 MHz	ISM
5.15-5.25 GHz	100 MHz	U-NII (U-NII-1)
5.25-5.35 GHz	100 MHz	U-NII (U-NII-2A)
5.470-5.725 GHz	255 MHz	U-NII (U-NII-2C)
5.725-5.850 GHz	100 MHz	U-NII (U-NII-3)

Table 1.1: Unlicensed (license-free) Bands used by IEEE 802.11

The license-free bands provide both a benefit and a disadvantage. The benefit comes from the fact that you are not required to obtain a license to communicate within these license-free bands. You can buy FCC-authorized equipment and install it in your environment without any required permits or fees. However, the disadvantage of using license-free bands is that others can also use them. You will have to deal with contention and interference issues and ensure that you have the RF bandwidth available for your intended purpose in the environment where you will be implementing the WLAN.

It would be nice if we could even say that the use of the license-free bands is on a "first-come-first-serve" basis, but it's not. You may have a WLAN installed for years only to have a nearby organization install another WLAN on the same frequencies (channels) you've been using; this can cause major contention on your network. As long as this neighboring network is within FCC regulations, very little can be done, aside from some careful negotiations on wireless device placement and channel usage.

Contention is the process used in 802.11 WLANs to access the medium (radio frequency) for transmission. It is a normal part of operations. As more devices access the same channel (radio frequency range), contention wait times become longer for each station (STA).

The FCC also regulates the output power levels of radio frequency devices within the allowed license-free bands. Table 1.2 gives a brief summary of the output power limits (in milliwatts) imposed by the FCC. More complex scenarios apply to the use of the ISM band that will be covered in later chapters. Remember, you will not be tested on specific output power constraints on the CWNA-108 exam. They are provided here for reference.

Finally, in the past, the FCC limited the 5.15-5.25 U-NII band (U-NII-1) to indoor usage only. This restriction was lifted in 2014 to allow indoor and outdoor use in this band. The other U-NII bands were allowed for use indoors or outdoors before this. However, the 5.725-5.850 band is especially well-suited for outdoor bridge links because of the high output power allowance of 1000 mW or 1 Watt. The area usage of the U-NII bands is summarized in Table 1.2. The point of this is to realize that regulatory constraints change over time and the CWNA should monitor for and be aware of these changes.

Band	Power Output Limits	Area Usage
U-NII-1 5.15-5.25 GHz	250 mW for client devices and 1 W for master devices (access points)	Indoor/outdoor.
U-NII-2A 5.25-5.35 GHz	250 mW	Indoor/outdoor.
U-NII-2-Extended (U-NII-2C) 5.470-5.725 GHz	250 mW	Indoor/outdoor.
U-NII-3 5.725-5.850 GHz	1000 mW	Higher output power assumes outdoor operations

Table 1.2: FCC Power Output Limits – U-NII Bands

When regulatory constraints change, it does not mean that WLAN hardware and software automatically changes. In some cases, software or firmware upgrades will allow you to take advantage of or comply with the new constraints. In other cases, you must replace the hardware to gain advantage of compliance.

The 2.4 GHz ISM band may be used indoors or outdoors and the output power at the intentional radiator cannot exceed 1 Watt (W). For indoor devices, the output power is usually well under 1 Watt and generally resides in a range from 10 to 300 milliwatts (mW). Special allowances are in place when higher gain antennas are used that ultimately result in more directional power for longer distance links, such as bridge links. In wireless transmissions, a power measurement called the Equivalent Isotropically Radiated Power (EIRP) is a measure of system output power after antenna gain. The 1 W output power constraint of the FCC assumes a 6 dBi antenna, resulting in 4 W EIRP. If a 9 dBi antenna is used, 1 dBm less power must be used, but the extra 3 dB in antenna gain results in an EIRP of approximately 6.3 W. The rule states that for every 3 dB in antenna gain, the output power must be lowered by 1 dBm.

A ***decibel (dB)*** is a relative measurement of power. It is covered in more detail in later chapters of this book. A dBm is an absolute measurement of power related to the Watt or milliWatt. It is also covered in more detail later. Chapter 3 provides extensive coverage of RF mathematics and measurements.

OfCom and ETSI

The Office of Communications (OfCom) is charged with ensuring optimal use of the electromagnetic spectrum, for radio communications, within the UK. OfCom provides documentation of and forums for discussion of valid frequency usage in radio communications. The regulations put forth by the OfCom are based on standards developed by the European Telecommunications Standards Institute (ETSI). These two organizations work together in much the same way the FCC and IEEE do in the United States.

MIC and ARIB

In Japan, the Ministry of Internal Affairs and Communications (MIC) is the governing body over radio communications. However, the Association of Radio Industries and Businesses (ARIB) was appointed to manage the efficient utilization of the radio spectrum by the MIC. In the end, ARIB is responsible for regulating which frequencies can be used and such factors as power output levels.

ACMA

The Australian Communications and Media Authority (ACMA) replaced the Australian Communications Authority in July of 2005 as the governing body over the regulatory domain of Australia for radio communications management. Like the FCC in the United States, the ACMA is charged with managing the electromagnetic spectrum to minimize interference. This is done by limiting output power in license-free frequencies, and by requiring licenses in some frequencies.

ITU-R

The International Telecommunications Union — Radiocommunication (ITU-R) is a Sector of the International Telecommunications Union (ITU). The ITU, after an extended history, was designated as a United Nations specialized agency on October 15, 1947. The constitution of the ITU has stated its purposes as:

- To maintain and extend international cooperation between all its Member States for the improvement and rational use of telecommunications of all kinds
- To promote and enhance participation of entities and organizations in the activities of the Union and foster fruitful cooperation and partnership between them and Member States for the fulfillment of the overall objectives embodied in the purposes of the Union
- To promote and to offer technical assistance to developing countries in the field of telecommunications, and also to promote the mobilization of the

material, human and financial resources needed for its implementation, as well as access to information

- To promote the development of technical facilities and their most efficient operation with a view to improving the efficiency of telecommunication services, increasing their usefulness and making them, so far as possible, generally available to the public
- To promote the extension of the benefits of new telecommunication technologies to all the world's inhabitants
- To promote the use of telecommunication services with the objective of facilitating peaceful relations
- To harmonize the actions of Member States and promote fruitful and constructive cooperation and partnership between Member States and Sector Members in the attainment of those ends
- In order to promote at the international level, the adoption of a broader approach to the issues of telecommunications in the global information economy and society cooperate with other world and regional intergovernmental organizations, and those non-governmental organizations concerned with telecommunications.

The ITU-R, specifically, maintains a database of the frequency assignments worldwide and helps coordinate electromagnetic spectrum management through five administrative regions. These five regions are:

- Region A: The Americas
- Region B: Western Europe
- Region C: Eastern Europe
- Region D: Africa
- Region E: Asia and Australia

Each region has one or more local regulatory groups such as the FCC in Region A for the United States or the ACMA in Region E for Australia. Ultimately, the ITU-R provides the service of maintaining the Master International Frequency Register of 1,265,000 terrestrial frequency assignments.

In the end, regulatory agencies typically control wireless use in unlicensed spaces in the following important areas:

- Allowed frequencies
- Channel bandwidth
- Area usage, such as indoor and outdoor
- Maximum transmission power at the radiator
- Maximum transmission power at the antenna

IEEE

The Institute of Electrical and Electronics Engineers (IEEE) states their mission as being the world's leading professional association for the advancement of technology. They provide standards and technical guidance for more than just the wireless industry. In this section, I focus on the specific standards developed by the IEEE that impact and benefit wireless networking. These standards include wireless-specific standards, as well as standards that have been implemented in the wired networking domain, which are now being utilized in the wireless networking domain. First, I provide you with a more detailed overview of the IEEE organization.

Overview of the IEEE organization

The IEEE is a global professional society with more than 423,000 members in 160 countries. The constitution of the IEEE defines the purpose of the organization as scientific and educational, directed toward the advancement of the theory and practice of electrical, electronics, communications and computer engineering, as well as computer science, the allied branches of engineering, and the related arts and sciences. Their mission is stated as promoting the engineering process of creating, developing, integrating, sharing, and applying knowledge about electro and information technologies and sciences for the benefit of humanity and the profession. Ultimately, the IEEE creates many standards for many niche disciplines within electronics and communications. In this book, the focus is on computer data networks and specifically wireless computer data networks. In this area, the IEEE has given us the 802 project and, specific to wireless, the IEEE 802.11 standard.

IEEE Standards Important to WLANs

Several standards are important to the implementation and support of WLANs, including the primary standard, which is 802.11. Table 1.3 provides an overview of important IEEE standards that will be referenced throughout this book in varying ways.

Standard	Description
802.11	The standard that defines WLAN networking, including the medium access control (MAC) and physical (PHY) specifications.
802.3	The Ethernet wired networking standard and the network type to which WLAN access points are most often connected.
802.1X	A port-based authentication standard that controls access to networks by devices. Used in enterprise-class WLAN authentication solutions.

Table 1.3: IEEE Standards Related to WLAN Implementations

The 802.11 standard itself has been updated many times since the first release in 1997. These updates are released in the form of amendments, which are discussed in more detail in the later section of this chapter titled *IEEE Standard Creation Process*. For example, 802.11n added the High Throughput (HT) MAC and PHY capabilities and 802.11ac added the Very High Throughput (VHT) MAC and PHY capabilities. Both of these amendments were ratified more than 10 years after the 1997 ratification of the 802.11 standard. At any moment, the

802.11 standard exists as "the standard as amended," which simply means that all amendments become part of the standard when they are ratified.

Wi-Fi Alliance

The Wi-Fi Alliance is a certification organization that provides testing and interoperability analysis for the wireless industry. While the FCC makes the rules and the IEEE determines how to abide by those rules according to a standard, the Wi-Fi Alliance ensures that devices are compatible with the IEEE standard and other devices compliant with the standard. Stated differently, the Wi-Fi Alliance has taken on the responsibility for testing Wi-Fi hardware to verify its compatibility with other Wi-Fi hardware.

Originally, the Wi-Fi Alliance was known as the Wireless Ethernet Compatibility Alliance (WECA). In October of 2002, the organization was re-branded as the Wi-Fi Alliance. This was done as a measure to improve brand awareness and make the name a more memorable and associative one (a creative way of saying: to make the name more marketable).

Only products of Wi-Fi Alliance members that have been tested successfully by the Wi-Fi Alliance can actually claim that they are Wi-Fi Certified. This is a subtle distinction as some vendors may say their equipment is Wi-Fi equipment and this equipment may or may not be Wi-Fi Certified. The result has been confusion for some consumers, but the good news is that the vast majority of WLAN devices today are truly Wi-Fi Certified. Ultimately, consumers should look for logos like the ones in Figure 1.3. If a logo like this is on the packaging, the product has been certified by the Wi-Fi Alliance.

Figure 1.3: The Wi-Fi Certified Logos

The Wi-Fi Alliance requests the use of the simple Wi-Fi Certified logo in the upper left of Figure 1.3 instead of the detailed logos in the lower half of the image. However, where the detailed logos are used, it is important to note the difference. For example, the left two logos on the bottom of Figure 1.3 indicate very different abilities. The leftmost logo is a 2.4 GHz-only device. This fact is known based on support for only 802.11b/g and 802.11n. The next logo to the right is a dual-band device supporting both 2.4 GHz and 5 GHz. This is known because 802.11a is supported as well, and it operates only in the 5 GHz band, while 802.11b/g operates in the 2.4 GHz band.

Wi-Fi doesn't stand for wireless fidelity as is often stated. It is simply a brand name. Legend has it that a Wi-Fi Alliance executive was speaking at a conference several years ago and indicated that it stood for wireless fidelity, and even though the Wi-Fi Alliance has clearly stated that the comment was incorrect the meaning has been resistant to its ultimate demise.

Table 1.4 provides a brief overview of the primary Wi-Fi Alliance certification programs of interest to the CWNA. When you see that a device or system states compliance with one of these certifications, you will now better understand the meaning.

Certification Program	Description
Wi-Fi Certified a	The device is certified as 802.11a compliant and operates in the 5 GHz band.
Wi-Fi Certified b/g	The device is certified as 802.11b and 802.11g compliant and operates in the 2.4 GHz band.
Wi-Fi Certified n	The device is compliant with 802.11n and may operate in 2.4 GHz, 5 GHz, or both. The certificate for a specific device will indicate if it is dual-band or single band and in what band(s) it operates.
Wi-Fi Certified ac	The device is certified as compliant with 802.11ac and operates in the 5 GHz band as an 802.11ac device. The specific features of an 802.11ac certified device are provided in the device certificate. Certificates are available in PDF form for each certified device.
Wi-Fi Certified WiGig	The device is certified as compliant with 802.11ad and operates in the 60 GHz band.
Wi-Fi Direct	A certification program that provides guidance for direct communications between two stations that are connected to an access point without requiring that the communications pass through the access point.
WPA2	A Personal and Enterprise version of this certifications exists. The Personal version tests pre-shared key authentication and the Enterprise version tests 802.1X/EAP authentication.
Wi-Fi Protected Setup (WPS)	Sometimes called push-button security, WPS allows for easy setup of security in 802.11 devices using a Personal Identification Number (PIN) or push-button activation.
Wi-Fi Multimedia (WMM)	Validates implementation of 802.11e Quality of Service (QoS) capabilities using access categories for wireless traffic so that higher-priority traffic can

	be transmitted earlier than lower-priority traffic in most situations.
Voice-Personal	Voice over IP (VoIP) over Wi-Fi provides interoperability testing and ensures good performance of voice communications on a single Wi-Fi link.
Voice-Enterprise	Takes Voice-Personal further to ensure good performance of voice communications on an enterprise WLAN even when transitioning from one access point to another during mobility.
Passpoint	Allows for mobile devices to discover, select and connect to WLANs with no user intervention. Devices can identify network ownership and authentication services available and connect to the network regardless of the SSID (network name) if it is partnering with a compatible backend authentication system of which the client device is aware and a member.

Table 1.4: Wi-Fi Alliance Certification Programs

Figures 1.4 and 1.5 show an example Wi-Fi Alliance certificate. The pictured certificate is for the Cisco 5508 WLAN controller and Cisco AP1852 access point. In the Summary of Certifications section, the classifications are listed, they include Connectivity, Optimization, and Access. Notice that the system is certified for 802.11a/b/g/n/ac in the Connectivity classification. However, you must inspect page 2 of the certificate to determine the bands supported and features of the device, such as the number of spatial streams and various communication forms. Don't worry, terms like spatial streams, MIMO, MU-MIMO and beamforming will become clearer as you study this book.

Page 1 of the certificate typically provides an overview of the system, including the data of certification, the company or vendor, the product tested with model numbers, hardware and firmware versions, operational bands, and general certifications acquired by the system. In order to acquire some Wi-Fi Alliance certifications (such as Voice-Enterprise) an access point must be certified as part of a larger system, such as the certificate shown in Figures 1.4 and 1.5. In this case, the AP and WLAN controller are certified together.

It is also important to note, in Figure 1.4, that the system was certified on a specific firmware version and operating system. While it is unlikely that a firmware update would remove a truly needed feature of the system, the certification is based on a specific release of software and hardware. At times, vendors will reapply for certification with newer firmware revisions, but this typically happens only when major changes have been made to the system and more often than not is not performed again with the same hardware.

Wi-Fi CERTIFIED™ Interoperability Certificate

This certificate lists the features that have successfully completed Wi-Fi Alliance interoperability testing. Learn more: www.wi-fi.org/certification/programs

CERTIFIED

Certification ID: WFA61335	Page 1 of 2
Date of Last Certification	March 21, 2018
Company	Cisco Systems
Product	Cisco 5508 WLAN Controller and Cisco AP1852 AP
Model Number	AIR-CT5508-K9 and AIR-CAP1852
Product Identifier(s)	AIR-CT5508-K9 and AIR-CAP1852 (SKU)
Category	Routers
Subcategory	Enterprise/Service Provider Access Point, Switch/Controller or Router
Hardware Version	Product: v1, Wi-Fi Component: v1
Firmware Version	Product: 8.5.124.30, Wi-Fi Component: 8.5.124.30
Operating System	Proprietary / Other: AireOS (controller) + Linux (...
Frequency Band(s)	2.4 GHz, 5 GHz - Concurrent

Summary of Certifications

CLASSIFICATION	PROGRAM
Connectivity	Wi-Fi CERTIFIED™ a, b, g, n, ac
	WPA2™ - Enterprise, Personal
Optimization	Wi-Fi Vantage™
	WMM®
	WMM®-Power Save
Access	Passpoint®

Figure 1.4: Wi-Fi Alliance Certificate for Cisco 802.11ac System (Page 1)

Wi-Fi CERTIFIED™ Interoperability Certificate

Certification ID: WFA61335 — **Page 2 of 2**

Security

WPA2™ - Enterprise, Personal
- EAP Type(s)
  - EAP-TLS
  - EAP-TTLS/MSCHAPv2
  - PEAPv0/EAP-MSCHAPv2
  - PEAPv1/EAP-GTC
  - EAP-SIM
  - EAP-AKA
  - EAP-AKA Prime
  - EAP-FAST
- Protected Management Frames

Wi-Fi CERTIFIED™ a

Wi-Fi CERTIFIED™ b

Wi-Fi CERTIFIED™ g

Wi-Fi CERTIFIED™ n

2.4 GHz, 5 GHz - Concurrent
3 Spatial Streams 2.4 GHz
3 Spatial Streams 5 GHz
Short Guard Interval
TX A-MPDU
STBC Transmit
40 MHz operation in 5 GHz
OBSS on Extension Channel
RIFS Test

Wi-Fi CERTIFIED™ ac

4 Spatial Streams 5 GHz
Rx MCS 8-9 (256-QAM)
Tx STBC 2x1
Rx A-MPDU of A-MSDU
Tx SU beamformer
Low Density Parity Check coding
Tx DL MU-MIMO

Wi-Fi Vantage™

WMM®

WMM®-Power Save

Passpoint®

Online Signup (OSU) and Policy Provisioning

Spectrum and Regulatory Features

802.11d
802.11h

Figure 1.5: Wi-Fi Alliance Certificate for Cisco 802.11ac System (Page 2)

Finally, consider that the Wi-Fi Alliance offers multiple classes of certification programs, including:

1. Connectivity (includes security certifications)
   1.1. Wi-Fi Certified a
   1.2. Wi-Fi Certified b/g
   1.3. Wi-Fi Certified n
   1.4. Wi-Fi Certified ac
   1.5. WPA2
      1.5.1. WPA2-Personal
      1.5.2. WPA2-Enterprise
   1.6. WPA – now defunct

2. Optimization
   2.1. Wi-Fi Agile Multiband
   2.2. Wi-Fi TimeSync
   2.3. Wi-Fi Vantage
   2.4. Wi-Fi Multimedia (WMM)
      2.4.1. WMM-Power Save (WMM-PS)

3. Access
   3.1. Passpoint
   3.2. Wi-Fi Protected Setup (WPS)

4. Applications and Services
   4.1. Miracast
   4.2. Voice-Personal
   4.3. Voice-Enterprise
   4.4. Wi-Fi Aware
   4.5. Wi-Fi Location

IETF

The Internet Engineering Task Force (IETF) is another standards development organization that has impacted the wireless networking industry. You can learn more about the IETF as an organization by visiting their website at ietf.org. The primary IETF standards impacting WLANs directly include RFC 3748, RFC 2865, and RFC 5415.

IETF RFC 3748

The IETF request for comments (RFC) 3748 details the functionality of the Extensible Authentication Protocol (EAP). EAP is used when IEEE 802.1X port-based authentication is implemented and is, therefore, an integral part of WLAN security. Several different EAP types are supported on WLAN systems, but the EAP standard provides the framework that drives the engine of every EAP type.

IETF RFC 2865

While EAP provides the authentication flow and specifications, RADIUS provides the highway on which EAP passes. In most implementations of 802.1X, EAP messages are passed to a RADIUS server where authentication is either approved or rejected. RADIUS is the remote access dial-in user service and it is used, obviously, for more than just dial-in connections today.

IETF RFC 5415

The Control and Provisioning of Wireless Access Points (CAPWAP) protocol is defined in RFC 5415 and is the newer such protocol replacing the Lightweight Access Point Protocol (LWAPP) that is defined in RFC 5412. CAPWAP (or LWAPP) is used for communications between lightweight access points (controller-based access points) and a WLAN controller. A tunnel is usually established between the access point and controller and communications traverse this tunnel. Not all vendors implement either CAPWAP or LWAPP. Some use Generic Routing Encapsulation (GRE – defined in RFC 1701) to tunnel traffic and proprietary (or other standards-based) protocols for management traffic. Vendors that use CAPWAP/LWAPP include Cisco and Ruckus Wireless. Aruba Networks uses GRE tunneling for user traffic and their own Protocol Application Programming Interface (PAPI) protocol for management traffic.

IEEE Standard Creation Process

IEEE projects, such as the IEEE 802 project, are divided into working groups. The Ethernet standard comes from the IEEE 802.3 working group and the WLAN standard comes from the IEEE 802.11 working group. The IEEE 802.11 working group was the 11th group formed under the IEEE 802 project. Figure 1.6 illustrates the hierarchy of the 802 project.

Other working groups of interest to wireless professionals include the IEEE 802.16 working group, which focuses on broadband wireless access commonly known as WiMAX, and the IEEE 802.20 working group, which focuses on mobile broadband wireless access. It's interesting to note that the majority of the newly developed working groups since IEEE 802.11, are wireless specific, which is more evidence of the importance of wireless technology in the digital future.

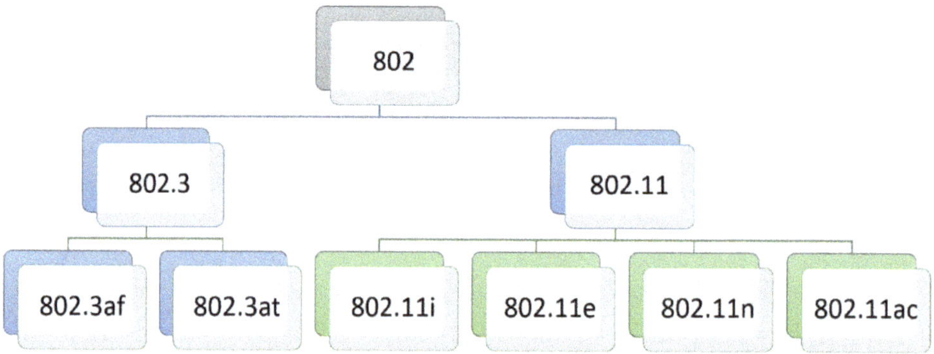

Figure 1.6: The 802 Project Hierarchy Showing 802.3 and 802.11

The original IEEE 802.11 standard was ratified in 1997 and is embodied in a document called IEEE 802.11-1997, today called 802.11-Prime by many wireless professionals. The standard was amended in 1999; the base document was altered slightly and replaced by a new base document called IEEE 802.11-1999, and several amendment documents were ratified. The base document was reaffirmed in 2003, when yet another amendment document was ratified and renamed IEEE 802.11-1999 Edition (R2003). Several more amendment documents have been ratified since, and a rollup was created in 2007 (802.11-2007), another in 2012 (802.11-2012), and finally the most recent in 2016 (802.11-2016).

The 802.11-2016 standard document includes 3,543 pages in the PDF document. This page count does not include additional information in the 802.11ah amendment or the other amendments ratified or soon-to-be-ratified since the 802.11-2016 amendment. The size of the document is, in part, because of the many different communication methods defined and the many different types of wireless networks accommodated (for example, ad-hoc networks between peers, infrastructure networks with access points, and mesh networks, to name a few).

The original IEEE 802.11-1997 standard specified three ways of implementing a physical communication layer (PHY). Frequency Hopping Spread Spectrum (FHSS) and Direct Sequence Spread Spectrum (DSSS) both use the 2.4 GHz band. The third PHY used Infrared but never saw the light of day (no pun intended) and is not covered by the CWNA certification for this reason. Each of these PHYs operates at a mandatory nominal data rate of 1 megabit per second (Mbps) and optionally at 2 Mbps.

BEYOND THE EXAM: Wireless Networking Growth
It is interesting to note the massive growth in WLAN technology since the initial release of the IEEE 802.11 standard. Before that time, WLAN devices used proprietary technologies for communications and this meant that hardware from one vendor could not usually communicate with hardware from another vendor. With the introduction of the 802.11 standard, a communications protocol was defined that vendors could implement and that would allow for interoperability between the various vendors' hardware solutions. Figure 1.2 illustrated the benefits of the regulations, standards and certifications combined, and we have seen growth in the industry because of this structure.

The next step was the release of the 802.11b amendment, which provided data rates of up to 11 Mbps. Once this milestone was reached, the WLAN market exploded with growth reminding us of a tremendous benefit standardization provides — compatibility. 802.11b was just the beginning. 802.11a, 802.11g, 802.11n and 802.11ac have all continued to expand the market and 802.11ax will soon take things even further.

We may have indeed learned some lessons from the past. The telephone industry was locked up in a monopolistic structure based around large organizations that developed the standards and owned the backbone. Eventually, an act of Congress put a stop to this and opened up the telephone industry to real competition. The IEEE stepped in and developed such standards for wireless communications before wireless networking became as popular as it is today, giving us the wonderful variety of vendors, while still providing compatibility.

The standards created by working groups are often updated by task groups and these updates are released as amendments. Amendments can be either ratified or in draft. When in draft mode, the amendment may still be modified and hardware or software development against the draft amendment is not usually recommended, though vendors will often release devices based on the draft. This action was seen quite often with 802.11n and somewhat with 802.11ac.

When an amendment is ratified, it has been stabilized and development of hardware and software is usually forthcoming. Table 1.5 provides a brief description of the amendments to the 802.11 standard that are important to the CWNA candidate. Those amendments with a date appended, were ratified in the year noted.

802.11 Amendment	Description
802.11a-1999	Uses Orthogonal Frequency Division Multiplexing (OFDM) instead of DSSS. Provides data rates up to 54 Mbps. Uses the 5 GHz bands. Not compatible with PHYs that use the 2.4 GHz band such as DSSS and HR/DSSS.
802.11b-1999 (amended slightly in 2001)	Uses High-Rate Direct Sequence Spread Spectrum (HR/DSSS) instead of the original DSSS. Provides data rates up to 11 Mbps. Uses the 2.4 GHz band. Backward compatible with DSSS.
802.11c-1998 (incorporated into the	Updates the IEEE 802.1D bridging standard for 802.11 operations.

802.1D-2004 standard, section 6.5.4)	
802.11d-2001	Provides specifications for the use of 802.11 in more regulatory domains (countries) than were originally specified.
802.11e-2005	Defines the layer 2 MAC controls used to meet the Quality of Service (QoS) requirements of multimedia and voice applications over 802.11 networks.
802.11g-2003	Supports DSSS and HR/DSSS and adapts OFDM modulation to 2.4 GHz band. Provides data rates up to 54 Mbps. Not compatible with the 5 GHz OFDM PHY due to the use of the 2.4 GHz band.
802.11h-2003	Enhances the 802.11 MAC and OFDM PHY with network management and control. Provides Dynamic Frequency Selection (DFS) and Transmit Power Control (TPC) mechanisms.
802.11i-2004	One of the most important enhancements to the 802.11 standard. Specifies the use of Counter Mode Cipher Block Chaining Message Authentication Code Protocol (CCMP), which is reliant on the Advanced Encryption Standard (AES) and allows for the use of the Temporal Key Integrity Protocol (TKIP). Requires the use of either IEEE 802.1X or pre-shared key (PSK) for authentication.
802.11j-2004	Extends the 802.11 MAC and OFDM PHY to operate in the newly available 4.9-5 GHz band in Japan (and the U.S.).
802.11k-2008	Specifies the use of transmit power control (TPC) in frequencies other than 5 GHz band, reporting of client statistics such as signal-to-noise ratio and frame transmissions, and channel statistics of channel management. General purpose is to provide radio resource measurements.
802.11n-2009	Defines modifications to the 802.11 physical and media access control layers that will allow for much higher throughputs and a maximum throughput of

	600 Mbps with 4 spatial streams in a 40 MHz channel. This is currently being accomplished with the use of MIMO (Multiple-Input/Multiple-Output) technology in conjunction with OFDM technology. Functions in both 2.4 GHz and 5 GHz bands.
802.11r-2008	Enhancements to the 802.11 MAC to improve Basic Service Set transitions within Extended Service Sets. Sometimes called the fast roaming amendment.
802.11s-2011	Specifies the interoperable formation and operation of an ESS Mesh network.
802.11u-2011	Provides amendments to the 802.11 PHY and MAC layers which enable InterWorking with other networks. May provide for handoffs between WiMAX and WLANs or between WLANs and cellular networks.
802.11v-2011	Enhancements to provide Wireless Network Management to the 802.11 MAC, and PHY, to extend prior work in radio measurement which results in a complete and coherent upper layer interface for managing 802.11 devices in wireless networks. Defines SNMP Management Information Bases that will allow for the configuration of a WLAN client device from a WLAN infrastructure device.
802.11w-2009	Improves security of 802.11 management frames like de-authentication frames. Provides data integrity, non-repudiation, confidentiality and replay protection to these management frames.
802.11-2007	A rollup project intended to roll the 802.11-1999 (R2003) base document and all its ratified amendments into a new expanded 802.11 standard base document. Also included are some specific definitions of behavior only hinted at in the original standard.
802.11-2012	A rollup project that includes all ratified amendments between 802.11-2007 and the time of the creation of 802.11-2012.

802.11ad-2012	A new specification for operations in 60 GHz called the Directional Multi-Gigabit (DMG) PHY.
802.11ae-2012	Provided prioritization of management frames.
802.11ac-2013	Provides a new PHY for the 5 GHz band providing gigabit and higher speeds. Operates only in the 5 GHz band.
802.11af-2013	A new PHY taking advantage of unused whitespace in low television frequencies. Named the Television Very High Throughput (TVHT) PHY.
802.11-2016	A rollup project that includes all ratified amendments between 802.11-2012 and the time of the creation of 802.11-2016.
802.11ah-2016	A sub-1 GHz PHY used for long-range, low-data-rate communications. May be popularized for IoT applications.
802.11ax	A PHY supporting 2.4 GHz, 5 GHz, and 6 GHz operations. High Efficiency (HE) PHY that implements OFDMA and other new features.

Table 1.5: 802.11 Amendments and Descriptions

In the real world, many vendors do leap onto amendments that are in the draft mode, as we have seen with 802.11n and 802.11ac. Figure 1.7 provides a flow diagram of the IEEE standards development process. As you can see in this figure, the process should take no more than four years, if the goal is accomplished. The standard must be reaffirmed or revised five years after publication, or it will be withdrawn.

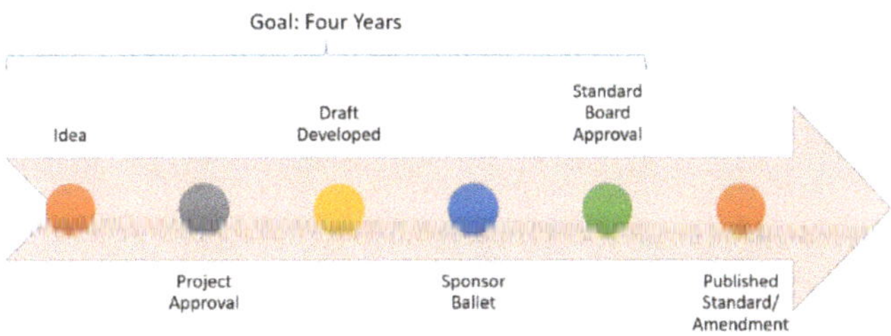

Figure 1.7: The IEEE Standards Development Process

IEEE Standards Impacting WLANs

In addition to the specific IEEE 802.11 standard and amendments, at least two other project 802 standards have a tremendous impact on IEEE 802.11. These two standards are IEEE 802.1X and 802.3-2015, Clause 33. Additional IEEE standards of importance to the CWNA are the 802.1D and 802.1Q standards.

IEEE 802.1X-2010

As Table 1.5 reveals, 802.1X is a referenced portion of the 802.11i amendment and, therefore, of 802.11-2016 wireless security. The 802.1X standard provides port-based authentication and control for your wireless networks in a similar way that it provides the same to wired networks. 802.1X is an important part of an 802.11 secure implementation and is an essential part of an enterprise-class implementation.

> You'll often see the ***X*** in ***802.1X*** typed as a lowercase x. It should be uppercase as shown in the 802.1X standards document. 802.1D and 802.1Q are also uppercase. If the document is a self-contained standard that operates on, or with, other standards, and not an amendment, it should use uppercase letters.

IEEE 802.3-2015, Clause 33

Commonly known as the IEEE 802.3af and 802.3at amendments, this is the standard that defines PoE (Power over Ethernet). Many wireless access points and bridges have support for PoE so that you can install them in locations where Ethernet cables exist, but power connections do not. This is particularly useful in closets, on towers and on rooftops and has become the most common way to power enterprise access points today. In fact, many access points ship with no wall outlet power supply provided and assume the use of PoE.

IEEE 802.1D-2004 and IEEE 802.1Q-2014

IEEE 802.1D is a standard that defines bridging and priority handling where IEEE 802.1Q focuses on priority tagging and VLAN handling for Quality of Service (QoS). 802.1D includes specifications for bridging, spanning tree protocol, and specifications for handling 802.11 MACs in the bridging process. The 802.1Q standard specifies the operation of bridges that support Virtual LANs (VLANs).

Wireless Network Types

Spread spectrum technology is used in multiple ways within modern organizations; however, these different ways can be organized within four primary categories: WLANs, Wireless PANs, Wireless MANs and Wireless WANs. Table 1.6 summarizes these uses and the following sections provide more detailed information.

Use	Examples	Range	Speeds
WLAN/Backhaul	IEEE 802.11	112 meters/375 feet to several miles	1 Mbps and higher
Wireless PAN	Bluetooth	1-3 meters	723 Kbps to 3 Mbps
Wireless MAN/Backhaul	WiMAX and EDGE	10 kilometers	Varies
Wireless WAN/Backhaul	AT&T microwave, Free Space Optics	Variable	Varies

Table 1.6: Spread Spectrum Uses and Examples

Wireless LANs

WLANs are the primary focus of the CWNA Certification. They provide mobility (moving around with active use during the move), nomadic ability (moving around without active use during the move) and unwired fixed connectivity (no movement). Mobility is provided because the user can move around within the coverage area of the access point, or even multiple access points, while still maintaining connectivity. Nomadic ability — the ability to move from place to place and use the network, although active communications do not take place while moving — is provided because you can power on a wireless client device from any location within a coverage area and use it for a temporary period of time as a fixed location device. It is a given that unwired fixed connectivity must exist if nomadic ability is provided.

Three primary roles exist, and WLANs play these roles in today's enterprise organizations:

- Access role
- Distribution role
- Core role

In the access role, the wireless network is used to provide wireless clients with access to wired resources. The access point remains fixed while the clients may move. The access point is usually connected to an Ethernet network where other resources, such as file servers, printers and remote network connections, reside. In this role, the access point provides access to the wireless medium first and then, when necessary, provides bridging to the wired medium or other wireless networks (such as in a mesh network implementation). Figure 1.8 illustrates the access role of a WLAN.

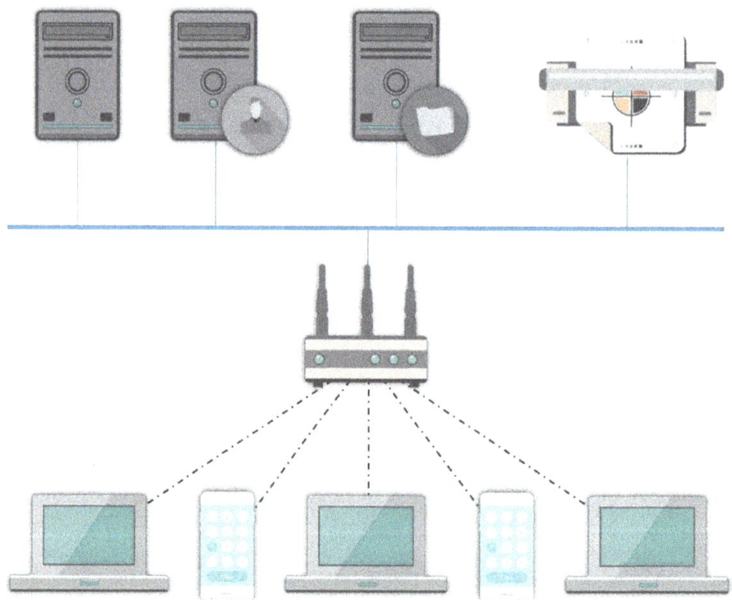

Figure 1.8: WLAN Access Role

In the distribution role, illustrated in Figure 1.9, wireless bridges provide a backhaul connection between disconnected wired networks. In this case, each network is connected to the Ethernet port of a wireless bridge and the wireless bridges communicate with each other using the 802.11 standard. Once these connections are made, network traffic can be passed across the bridge link so that the two previously disconnected networks may act as one.

The final role is the core role. In the core role, the WLAN is the network. This may be suitable for small networks built on-the-fly, such as those built at construction sites or in disaster areas; however, the limited data throughput will prohibit the WLAN from being the core of the network in a large enterprise installation. Future technologies may change this, but for now WLAN technologies play the access and distribution roles most often.

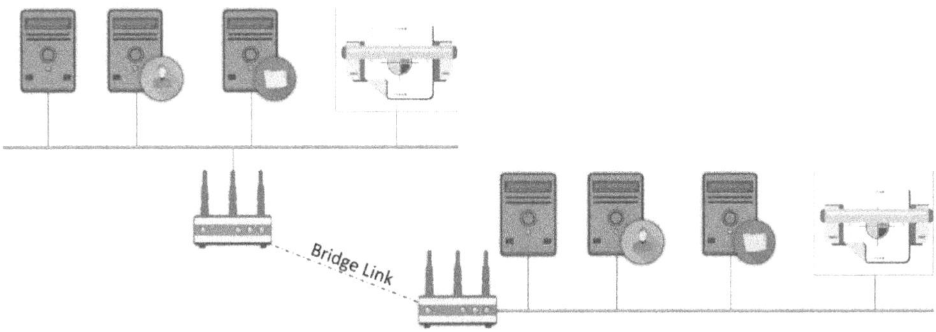

Figure 1.9: Wireless APs in a Distribution Role Creating a Bridge Link

Wireless PANs

A *WPAN* (Wireless Personal Area Network) provides hands-free connectivity and communications within a confined range and limited throughput capacity. They also provide for small-scale mesh-type wireless networks, like those implemented with ZigBee technology. In addition, RFID systems are frequently categorized as WPAN technologies because they have a short communications range. Bluetooth is also a good example of a WPAN technology that is both beneficial and in widespread use. Everything from Bluetooth mice to headsets and speakers are being used daily throughout the world.

Operating in the 2.4 GHz band, Bluetooth technologies can cause interference with WLAN technologies like DSSS (802.11-Prime), HR/DSSS (802.11b), ERP (802.11g), and HT (802.11n) when operating in the 2.4 GHz band. However, the newer adaptive frequency hopping technology helps to reduce this interference if not completely removing it. Adaptive frequency hopping is a newer feature found in Bluetooth 1.2 devices and higher.

Wireless MANs

WMANs (Wireless Metropolitan Area Networks) differ from WLANs and WPANs in that they are not usually implemented by the organization that wishes to use the network. Instead, they are generally implemented by a service provider and then access to the network is leased by each subscribing organization. However, unlike wireless Wide Area Networks, this does not have to be the case. For example, 802.16-compliant hardware could be purchased, and frequency licenses could be acquired to implement a private WMAN, but the expense is usually prohibitive.

WiMAX is the most commonly referenced WMAN technology. WiMAX is based on the IEEE 802.16 standard and provides expected throughput of approximately 130 Mbps in the latest specifications. In addition to the throughput speeds, WiMAX incorporates QoS mechanisms that help to provide greater throughput for all users and important applications using the network.

Wireless WANs

Wide Area Networks (*WANs*) are usually used to connect Local Area Networks (LANs) together. If the LANs are separated by a large distance, WAN technologies may be employed to connect them. These technologies include Frame Relay, analog dial-up lines, DSL, ISDN and others. What they have traditionally had in common is a physical wire connected to some device that is connected to some other device (usually across a leased line) that is eventually connected to the remote LAN. The wireless WAN (WWAN) is completely different because there is no wire needed from your local LAN to the backbone network or from the backbone network to your remote LAN. Wireless connections are made from each of your LANs to the backbone network.

Examples of WWAN technologies include Free Space Optics, licensed and unlicensed radio, and hybrids of the two. For WAN links that span hundreds of miles, you may need a service provider such as AT&T microwave. For shorter links of a few miles, you may be able to license frequency bands or use unlicensed technology to create the links. The key differentiator of WWAN technologies from WLAN, WPAN, and WMAN is that the WWAN link is aggregating multiple communication channels together (multiplexing) and passing them across the single WAN link.

Chapter Summary

In this chapter, you were introduced to the world of wireless from the perspectives of regulatory agencies, standards development organizations, and product certification organizations. You learned about the standards development process and some common use cases for wireless networking. In the next chapter, you will begin exploring the technical details of wireless network functionality as you learn about radio frequency characteristics and behaviors.

Points to Remember

Remember the following important points:

- The IEEE creates the standards used to allow interoperable communications among Wi-Fi devices.
- The 802.11-2016 rollup is the most recent complete version of the standard.
- Amendments modify the standard as they are ratified, giving rise to the phrase 802.11 as amended.
- The regulatory agencies control output power levels, frequencies available and areas of use, such as indoor and outdoor use.
- The most popular bands used by Wi-Fi networks are the 2.4 GHz and 5 GHz bands.
- The Wi-Fi Alliance created certification programs that verify device compatibility with the 802.11 standard under test conditions.
- When a device is 802.11b/g/n-certified, this indicates that it is a 2.4 GHz-only device.
- When a device is 802.11a/b/g/n/ac-certified, you must evaluate the certificate for the device to determine if it supports one band at a time or is dual-band concurrent.
- The IETF also created standards or recommendations that are useful in WLAN deployments, including RADIUS, EAP and CAPWAP.
- Access points used in access mode are used by clients to gain access to the networks to which the access points are connected.
- Access points used in distribution mode are used as bridges to interconnect multiple networks.

Review Questions

1. Which one of the following organizations performs compatibility testing of 802.11 hardware?
   a. IETF
   b. Wi-Fi Alliance
   c. FCC
   d. MIC

2. You must locate information within the 802.11 standard to understand the details of Beacon frame options. What organization provides a website that you can use to gain access to the 802.11 standard?
   a. IEEE
   b. FCC
   c. ITU-R
   d. Wi-Fi Alliance

3. Your organization has decided to install a WLAN in their warehouse facility. The warehouse is approximately 400 meters from the main building and it has no Ethernet or other wired network connection run between the main building and the warehouse. Which one of the following wireless technologies will most likely be used to provide a link between the main building and the warehouse?
   a. WPAN
   b. WMAN
   c. WWAN
   d. WLAN

4. What IEEE standard specified a port-based authentication solution?
   a. 802.3
   b. 802.1X
   c. 802.1Q
   d. 802.1D

5. What frequency bands are most commonly used by WLAN hardware?
   a. 2.4 GHz and 5 GHz
   b. 2.4 GHz and 100 GHz
   c. 5 GHz and 10 MHz
   d. None of these

6. What is indicated when a device is certified by the Wi-Fi Alliance as 802.11b/g/n-certified?
   a. It is a 5 GHz-only device
   b. It is a 2.4 GHz-only device
   c. It supports WPA2 Enterprise

d. It supports WPA Personal

7. What is based on an IETF document and defines communications between a lightweight access point and a controller?
   a. CAPWAP
   b. 802.1Q
   c. EAP
   d. RADIUS

8. What Wi-Fi Alliance certification ensures proper VoIP communications even when moving from one access point to another?
   a. WMM
   b. Voice-Personal
   c. Voice-Enterprise
   d. WPS

9. You have purchased an AP for a specific regulatory domain. You know that you cannot turn up the transmit power beyond a specific level. What is the likely cause of this behavior?
   a. Firmware problems
   b. Radio failure
   c. Regulatory constraints
   d. IEEE constraints

10. In what mode is an access point operating when clients connect to it in order to gain access to a printer?
   a. Distribution
   b. Core
   c. Access
   d. None of these

Review Answers

1. **B is correct.** The Wi-Fi Alliance performs compatibility testing against 802.11 wireless devices based on 802.11 standards or subsets of those standards.

2. **A is correct.** The IEEE website can be used to access 802.11 standard documents. They are free after a period of six months passes from their ratification.

3. **D is correct.** WLAN bridges will most likely be used in this scenario as it is a shorter distance link than required by WWAN solutions.

4. **B is correct.** The 802.1X standard specified port-based authentication and is used in enterprise WLAN security solutions.

5. **A is correct.** While the sub-1 GHz, TV whitespace, and 60 GHz bands are used, by far the most commonly utilized bands are 2.4 GHz and 5 Ghz.

6. **B is correct.** When the certificate indicates only 802.11b/g/n, it indicates 2.4 GHz-only support as these amendments are all operable in 2.4 GHz and, if 5 GHz were supported, 802.11a would be listed as well.

7. **A is correct.** CAPWAP and LWAPP are defined in IETF RFCs. They both specify methods of communications between an access point and a controller.

8. **C is correct.** Voice-Enterprise is a certification that ensures proper communications for VoIP across a WLAN, even when roaming between access points. Voice-Personal only ensures such communications when connected to a single AP.

9. **C is correct.** One area of constraint imposed by regulatory agencies is output power. When an AP ships for a specific regulatory domain, those constraints are enforced.

10. **C is correct.** When an access point is in access mode, it is providing access to the resources to which it is connected for clients connected to the access point.

Chapter 2 — RF Characteristics and Behaviors

In the beginning, there was electromagnetic radiation, at least as far as we know. Electromagnetic radiation has been with us as long as we've been on this planet and it is a recent development, in the grand scheme of history, to harness and use this phenomenon for short and long-distance communications. In this chapter, you will learn about electromagnetic waves (at a basic level), and specific electromagnetic waves, known as Radio Frequency (RF) waves, which are used in wireless communications today. In the process, you will explore the objective 1.1 of the CWNA-108 exam. Given that light waves (a portion of the electromagnetic spectrum) travel at approximately 300,000,000 meters per second and you are likely less than one meter from this page, we should get started.

Electromagnetic Waves

You will not have to know a great deal about the physics behind electromagnetic waves to pass the CWNA exam, or to implement enterprise-class wireless networks. I do, however, hope this overview gives you a desire to learn more. This summary of the fundamentals will also help you to better understand the RF Behavior section later in this chapter.

Waves

The first thing we must define is a wave. A *wave,* in the realm of physics, can be defined as a motion traveling through matter or space. Notice that the wave is not necessarily a movement of matter, but it is a motion — such as oscillation — traveling through matter or space. Think of the waves in the ocean bobbing up and down. Now imagine a ball placed on top of the waves. The waves pass by and the ball bobs up and down as they pass by, but the ball does not move with the waves. If you were to investigate even more closely, you would see that the water does not travel with the waves either, but the waves pass through the matter (water).

An electromagnetic wave is an oscillation traveling through space. In the early days of electromagnetic wave study, some thought an invisible medium existed through which the waves traveled. This invisible medium was called the Ether. You may recognize this term as it is used today, in the word Ethernet, paying homage to this earlier thinking. In fact, electromagnetic waves can travel in a vacuum where all matter has been removed and, because of this, we theorize that they need no material medium to travel from one place to another. How then do they propagate through space? It is through an interesting relationship between electric and magnetic fields.

Electric Fields

An *electric field* is the distribution in space of the strength and direction of forces that would be exerted on an electric charge at any point in that space, according to the American Heritage Science Dictionary. Stated more simply, the electric field is the space within which an electrically charged object will feel a pull or a push, depending on whether the charge is unlike (pull) or like (push) that of the pulling or pushing source — yes, that was as simple as it gets when talking physics. Positively charged objects attract negatively charged objects, and negatively charged objects attract positively charged objects. The attraction is greater when the objects are closer together, and lesser when they are farther apart. The electric field represents the space within which this attraction can be detected, although, theoretically, the attraction extends infinitely; though it cannot be measured by today's measuring tools, we'll see what the future holds.

Electric fields result from other electric charges, or from changing magnetic fields. Electric field strength is a measurement of the strength of an electric field at a given point in space and is equal to the force induced on a unit of electric charge at that point.

Magnetic Fields

A *magnetic field* is a force produced by a moving electric charge that exists around a magnet or in free space. Magnetic fields extend out from the attracting center. The space in which it can affect objects is considered the extent of the magnetic field. A changing magnetic field generates an electric field and this is important for radio frequency propagation.

> ***Radio frequency propagation*** is a phrase used to describe how RF waves move through space. They move in the direction of propagation, as electric fields give rise to magnetic fields and magnetic fields give rise to electric fields. They also respond to the environment in reflections, refractions, scattering, diffraction, and absorption.

Electromagnetic Waves

Now that you have definitions of electric fields and magnetic fields, you are ready to investigate electromagnetic waves. An *electromagnetic wave* is a propagating combination of electric and magnetic fields. Remember that a magnetic field can generate an electric field and an electric field can generate a magnetic field. While the analogy is not perfect, consider that a chicken creates an egg that creates a chicken that creates an egg, ad infinitum. The alternating current (AC) in the antenna generates a magnetic field around the antenna that generates an electric field that generates a magnetic field into space, ad infinitum.

The electric and magnetic fields are oscillating perpendicular to each other and they are both perpendicular to the direction of propagation as is shown in Figure 2.1. You can see that the electric field is parallel to the generating wire (antenna) and the magnetic field is perpendicular to the generating wire. The wave is traveling out from the generating wire.

> Deeper physics would address flux lines of the electric and magnetic fields, but this knowledge is beyond that required for the CWNA and is not covered here. However, if you would like to learn more, the book "Physics Demystified" is an excellent beginning resource.

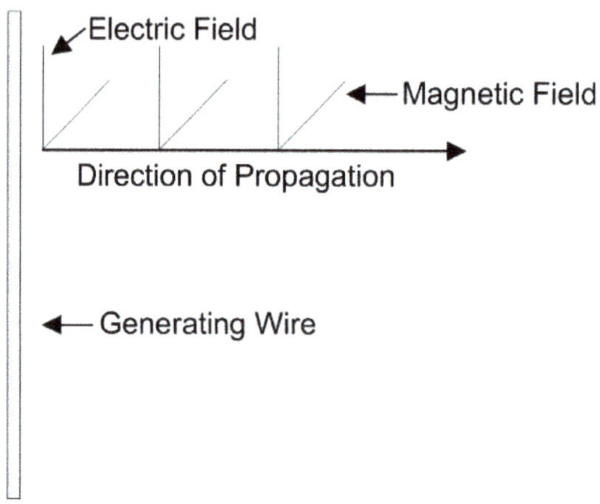

Figure 2.1: Electromagnetic Wave Propagation Out from an Antenna Wire

A very specific form (wavelength and frequency) of these electromagnetic waves is used to communicate wirelessly in 802.11 networks. This form of wave is a radio frequency wave and is often shortened to RF. An RF-based system, then, is a system that relies on the phenomenon of electromagnetic wave theory to provide data and voice communications wirelessly.

The sine wave is a useful tool in the learning process of wave theory. Figure 2.2 shows an example sine wave. Such a wave can be manipulated to represent data, and in wireless networks, this is what we call modulation. Representative sine waves will be used throughout this chapter to help you understand basic RF characteristics.

RF Characteristics and Behaviors

RF waves have specific characteristics and behaviors that must be understood by network administrators to effectively implement and troubleshoot WLANs. The characteristics define the properties of the RF waves. The behaviors define the way in which these RF waves propagate throughout an area.

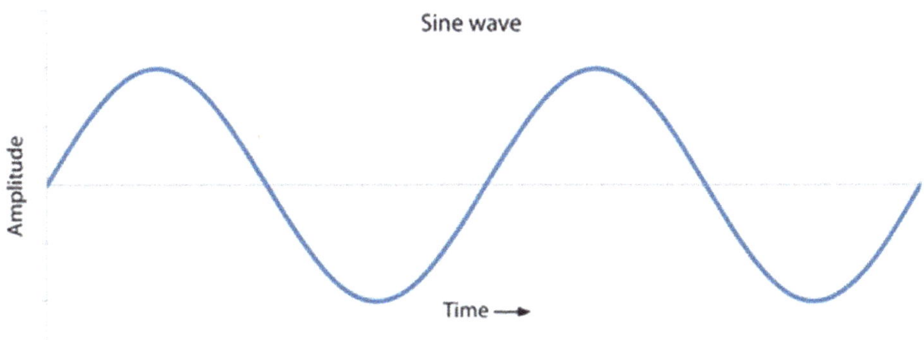

Figure 2.2: The Standard Sine Wave

RF Characteristics

All RF waves have characteristics or properties that vary to define the wave. Some of these properties can be manipulated or modified to allow modulation of information onto the wave. The most important characteristics for the CWNA are wavelength, frequency, amplitude and phase. These four properties are also very important when considering antennas for your network installation because antennas can be designed to work best with specific frequencies. This section explores and explains these characteristics.

Wavelength

The *wavelength* of an RF wave is calculated as the distance between two adjacent identical points on the wave. Figure 2.3 shows a standard sine wave. Point A and Point B mark two identical points on the wave, and the distance between them is defined as the wavelength. Notice the other marker also defines the wavelength. It is, again, simply the distance between two recurring points in the wave. The wavelength is frequently measured as the distance from one crest of the wave to the next.

The wavelength is an important factor in wireless networking. The wavelength dictates the optimum size of the receiving antenna, and it determines how the RF wave will interact with its environment. An RF wave will react differently when it strikes an object that is large in comparison to the wavelength than when it strikes an object that is small in comparison to the wavelength (reflection versus scattering, respectively).

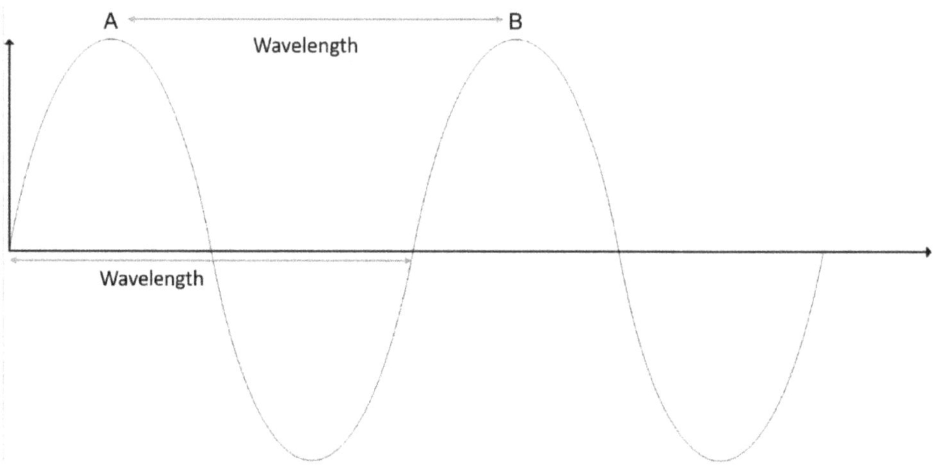

Figure 2.3: Wavelength Measurement Examples

Frequency is covered in more detail later in this section, but it is important that you understand that the wavelength and the frequency are interrelated. For a given medium, if you know the wavelength you can calculate the frequency, and if you know the frequency you can calculate the wavelength.

The wavelength is related to the frequency and the speed of light. If you know the frequency, you can calculate the wavelength. If you know the wavelength, you can calculate the frequency. This is because the speed of the wave is constant (roughly the speed of light).

One of the great discoveries in the history of electromagnetism is that electromagnetic waves travel at the speed of light. Since we know the speed of light to be 299,792,458 meters per second (or the simple 300,000,00 meters per second, if you prefer rounding), we also know that this is the speed at which electromagnetic waves travel in a vacuum. This phenomenon was theorized by James Clerk Maxwell and proved through experimentation by Heinrich Hertz.

You have probably heard of measurements such as 100 megahertz (MHz) and 3.6 gigahertz (GHz). These measurements refer to the number of cycles that a wave generates per second. When we say that the access point is using the 2.4 GHz spectrum, we are saying it is using the spectrum that uses a wave cycle rate of 2,400,000,000 (yes, that's 2 billion, 400 million) times per second. This measurement is named for Heinrich Hertz and his research in electricity and magnetism. A kilohertz (kHz) is 1,000 hertz or cycles per second. A megahertz (MHz) is 1,000,000 hertz and a gigahertz (GHz) is 1,000,000,000 hertz.

Because we know that RF waves travel at the speed of light, we can calculate the frequency when we know the wavelength or the wavelength when we know the frequency. The following formula can be used to calculate the wavelength (λ) in meters when the frequency is known.

$$\lambda = 299{,}792{,}458 / f$$

Where λ is the wavelength in meters and f is the frequency in hertz and the medium is a vacuum. Therefore, the 2.45 GHz frequency would have a wavelength that is calculated with the following formula:

$$\lambda = 299{,}792{,}458 / 2{,}450{,}000{,}000$$

The result is .123 meters or approximately 12.3 centimeters in length. This translates to about 4.8 inches. To calculate inches from centimeters, just multiple the number of centimeters times 0.3937. The formal character used to represent a wavelength is the Greek lambda (λ), and the symbol for the speed of light is c. Therefore, the formal representation of the previous formula would be:

$$\lambda = c / f$$

The calculation for frequency is just the opposite. You will divide the speed of light by the wavelength in meters to discover the frequency. Keep in mind that the numbers we've been using have been rounded and that impacts the results of the following formula; however, the results are close enough to recognize that a wavelength of .123 meters would indicate an RF wave in the 2.45 GHz frequency:

$$f = 299{,}792{,}458 / .123$$

$$f = 2437337056.91$$

Due to the complex measurement number that is the speed of light, this number is often rounded to 300 million meters per second. While this will change formula results, the findings are close enough for understanding the behavior of RF waves; however, engineers developing RF systems must use more precise measurements. Wireless network administrators need not be as precise as they are implementing gear (access points, clients, antennas, etc.) created by precise engineers. Additionally, formulas like the following simplify matters:

$$\text{wavelength in inches } (\lambda) = 11.811 / f \text{ (in GHz)}$$

$$\text{wavelength in centimeters } (\lambda) = 30 / f \text{ (in GHz)}$$

Because wireless networks use such high frequency ranges, formulas like this make the calculations easier. Exercise 2.1 steps you through the calculation of wavelengths for different frequencies in both the 2.4 GHz and 5 GHz bands.

While I provide formulas like these for wavelength calculations, for your reference and use as a WLAN administrator, you will not see these formulas on the CWNA exam. However, my goal is to help you fully understand wireless networking as you journey toward your CWNA and further CWNP certifications. For this reason, to make sure you understand the concepts, I will frequently go deeper than the exam requires. I will also point out these areas to you so that you will not have to spend time memorizing facts that you can reference in this book as you go about your administration tasks. At the same time, working with calculations against wavelengths, does help you to remember the importance of the wavelength on antenna design and propagation behaviors.

Exercise 2.1: Calculating Wavelengths

In this exercise, you will calculate the wavelengths in inches and centimeters using Microsoft Excel. You will first calculate the wavelength of a wave on the 2.4 GHz channel 1 center frequency of 2.412 GHz and then on the 5 GHz channel 157 center frequency of 5.785 GHz. The purpose of this exercise is to enforce the fact that wavelengths are related to frequency and the fixed wave speed.

1. Open Microsoft Excel and create a new Worksheet or Workbook.
2. Format the worksheet like that shown in Graphic 2.1. The values for the Variable for Centimeters and Variable for inches are very important and should be 30 and 11.811 respectively.

	A	B	C	D	E
1	Frequency in GHz	Variable for Centimeters	Variable for Inches	Wavelength (cm)	Wavelength (in)
2	2.412	30	11.811		

Graphic 2.1

3. In the cell D2, enter the formula: =B2/A2
4. In the cell E2, enter the formula: =C2/A2
5. You should see results like those in Graphic 2.2.

	A	B	C	D	E
1	Frequency in GHz	Variable for Centimeters	Variable for Inches	Wavelength (cm)	Wavelength (in)
2	2.412	30	11.811	12.43781095	4.896766169

Graphic 2.2

6. Now, enter the values 5.785, 30, and 11.811 in cells A3, B3, and C3 respectively.
7. In cell D3, enter the formula: =B3/A3
8. In cell E3, enter the formula: =C3/A3

9. Note the results and notice that the 5 GHz wavelength is significantly shorter than the 2.4 GHz wavelength. This fact has an impact on receptivity of 5 GHz signals at greater distances. Graphic 2.3 shows the final results.

	A	B	C	D	E
1	Frequency in GHz	Variable for Centimeters	Variable for Inches	Wavelength (cm)	Wavelength (in)
2	2.412	30	11.811	12.43781095	4.896766169
3	5.785	30	11.811	5.185825411	2.041659464

Graphic 2.3

Frequency

Frequency refers to the number of wave cycles that occur in a given window of time. Usually measured in one-second intervals, a frequency of 1 kilohertz (KHz) would represent 1,000 cycles of the wave in one second. To remember this, just keep in mind that a wave cycles frequently and just how frequently it cycles determines its frequency.

Because all electromagnetic waves, including radio waves, move at the speed of light, the frequency is related to the wavelength. We observe that wavelength, frequency and the medium are interdependent. Higher frequencies have shorter wavelengths and lower frequencies have longer wavelengths. This fact means that the 802.11n networks using the 2.4 GHz band have longer wavelengths than 802.11ac networks using the 5 GHz bands, as you learned in the preceding wavelength section. Figure 2.4 shows two waves with differing frequencies.

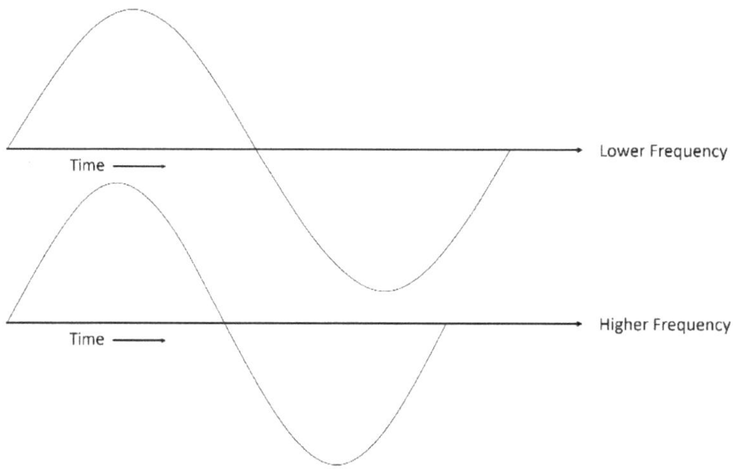

Figure 2.4: Frequency Illustrated

The concept of frequency is used in sound engineering, as well as RF engineering. Figure 2.5 shows a piano keyboard and the sound frequencies to which the keys are traditionally tuned. Knowing this can help establish your understanding of frequencies in RF communications; however, you must be clear in your thinking about the differences between sound waves and electromagnetic waves. The two wave types are not the same phenomenon, but share similar characteristics, such as amplitude, frequency and wavelength. As most people are already somewhat familiar with the behavior of sound waves through life experience, they work as a good analogy and a starting point for your understanding of electromagnetic wave frequencies, amplitude and wavelengths.

With sound waves, the proper string that is tightened to the appropriate tension will emit a sound of the desired frequency. Sound waves travel much more slowly than electromagnetic waves — at a rate of approximately 344 meters per second or 1,100 feet per second through the air. If you are standing 100 feet (30.5 meters) from the source of the sound, it will take that sound approximately 1/10 of a second to reach you.

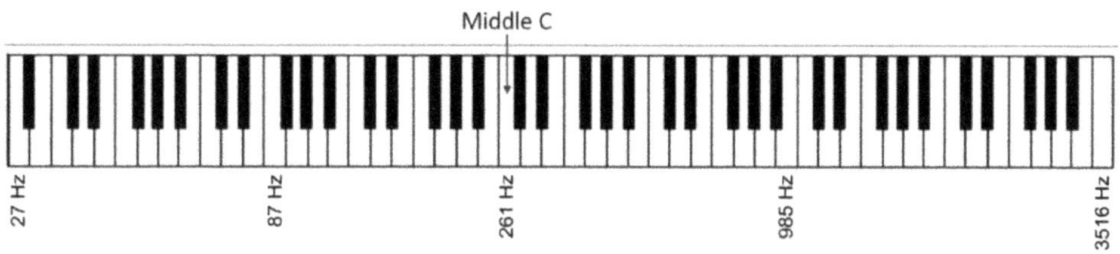

Figure 2.5: Sound Frequencies on a Piano Keyboard

Looking again at the piano keyboard in Figure 2.5, you can see that Middle C has a frequency of 261 Hz. From this, the wavelength can be calculated by dividing 344 meters by 261 Hz for a wavelength of 1.32 meters or 4.33 feet. In effect, we are saying that 261 waves are generated in a second, and in any given second each existing wave travels 344 meters. Now, it is important to note that the lower frequencies still travel at a rate of 344 meters per second as well; however, fewer — though longer — waves are in each second. Lower frequency sound waves can be perceived at a greater distance due to the working of the human ear. To show that the other sound waves still exist at the greater distance, you can use an amplifier like that commonly seen along the sidelines at American football games. This device has a larger "receive" space than the human ear so it is able to "pick up" sound waves that would otherwise be missed. RF waves are similar in that the lower frequency waves are easier to pick up at greater distances due to constraints in antenna engineering that make it more difficult to pick up higher frequencies at the same range.

The impact of frequency usage on WLANs is tremendous. By using different frequencies, you can enable distinct connections or RF links in a given coverage area or cell. For example, an 802.11n network using channel 1 can exist in the same coverage area as an 802.11n network using channel 11. Because these channels use different frequencies sufficiently spaced apart, they do not cancel or interfere with each other. To be clear, you can place two 802.11n networks in the same coverage area on the same channel, but you increase something called co-channel interference (CCI) and greatly reduce the performance of both networks. But you will learn more about CCI in later chapters.

Think of it like a beautiful orchestra. Many instruments are playing on many different frequencies, but together they make wonderful music. Now, consider the sound you get when you walk up to a piano and press the palm of your hand down on 6 or 7 keys simultaneously. Few people call that pleasant music. The sound frequencies are so close together that they just sound like noise. In a similar way, overlapping radio frequency waves will be very difficult, if not impossible, to distinguish from one another. However, consider the melodious sound of the C-major chord or the D-minor chord. In the same way, multiple 802.11n networks can work side by side when they are configured to channels 1, 6 and 11 in a cell.

In some wireless systems, the frequency is used for modulation. That is, by altering the frequency, we can represent data. This modulation type is called Frequency Shift Keying (FSK). It is not used in WLANs, but it is a modulation type available in other systems.

Modulation is the process of manipulating a carrier so that it represents meaningful information. For example, you can manipulate sound waves so that data can be transmitted using sound. This function is effectively what telephone modems do. They modulate/demodulate (modem) digital data into and from analog audio data. Wireless networks use modulation that manipulates RF waves to represent meaningful information.

Amplitude

With the explanation in the previous section, you might be tempted to think that the volume of sound waves is dependent on the frequency since lower frequency waves are heard at a greater distance; however, there is actually another characteristic of sound waves that impacts the volume. Remember, at greater distances shorter wavelength waves are more difficult to detect, as the waveform spreads ever wider (though this may be more a factor of the receiving device used than of the waveform itself). The characteristic that defines the volume is known as *amplitude*. In sound wave engineering, an increase in amplitude is equivalent to an increase in volume, hence, an amplifier adds to the volume or makes the sound louder. While the frequency affects the distance at which a sound wave can be effectively received, the amplitude affects the volume of the sound wave at that distance. This behavior is why you can turn up the volume (increase the amplitude) of a sound system and hear it at a greater distance. RF waves are similar.

An RF wave with greater amplitude is easier to detect than an RF wave with lesser amplitude, assuming all other factors are equal. In a vacuum, an RF wave will be said to have better quality at a distance if it has greater amplitude. Realize, that RF waves travel, theoretically, forever. This reality being the case, the detectability of the wave is greater at a specific distance when the wave starts with a greater amplitude. A wave with a lesser amplitude may not be detectable due to the noise floor at the point of reception, or other interfering signals present at the time of reception.

Noise floor is a term used to define the continual RF noise in an area. For 2.4 GHz, in a given area, it will likely be higher than for 5 GHz due to the multitude of devices that generate signals and noise in the 2.4 GHz band. A high noise floor impedes the ability to achieve higher data rates in WLANs.

There is a point in space where an RF wave still exists, but it cannot be distinguished from the electromagnetic noise in the environment and is no longer useful. Two waves having been modulated to carry data (now called signals) of differing amplitudes can arrive at the same receiver area, one can be processed properly while the other cannot. In effect, both the high amplitude and low amplitude waves exist at that point, but only the high amplitude wave can be detected and processed. Both waves have traveled the distance, but only the high amplitude wave is useful. For this reason, in common usage, engineers often say that an increase in amplitude will extend the range of the RF wave. What is meant by this is that the RF wave's useful range has been extended. However, a CWNA should not rely on increased output power in the access points alone to accomplish improved received signal capabilities. If the clients cannot match this output power, they may not be able to transmit a successful signal back to the access point. Using higher gain antennas or more access points is usually the better solution.

Figure 2.6 shows an RF signal with original, increased and decreased amplitudes.

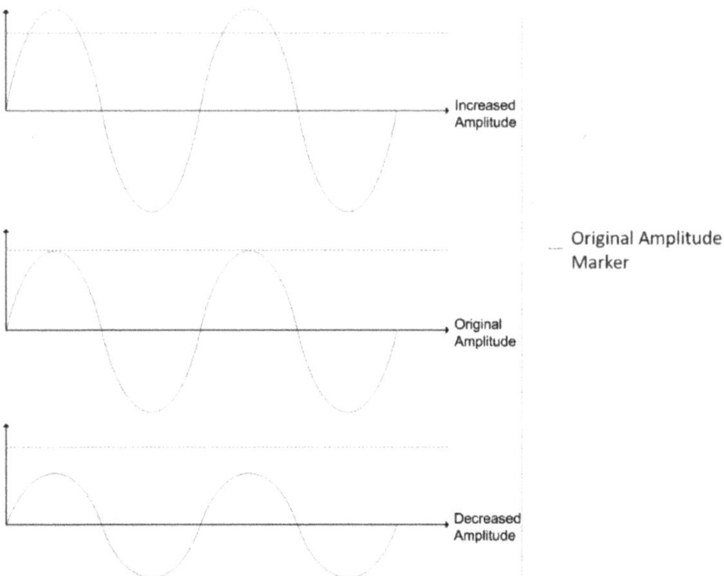

Figure 2.6: RF Waves at Varying Amplitudes

Like frequency, amplitude can be used in modulation techniques. Changing or shifting the amplitude can represent information. This modulation technique is known as Amplitude Shift Keying (ASK). ASK, unlike FSK, is used in WLAN implementations. For example, Quadrature Amplitude Modulation (QAM), as its name implies, uses ASK as well as other modulation techniques.

Phase

Unlike wavelength, frequency and amplitude, *phase* is not a characteristic of a single RF wave but is instead a comparison between two RF waves. If two copies of the same RF wave arrive at a receiving antenna at the same time, their phase state will impact how the composite wave is able to be used. When the waves are in phase, they strengthen each other and when the waves are out of phase, they sometimes strengthen and sometimes cancel each other. In specific out-of-phase cases, they only cancel each other, such as when they are 180 degrees out of phase.

Phase is measured in degrees, though real-world CWNA tasks usually benefits only from the knowledge of whether the waves are in phase or out of phase. Two waves that are completely out of phase would be 180 degrees out of phase, while two waves that are completely in phase would be 0 degrees out of phase. Figure 2.7 shows a main wave signal, another in-phase signal and an out-of-phase signal.

> Two arriving copies of a wave that happen to arrive in-phase will increase the strength of the received signal; however, they will never make the signal stronger than the transmitted signal. Signals weaken as they travel through space and the energy is spread over an ever-increasing area. However, the two copies can make the received signal stronger than it would have been otherwise.

When troubleshooting wireless networks, the phase of duplicate RF signals is mostly an implication of reflection or scattering in an area that may cause dead zones due to the out-of-phase signals. However, in an interesting twist on wireless communications, Multiple-Input/Multiple-Output (MIMO) systems (like 802.11n and 802.11ac) can take advantage of the multiple paths a signal may travel, and slightly different antenna locations in the transmitter and receiver to process multiple concurrent streams of modulated data.

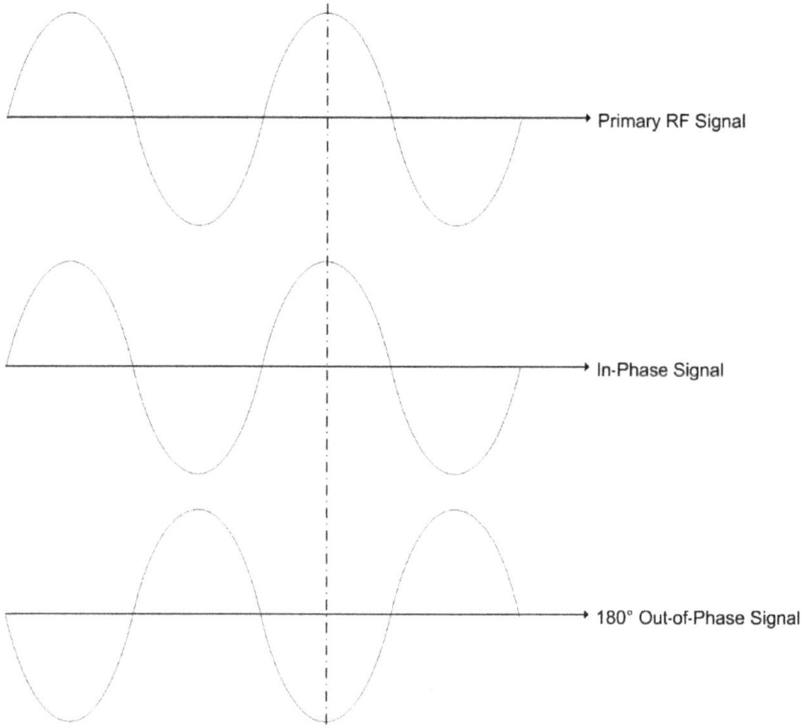

Figure 2.7: RF Wave Phases

Like amplitude, phase shifting can be used as a modulation technique. In fact, 802.11 networks use Phase Shift Keying (PSK) as well as ASK for modulation.

RF Behavior

RF waves that have been modulated to contain information are called RF signals. These RF signals have behaviors that can be predicted and detected. They become stronger and they become weaker. They react to different materials in varying ways and they can interfere with other signals. The following sections introduce you to the major RF signal behaviors and their implications including:

- Gain
- Loss
- Reflection
- Refraction
- Diffraction

- Scattering
- Absorption
- VSWR
- Return Loss
- Amplification and Attenuation
- Wave Propagation
- Free Space Path Loss
- Delay Spread

Gain

Gain is defined as the positive amplitude difference between two RF wave signals (I'll just call these RF signals from now on). Amplification is an active process used to increase an RF signal's amplitude and, therefore, results in gain. Two basic types of gain exist: active and passive. Both types can be intentional and passive gain can also be unintentional. Figure 2.8 shows an example of a signal that demonstrates both gain and loss.

Active Gain

Active gain is achieved by placing an amplifier in-line between the RF signal generator (such as an access point) and the propagating antenna. These amplifiers usually specify the gain they provide in dB (covered as a metric in the next chapter). For example, an amplifier may provide 6 dB of gain to the incoming RF signal. To determine the actual power of the signal after passing through the amplifier, you will have to know the original power of the signal from the RF generator, and then process the appropriate RF math algorithms. RF math is covered in the next chapter.

When using any type of intentional gain, you must be careful not to exceed the legal output constraints within your regulatory domain. For example, the FCC in the United States limits the output power at the intentional radiator to 1 watt and at the antenna to 4 watts, for point-to-multipoint applications in the unlicensed 2.4 GHz band. An access point with clients is an example of a point-to-multipoint application.

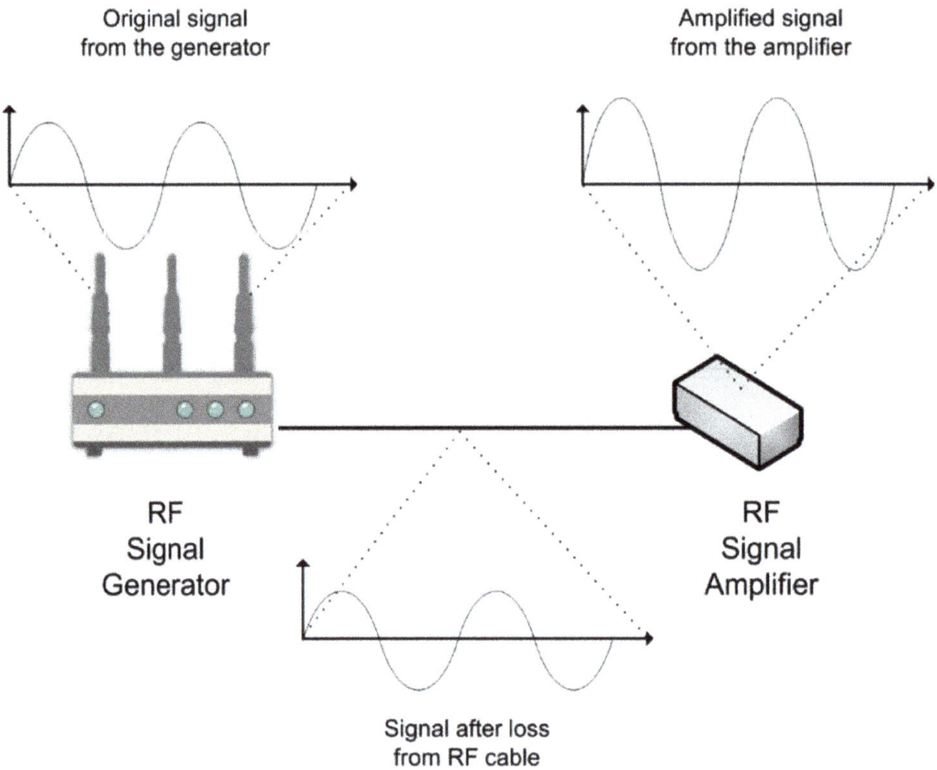

Figure 2.8: RF Signal Gains and Losses in a System

> While the concept of the intentional radiator is covered in greater depth in Chapter 3, it is mentioned periodically throughout the book. For now, consider the following definition: The ***intentional radiator*** is the point in the radio system where the system is connected to the antenna. Restrictions exist on the output power at the intentional radiator and restrictions exist on the output power of the antenna after passive gain.

Passive Gain

Passive gain is not an actual increase in the amplitude of the signal delivered to the intentional radiator, but it is an increase in the amplitude of the signal, in a favored direction, by focusing or directing the output power. Passive gain can be either intentional or unintentional.

Intentional passive gain is like cupping your hands around your mouth as you yell to someone at a distance. You are directing the sound waves, intentionally, toward that targeted location. You are not necessarily increasing your volume or your ability to yell louder. If you yell at your loudest without cupped hands, it will not be as detectable at a greater distance as it would with cupped hands. This is intentional passive gain. To experience this, read this paragraph out loud. As you are reading, cup your hands around your mouth and notice how the sound changes (becomes muffled and seems to change tonality). This is because more of the sound waves are traveling out from you, and your ears detect the difference. Of course, anyone else in the room with you can tell a difference as well, and they might even think you're a little strange — so make sure you are alone when you perform this test.

Antennas are used to provide intentional passive gain in wireless networks using RF signals. The antenna propagates more of the RF signal's energy in a desired direction than in other directions. The RF signal is said to have gain in that direction.

Antennas provide passive gain. They can in no way increase the total power, but they can focus the power in a desired direction. Amplifiers provide active gain and they may be included in the access point or in line between the access point and the antenna.

Unintentional passive gain happens because of reflection and scattering (defined in detail later in this chapter) in a coverage area. When the RF signal leaves the transmitting antenna, the primary signal travels out from the antenna according to the propagation patterns for which the antenna is designed. However, this signal may encounter objects that cause reflection and scattering, resulting in multiple copies of the same signal arriving at the receiving antenna. If these signals arrive in phase, they can cause the signal strength to actually increase, and this would be a form of unintentional passive gain.

> Some RF engineers doubt that RF energy, once scattered, is ever joined with other signal paths to produce passive gain of any measurable value. This concept of unintentional passive gain is not universally supported among theoreticians (smart people who think about how things might work).

Loss

Loss is defined as the negative amplitude difference between two RF signals. Like gain, loss can be either intentional or unintentional (referenced as natural in this section).

Intentional Loss

Due to FCC regulations and the regulations of other regulatory domains, you will have to ensure that the output power of your wireless devices are within specified constraints. Depending on the radios, amplifiers, cables and antennas you're using, you may have to intentionally cause loss in the RF signal. This action indicates that you're reducing the RF signal amplitude. Attenuators are used to cause intentional loss in an RF signal and are placed in line between the access point or bridge and the antenna.

Attenuators may also be used to comply with design specifications. For example, if the design specification requires 10 mW of output power from the AP, but the AP supports only a low output power setting of 20 mW, a 3-dB attenuator could be used between the AP and external antennas to meet design specification. This action is not frequently taken but may be available in some scenarios.

Natural Loss

In addition to the intentional loss that is imposed on an RF signal to comply with regulatory demands or design requirements, natural or unintentional losses can occur. The natural kind of loss happens because of the normal process of RF propagation, which involves spreading, reflection, refraction, scattering, diffraction and absorption.

Reflection

When an RF signal bounces off a smooth non-absorptive surface, changing the direction of the signal it is said to reflect, and the process is known as *reflection.* This is probably the easiest RF behavior to understand simply because we see it frequently in our everyday lives. You can shine a light on a mirror at an angle and see that it reflects off that mirror in relation to the angle. In fact, when you look in the mirror, you are experiencing the concept of electromagnetic reflection, which is the same as RF reflection.

Figure 2.9 illustrates this concept. As you can see, the light waves, which are electromagnetic waves are similar to RF signals first reflecting off the object and travel toward the mirror. Next, the light waves reflect off the mirror and travel toward your eye. Finally, your eye acts as a focusing device and brings the light waves together at the back of the eye, giving you the sense of sight. However, the important thing to note is that what you are "seeing" is the light reflected off the object onto the mirror, and off the mirror into your eyes. The ability to see objects all around us is driven by the reflective properties of the materials and the light waves striking against them.

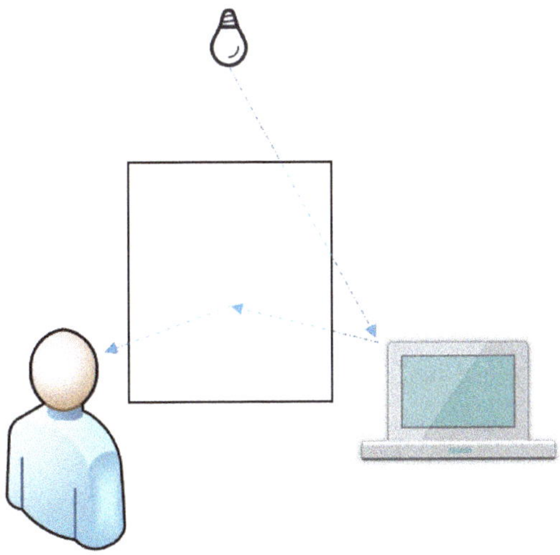

Figure 2.9: Reflection Illustrated with a Mirror

RF signals also reflect off objects that are smooth and larger than the waves that carry the signals. Earlier, it was noted that the wavelength impacts the behavior of the RF wave as it propagates through space. This is the first example of the relationship of the wavelength and the space through which the wave travels. If the space were empty, there would be no reflection, but since all space we operate in (earth and its atmosphere) contains some elements of matter, reflection, refraction, scattering, diffraction and absorption are expected.

Because the object that causes reflection will normally be smooth and larger than the wavelength, and because waves used by 802.11-compliant radios are roughly between 5 and 13 centimeters, it follows that the objects will be greater than 5 centimeters in size (for 5 GHz bands) or 13 centimeters in size (for the 2.4 GHz band) and smooth. Such objects include metal roofs, metal or aluminum wall coverings, elevators and other larger smooth objects. Figure 2.10 shows the traditional diagram of RF signal reflection. It is important to remember that reflected signals are usually weaker after reflection. This weakening is because some of the RF energy is typically absorbed by the reflecting material.

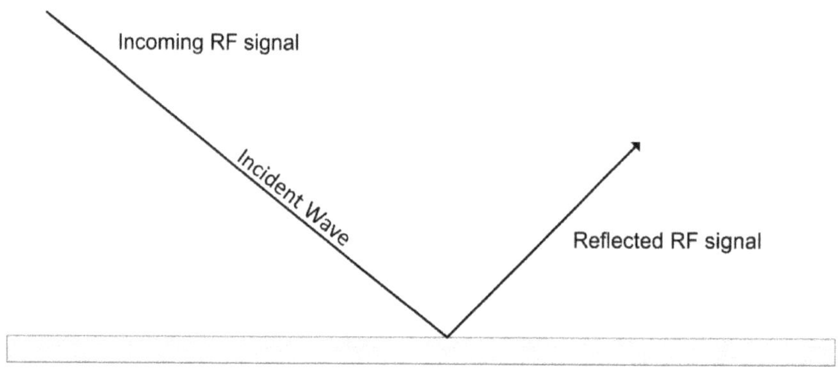

Figure 2.10: RF Reflection

An RF phenomenon that results from reflection (and scattering) is multipath. Multipath occurs when signals traverse different paths to the same receiver. In single-input/single-output (SISO) systems, this was commonly a problem. 802.11a/b/g are all SISO systems. To address it, antenna diversity was used. Antenna diversity involves using more than one antenna connected to a single radio receiver. The antenna perceiving the best copy of an incoming signal is used to receive the communication. A small amount of space difference can allow a SISO system to function in the presence of multipath.

MIMO systems thrive on multipath. They need multipath to function. If multipath does not exist, they cannot transmit and receive multiple streams of data concurrently. By engineering the antennas and antenna placement for optimum use, we can take advantage of technologies in 802.11n and 802.11ac (and the future 802.11ax) to use multiple spatial streams in communications. Each additional spatial stream increases the overall data rate. More information on MIMO systems will follow in later chapters.

While this book mentions 802.11ax on various occasions. The CWNA-108 exam in no way tests your knowledge of this currently unratified amendment to the standard. Future versions of the exam will test your knowledge of 802.11ax, after it has been ratified.

Refraction

Refraction occurs when an RF signal changes speed and is bent while moving through media of different densities. Different mediums, such as drywall, wood or plastic, will have different refraction indexes. The refraction index helps in determining how much refraction will occur.

Let's go back to the light analogy for a moment. If you wear glasses, you are wearing a refraction device. The lens refracts, or bends the light, to make up for the imperfect lens in your eye. This allows you to see clearly again because the lacking focus of the eye is corrected by the refraction caused in the lens of the glasses.

Figure 2.11 shows an RF signal being refracted. As you can see, when refraction occurs with RF signals, some of the signal is reflected and some is refracted as it passes through the medium. Of course, as with all mediums, some of the signal will be absorbed as well.

RF signal refraction is usually the result of a change in atmospheric conditions. For this reason, refraction is not usually an issue within a building, but it may introduce problems in wireless site-to-site links outdoors. Common causes of refraction include changes in temperature, changes in air pressure or the existence of water vapor.

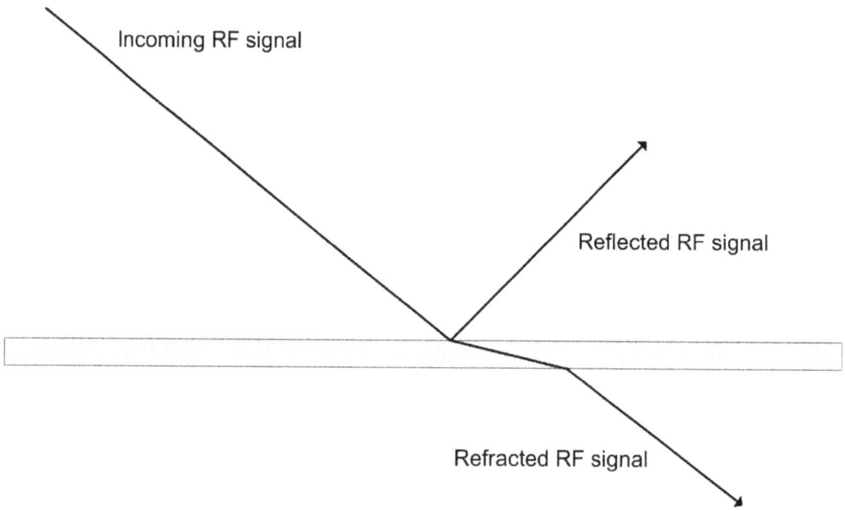

Figure 2.11: RF Refraction

The issue here is simple: if the RF signal changes from the intended direction as it's traveling from the transmitter to the receiver, the receiver may not be able to detect and process the signal. The result can be a broken connection or an increase in error rates if the refraction is temporary, or sporadic due to fluctuations in the weather around the area of the link.

An excellent experiment can be performed easily that demonstrates the concept of refraction. Take a large clear bowl and fill it with water. Now, place a large butter knife into the water at an angle and look through the clear side of the bowl at the knife. What did the knife do? Well, nothing other than enter the water; but what did it appear to do? It appears to bend. This is because the light waves are traveling slower in the water medium and this causes refraction of the light waves.

It's not the knife that's bending — because it's not the knife you see. It's the light that's bending, because it's the light that you see.

Diffraction

Diffraction is defined as a change in the direction and/or intensity of a wave as it passes by the edge of an obstacle. As seen in Figure 2.12, this can cause the signal's direction to change and it can also result in areas of RF shadow. Instead of bending as it passes into or out of an obstacle, like refraction, diffraction describes what happens as light travels around the obstacle.

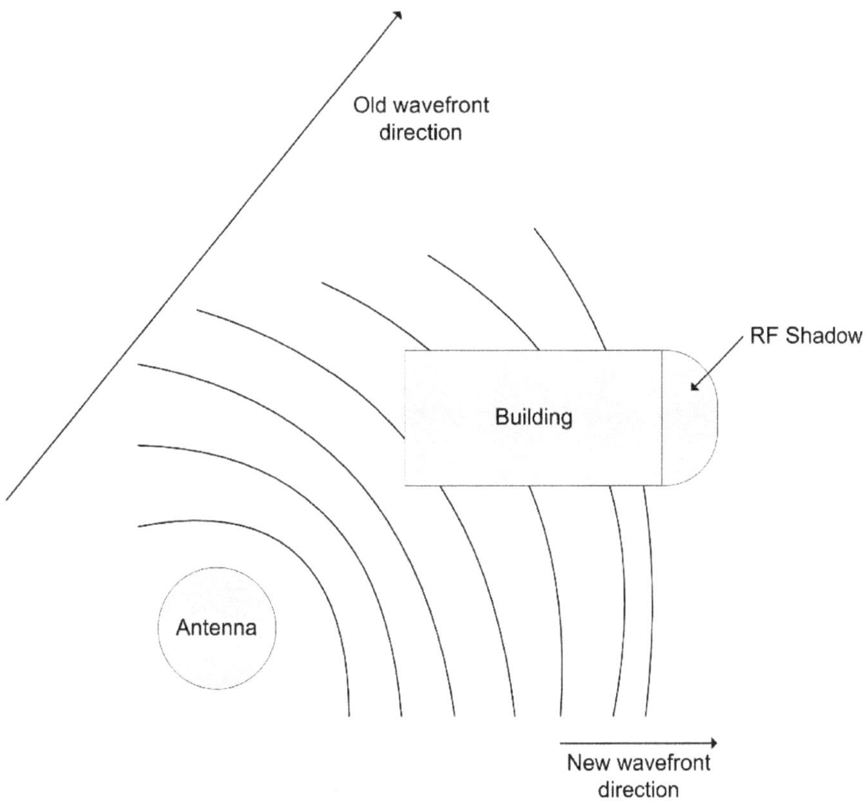

Figure 2.12: RF Diffraction

Diffraction occurs because the RF signal slows down as it encounters the obstacle, and this causes the wave front to change directions. Consider the analogy of a rock dropped into a pool and the ripples it creates. Think of the ripples as analogous to RF signals. Now, imagine there is a stick being held upright in the water. When the ripples encounter the stick, they will bend around it, since they cannot pass through it. A larger stick has a greater visible impact on the ripples, and a smaller stick has a lesser impact. Diffraction is often caused by buildings, small hills and other larger objects in the path of the propagating RF signal.

The RF shadow caused by diffraction can result in areas without proper RF coverage. If you are in an RF shadow area, you will not be able to receive communications from the wireless network. An example of this phenomenon indoors is an elevator shaft. Often, when the access point is on one side of the elevator, and the client is on the opposite side, the signal will be insufficient for communications in that location. Many times, RF shadow problems can be resolved with very slight adjustments in the location of the antennas used on the access point or wireless router, or by installing additional access points. For example, if you install access points in the areas on both sides of the elevator shaft, one access point can serve one side and the other can serve the remaining side.

Scattering

Scattering happens when an RF signal strikes an uneven surface (a surface with inhomogeneities — there's a word you can use with your kids to sound smart) causing the signal to be scattered instead of absorbed, so that the resulting signals are less significant than the original signal. Another way to define scattering is to say that it is simply multiple reflections. Figure 2.13 illustrates this.

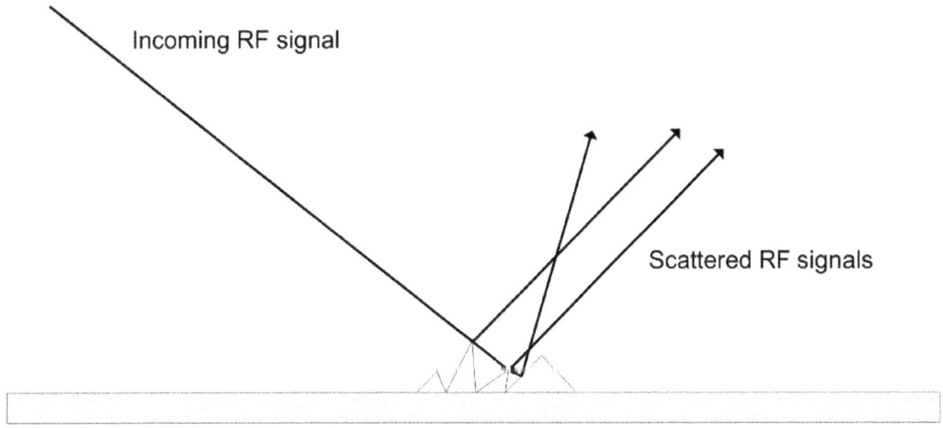

Figure 2.13: RF Scattering

Scattering can happen in a minor, almost undetectable way, when an RF signal passes through a medium that contains small particles. These small particles cause the scattering. Smog is an example of such a medium. The more common and more impacting occurrence is that caused when RF signals encounter things like rocky terrain, leafy trees or chain link fencing. Rain and dust can cause scattering as well.

Absorption

Absorption is the conversion of the RF signal energy into heat. This conversion happens because the molecules in the medium through which the RF signal is passing cannot move fast enough to "keep up" with the RF waves. Many materials absorb RF signals in the 2.4 GHz ISM spectrum. These include water, drywall, wood and even humans. Figure 2.14 shows RF signal absorption.

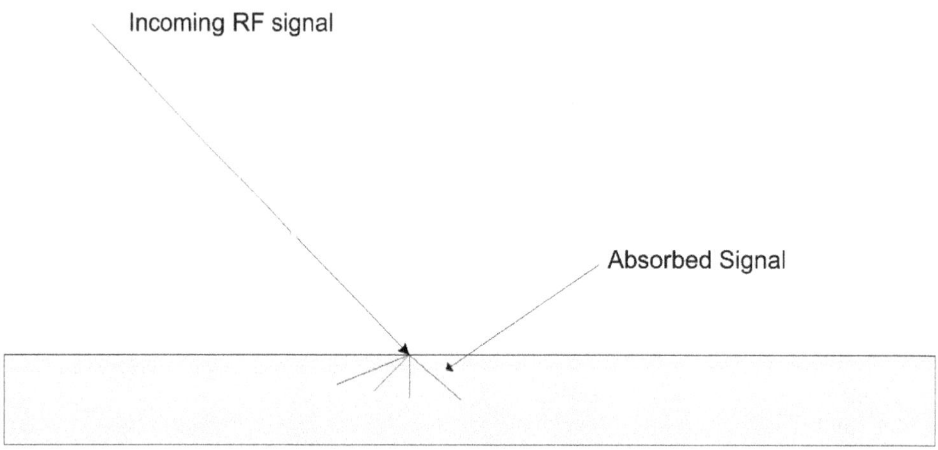

Figure 2.14: RF Absorption

Microwave ovens use the 2.4 GHz frequency range to heat food. While your WLAN devices have output power levels from 30 milliwatts (mW) to 4 watts (W), microwave ovens usually have an output power between 700 and 1400 W. What does this have to do with WLAN engineering? The microwave oven works because RF waves are absorbed well by materials that have moisture (molecular electric dipoles) in them. The absorption converts the RF wave energy into heat energy, and therefore heats your food.

If you've ever set up a wireless network in a large auditorium, only to notice that the coverage was less acceptable after the room filled with hundreds or thousands of people, you've experienced the phenomenon of absorption firsthand. Before the people were in the room, most of the items were reflecting, refracting, scattering or diffracting the RF signals. People tend to absorb the RF signals instead of reflecting them, causing a reduction in the available signal strength within the coverage area.

Different materials have different absorption rates. Table 2.1 provides a listing of some of the more common types of materials, and the absorption rates associated with them. When performing a site survey or troubleshooting a communications problem, these types of materials should certainly be considered.

Material	Absorption Rate
Plasterboard/Drywall	3 – 5 dB
Glass wall and metal frame	6 dB
Metal door	6 – 10 dB
Window	3 dB
Concrete wall	6 – 15 dB
Block wall	4 – 6 dB

Table 2.1: RF Absorption Rates of Materials

Earlier, I suggested that you cup your hands in front of your mouth to see the impact this has on sound waves, and I used this as an analogy of intentional passive gain. The total output power was not increased, but it was focused in a specific direction. Let's do another experiment. Begin reading this text aloud. As you continue to read, place your hand over your mouth so that it completely covers it and continue reading. If you are reading this with your hand over your mouth, you're experiencing the results of absorption in relation to sound waves, and you're looking a bit odd. The sound disturbance has great difficulty passing through your hand and so the sound is muffled. RF signals can be absorbed by materials in a similar manner.

BEYOND THE EXAM: Popcorn and Electromagnetic Waves

I don't know about you, but I like popcorn with a great movie. Of course, today we can get freshly popped popcorn in under four minutes, thanks to the power of the microwave. Microwave ovens use the 2.4 GHz frequency range (some use it all and some use just a portion) to pop that popcorn. The popcorn absorbs the RF energy by converting it to heat. Eventually, the heat builds up pressure and you hear that wonderful poppety-pop-pop sound coming from the microwave, which means you'll be enjoying your movie and your popcorn in just a few minutes.

So, what does a microwave oven have to do with WLANs? Well, the answer to that question is twofold. First, it can be used as a teaching tool to understand concepts like absorption and reflection. Second, microwave ovens can cause interference with your WLAN in many scenarios.

As a teaching tool, the microwave oven can help you understand both absorption and reflection. When you put a glass of cold water in the microwave and turn the microwave on, the water heats up. Why? Because absorption occurs. Absorption, remember, is the conversion of RF energy to heat. Now, you can take out that glass of water and dip a tea bag in it to get some soothing hot tea.

Reflection is seen in the fact that very little of the output energy escapes from the microwave. Why? It is being reflected inward by the design of the internal unit. Place your cellphone in a microwave (without turning the microwave oven on, of course) and close the door. After a few seconds, open the door again — you'll likely see that your phone is looking for service. Why? The design of the microwave keeps as much of the RF energy in as possible, and that results in keeping the cell tower's energy out.

Microwave ovens can cause interference simply because they operate in the same frequency space as 802.11, 802.11b, 802.11g and 802.11n devices. While the microwave ovens do a good job of protecting you as a human (by keeping dangerous levels of RF energy inside the microwave), they certainly let plenty of the energy escape from the perspective of a nearby WLAN. Always test the microwaves in the area where you are installing a WLAN. It may dictate the channel you have to use.

VSWR

Before the RF signal is radiated through space by the antenna, it exists as an alternating current (AC) within the transmission system. Within this hardware, RF signal degradation occurs. All cables, connectors and device have some level of inherent loss. Even in a properly designed system, this loss by attenuation is unavoidable. However, the situation can be even worse if all the cables and connectors do not share the same impedance level.

Impedance is defined as the resistance in an electric circuit and is typically rated in ohms (Ω). Most WLAN equipment uses 50-ohm components. Mismatched components may result in damage to equipment.

If all cables, connectors and devices in the chain from the RF signal generator to the antenna do not have the same impedance rating, there is said to be an impedance mismatch. For example, you would not want to use cable rated at 50 ohms with connectors rated at 75 ohms. This would cause an impedance mismatch. Maximum power output and transfer can only be achieved when the impedance of all devices is the same.

Voltage Standing Wave Ratio (*VSWR*) is a measurement of mismatched impedance in an RF system and is stated as an X:1 (read as "X to one") ratio. Table 2.2 provides a reference for different common VSWR ratings and their meanings.

In a VSWR rating, a lower first number means a better impedance match. Therefore, 1.5:1 is better than 2.0:1. To help with your understanding, think of a series of pipes connected to a water pump, as depicted in Figure 2.15. The water pump is analogous to the RF transmitter and the pipes are analogous to the cables and connectors leading up to the antenna.

VSWR	Definition
1.0:1	One to one. Exact match. An ideal that cannot be accomplished with current technology.
1.5:1	One point five to one. Good match. Only 4% loss in power.
2.0:1	Two to one. Acceptable match. Approximately 11% loss in power.
6.0:1	Six to one. Poor match. Approximately 50% loss in power.
10:1	Ten to one. Unacceptable match. Most of the power is lost.
∞:1	Infinity to one. Useless to measure as the mismatch is so great.

Table 2.2: VSWR Ratings

Assuming the water pump can pump water at a rate and force equal to pipe A, pipe B will cause a mismatch in impedance because it is a smaller pipe. Pipe B cannot handle the amount of water at the level of pressure that pipe A and the pump can handle. This results in a buildup of pressure in pipe B and the pressure is pushed back into pipe A and within the pump. At this point two things can happen: the water flowing out of the end of pipe B will be less than the original potential of the water pump, or the pipe A and the water pump may be destroyed in some way. Pipe A or B could burst, or the seals around the connectors between the water pump and pipe A and between pipe A and pipe B could leak. The water pump itself could begin leaking internally, or even overheat and malfunction. As you can see, the least impacting result would be that the water flow is less than that of which the pump and pipe A are capable. RF systems have similar potentials, including potential for damage, as you will see in the next section on return loss.

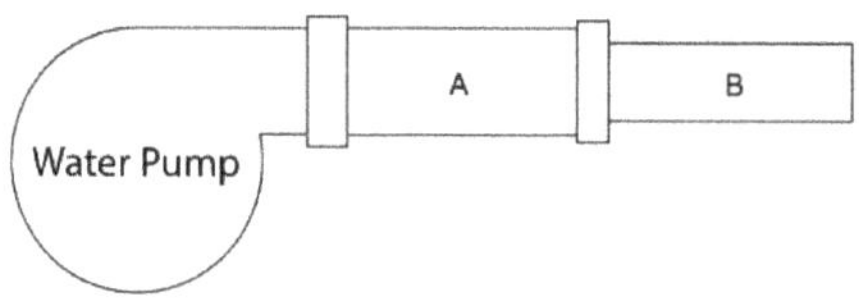

Figure 2.15: VSWR Analogy Using a Water Pump and Pipe Connections

Return Loss

When there is VSWR greater than 1.0:1, which will always be the case, there is some level of power loss due to backward reflection of the RF signal within the system. This fact is always important to keep in mind when building bridge installations using multiple cables and connectors between the WLAN bridges and the antennas. The energy that is reflected toward the RF generator or transmitter causes return loss. *Return loss* is a measurement, usually expressed in decibels (dB), of the ratio between the forward current (incident wave) and the reflected current (reflected wave). The results of this return loss will be similar to those in the water pump analogy presented previously. The RF transmitter may be destroyed, as may other components in the RF system, if the reflected energy is too high, but this would be a worst-case scenario. It is more common that the output power at the intentional radiator is simply less than the original potential generated by the RF transmitter.

To minimize VSWR and return loss, you must avoid impedance mismatches in the entire link chain from the radio to the antenna. You will want to use all equipment (RF transmitters, cables and connectors) with the same ohm rating. The rating is usually 50 ohms when considering RF systems used in 802.11 wireless networks. If you purchase an entire RF system as a unit from a manufacturer, all the components should have the same ohm rating already. If you build an RF system from scratch, you will have to take the responsibility of ensuring there is no impedance mismatch and you must also ensure this custom system is following local regulatory agency requirements.

When considering VSWR, there are two extreme scenarios that create the infinite rating listed in Table 2.2: perfect open and perfect short. Perfect open would mean that someone failed to connect the end of the cabling to an antenna and perfect short would occur if someone shorted out a perfect open with something. In these cases, most of the RF energy is reflected and the VSWR leans toward ∞:1. This, of course, should be avoided if you value your RF equipment.

Amplification

Amplification is an increase of the amplitude of an RF signal. Passive gain, as discussed earlier, is not an amplification of an RF signal up to the point of the intentional radiator. Passive gain is a focusing or directing of an RF signal through the use of antennas. Active gain is an increase in the amplification of an RF signal somewhere in the path up to the point of the intentional radiator. Amplification is achieved through active gain and is accomplished with an amplifier.

Many access points contain variable power output settings, while this capability is not technically an amplifier separate from the access point, these settings will impact the amplitude of the RF signal that is generated. Therefore, the changing of this setting to a higher setting results in a stronger RF signal from the access point.

Attenuation

Attenuation is the process of reducing an RF signal's amplitude. This is occasionally done intentionally with attenuators to reduce a signal's strength within a regulatory domain's-imposed constraint. Loss is the result of attenuation and gain is the result of amplification. RF cables, connectors and devices may have some level of imposed attenuation, and this attenuation is usually stated in dB, and is often stated as loss in dB per foot — this is also known as insertion loss. Insertion loss is the loss incurred by simply inserting the object (cable, connector, etc.) into the path of the RF signal, between the source and the intentional radiator. For example, if an RF splitter is used to split a signal across two antennas, for some reason, the loss incurred by the splitter is called insertion loss.

Wave Propagation

The way RF waves move through an environment is known as wave propagation. Attenuation occurs as RF signals propagate through an environment. When the RF signal leaves the transmitting antenna, it will begin propagation through the local environment and continue, theoretically, forever. The signal cannot be detected after a certain distance, and so the distance in which it can be detected becomes the usable range of the signal. Given that the signal could theoretically propagate forever, why is there a point at which it can no longer be detected? It is because attenuation occurs as the signal is propagating through the environment. Some of the signal strength is lost through absorption by materials encountered by the RF signal; however, even without any materials in the path of the signal, the amplitude will be lessened. This is due to a phenomenon known as free space path loss.

As the signal strength becomes weaker and weaker, it gets ever closer to the RF noise floor. Eventually, it disappears into the noise floor and can no longer be detected, though we theorize that it is still there. I suppose it is kind of like the Borg — the RF signal becomes part of the collective and is still there (else there would be no collective), but it is no longer useful on its own. If you are not a Star Trek fan, I apologize for that last analogy.

Free Space Path Loss

Free space path loss (FSPL), sometimes simply called free space loss (FSL) or just path loss, is a weakening of the RF signal due to a broadening of the wave front. The broadening of the wave front is known as signal dispersion. Consider the concentric circles in Figure 2.16 as representing an RF signal propagating out from an omni-directional antenna (antenna types are covered in detail in later chapters). Notice how that the wave front becomes larger as the wave moves out from the antenna. The broadening of the wave front causes a loss in amplitude of the signal at any specific point in space, because the energy is spread over a larger area.

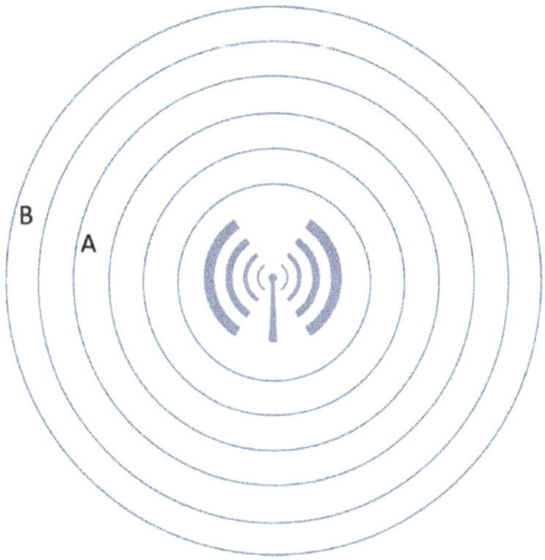

Figure 2.16: FSPL Illustrated

If you place a receiving antenna at point B in Figure 2.16, you will detect a weaker signal than if you place a receiving antenna at point A. The broadening of the wave is also sometimes called beam divergence. Beam divergence can be calculated by subtracting the beam diameter (D1) at a greater distance from the beam diameter (D2) closer to the antenna and then dividing by the distance between these two points (L). The following formula illustrates this:

Divergence = (D1 – D2) / L

Free space path loss can be understood by thinking of the results you get when blowing bubbles with bubble gum. When blowing a bubble, you may notice that the outer shell that forms the bubble boundary becomes thinner as the bubble grows larger. Similarly, RF signals grow weaker as the cell grows larger or the distance becomes greater. The reduction in signal strength is logarithmic rather than linear. For example, a 2.4 GHz signal, such as that used by many 802.11 devices, will attenuate by approximately 80 dB in the first 100 meters and then by another 6 dB in the second 100 meters. As you can see, the attenuation becomes much less in the second 100 meters than in the first, and this is due to logarithmic attenuation.

The following formulas are used to calculate free space path loss in dB (the CWNA exam will not test your knowledge of any FSL formulas):

$$FSPL = 36.6 + (20 * \log_{10}(F)) + (20 * \log_{10}(D))$$

Where FSPL is the free space path loss, F is the frequency in MHz and D is the path length in miles. The result is based on a distance measurement in miles. To get the results based on a distance measurement in kilometers (for example, D is the path length in kilometers) change 36.6 to 32.4, giving you the following formula:

$$FSPL = 32.4 + (20 * \log_{10}(F)) + (20 * \log_{10}(D))$$

For example, assuming you are using the 2.4 GHz spectrum (we'll say 2,450 MHz), and the distance you want to evaluate is 2.5 miles, the following equation will result in the free space path loss:

$$36.6 + (20 * \log_{10}(2450)) + (20 * \log_{10}(2.5))$$

or

$$36.6 + 67.78 + 8 = 112.38$$

The result is a loss of roughly 112 dB at 2.5 miles. I rounded the numbers in this case. More accurate numbers can be found in Table 2.3, which provides a breakdown of free space path loss attenuation in dB for different distances with both the 2.4 GHz spectrum and the 5 GHz spectrum. The next chapter on RF mathematics will give you the knowledge you need to calculate an estimate of signal strength in dB after the signal travels this 2.5 miles through free space. There is an element not considered in the free space path loss calculations that will be added at that time: output power. When you know the free space path loss calculation formula, and the output power, you can estimate the signal power, in dBm (decibels related to milliwatts), at any point in space. The result will be an ideal estimate because weather and other factors can worsen the signal strength in reality.

Distance (miles)	2.4 GHz (attenuation in dB)	5 GHz (attenuation in dB)
0.5	98.36	104.56
1	104.38	110.58
1.5	107.91	114.10
2	110.40	116.60
2.5	112.34	118.54
3	113.93	120.12
4	116.42	122.62
5	118.36	124.56
10	124.38	130.58

Table 2.3: FSPL in dB for 2.4 and 5 GHz Spectrums

Another method that is simpler to use is the 6-dB rule. This is an estimation method that is less accurate than the free space path loss formula we've covered, but it provides a quick calculation that is very close to the results that would be provided by the formula. If you look at the 2.4 GHz column in Table 2.3, you will see a pattern that may not stand out at first. Paying close attention to the 1, 2, and 4-mile distances, you can see that there is an increase in dB loss of approximately 6 dB at each of these intervals. You'll also notice that each of these intervals represents a doubling of the distance. Therein lies the 6-dB rule: For every doubling of distance there is an amplitude loss of approximately 6-dB. Even in the 5 GHz column, you can see that this is true. Though the 5 GHz frequencies attenuate more quickly in the first mile, they follow the 6-dB rule thereafter.

The ***6-dB rule*** is based on the *inverse square law*. They reveal that an increase of 6 dB in the signal doubles the distance at which the signal is usable. A decrease in 6 dB halves the usable distance.

While the general understanding of free space path loss is usually stated as seen here, it is equally valid to consider a different perspective. This alternate perspective states that the RF signal still travels the farther distance, but that the higher frequencies have shorter wavelengths, and therefore shorter optimum antenna sizes. The result is that the smaller antenna has a greater difficulty gathering sufficient RF energy because of its smaller receiving surface. Think of it like the small receiving surface of the human ear compared to the listening devices used on American football sidelines mentioned earlier. The argument is that the RF signal may not be attenuating any "faster," but that it attenuated the same and the receiver is the actual point of problem, rather than the attenuated signal strength. Either way, as a WLAN administrator, you must simply remember that FSPL causes loss in signal strength for all wireless systems.

Multipath and Delay Spread

When signals bounce around in an environment through reflection, refraction, diffraction and scattering, they create an effect known as multipath. *Multipath* occurs when multiple paths of the signal, understood as multiple signals, arrive at the receiving antenna at the same time or within a small fraction of a second (nanoseconds) of each other. Multipath can also occur outdoors when signals reflect off large objects in the RF link path, such as in Figure 2.17.

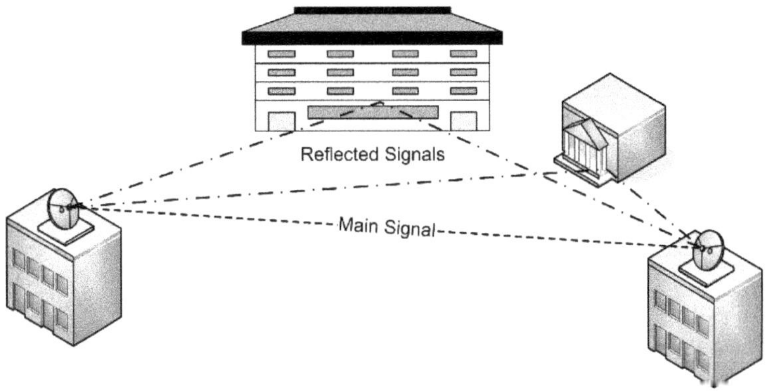

Figure 2.17: Multipath in Outdoor Spaces

Multipath occurs very frequently indoors and is so common an occurrence that many access point vendors included multiple antennas for dealing with this phenomenon in the early days of WLANs. Today with MIMO technology, the multipath behavior indoors is an advantage, and this applies to 802.11n and 802.11ac devices using multiple radio chains (radios and antennas). Indoors, file cabinets, walls, desks and doors —among other things — can cause RF propagation patterns that result in multiple paths arriving at the receiving antenna. In an indoor environment there is often no direct signal path between the transmitter and the receiver (or the access point and the client station). All signals reaching the client station will have arrived due to the RF propagation patterns similar to those in Figure 2.17, only based on indoor reflections.

The difference in time between the first and second signal arriving at the receiver in a multipath occurrence is known as the *delay spread*. Earlier in this chapter, you learned that signals can be in phase or out of phase. These signals arriving at the receiver with a delay spread of nearly 0 will complement each other and cause signal up-fade, which is a good thing. In other words, the received signal will be stronger at the receiver than it would have been without the multipath occurrence. When the delay spread is greater, so that the signals arrive out of phase, the signal will either be down-faded, corrupted or nullified. In such cases, antenna diversity can be used to resolve problems in SISO systems. Remember, in MIMO systems, this multipath phenomenon is used to advantage.

Chapter Summary

In this chapter, you learned about electromagnetic waves and a specific type of electromagnetic wave called a radio frequency (RF) wave. You also learned about RF characteristics, including wavelength, frequency, amplitude and phase. You discovered that amplitude and phase are used for modulation in 802.11 wireless networks as ASK and PSK respectively. Next, you explored the many facets of RF signal behaviors, including reflection, refraction, scattering, diffraction, FSPL, VSWR, gain, loss, attenuation and absorption. In the next chapter, you will begin the learning process to understand how to calculate signal strengths in an RF environment.

Points to Remember

Remember the following important points:

- RF waves are electromagnetic waves comprised of an electric field and a magnetic field.
- The wavelength of the RF waves is the distance between two instances of the same point in the waveform.
- The wavelength is related to the frequency and the speed of the wave.
- The speed of a wave is constant (the speed of light) so when you know the frequency, you can determine the wavelength.
- The frequency is the rate at which the wave cycles per second and is measured in Hertz.
- The amplitude of the wave is the strength or power of the wave.
- Amplitude is used in 802.11 networks as Amplitude Shift Keying (ASK) modulation.
- The phase is a comparison between two waves that are either in phase with each other or some degree out of phase.
- Phase is used in 802.11 networks as Phase Shift Keying (PSK) modulation.

- Reflection occurs when an RF wave encounters a large reflective object in relation to the wavelength.
- Scattering occurs when an RF wave encounters small reflective objects in relation to the wavelength.
- Free Space Path Loss (FSPL) occurs as an RF wave front spreads while moving through space.
- Every 6 dB of gain doubles the usable distance of an RF signal and every -6 dB of loss halves the usable distance of an RF signal.
- Voltage Standing Wave Ratio (VSWR) is a measurement of the return loss in an RF system.
- Multipath occurs when RF signals travel multiple paths to the receiver and arrive at slightly different times.

Review Questions

1. What denotes the strength of an RF wave or signal?
    a. Wavelength
    b. Phase
    c. Amplitude
    d. Modulation

2. What is decreased as the frequency of a wave increases?
    a. Wavelength
    b. Phase
    c. Modulation
    d. Coding

3. What RF behavior is similar to light waves as seen in a mirror?
    a. Absorption
    b. Attenuation
    c. Reflection
    d. Phase

4. What kind of modulation is not used in WLANs?
    a. PSK
    b. FSK
    c. ASK
    d. BPSK

5. What is the common ohm rating in equipment designed for use in WLANs?
    a. 75
    b. 50
    c. 25
    d. 100

6. What happens when RF energy is converted to heat when passing through some materials?
    a. Absorption
    b. Reflection
    c. Phase shifting
    d. Frequency shifting

7. What phenomenon is good for MIMO systems, but must be handled with something like diversity in SISO systems?
    a. Modulation
    b. Multipath
    c. FSPL
    d. Refraction

8. What amount of additional signal strength doubles the usable range of an RF signal in free space?
   a. 2 dB
   b. 3 dB
   c. 6 dB
   d. 10 dB

9. What comparative attribute relates an RF wave to another wave?
   a. Wavelength
   b. Frequency
   c. Amplitude
   d. Phase

10. What occurs as an RF signal moves through space and spreads to cover an even larger area?
    a. Amplification
    b. FSPL
    c. VSWR
    d. Diffraction

Review Answers

1. **C is correct.** The amplitude is the strength or power of the RF wave.
2. **A is correct.** As the frequency increases, the wavelength decreases. Alternatively, as the wavelength increases, the frequency decreases.
3. **C is correct.** Reflection occurs in RF waves as well as light waves and is analogous to light in a mirror.
4. **B is correct.** Frequency Shift Keying (FSK) is not used in WLANs. Both ASK and PSK are used, and BPSK is a form of PSK modulation.
5. **B is correct.** In most cases, WLAN equipment uses a 50-ohm rating for resistance.
6. **A is correct.** Absorption occurs when RF energy is converted to heat, and the result is an attenuation of the RF signal as it exits the absorbing medium.
7. **B is correct.** Multipath is good for MIMO systems because it allows for the transmission of multiple spatial streams. In SISO systems, diversity antennas are most often used to address the issues presented by multipath.
8. **C is correct.** The inverse square law results in the reality that an increase of 6 dB doubles the distance at which a signal is usable and a reduction of -6 dB halves the usable distance.
9. **D is correct.** The phase of an RF wave is an attribute that compares one wave with another wave. They can either be in phase or, to some degree, out of phase.
10. **B is correct.** FSPL occurs naturally as the RF signal moves through space and the wave front spreads to cover an even larger area.

Chapter 3 — RF Mathematics and Measurements

Mathematics, a topic we all loved in primary school. Or maybe not, but either way, I will attempt to make RF math as basic as I can without diminishing the learning experience in this chapter. To understand signal strengths at the transmitter and the receiver, you must understand the basics of RF math. You will need to know about terms like watts, milliwatts, dB, dBm, dBi, RSSI, SNR, SINR and more. Yes, it's an alphabet soup of terminology, but the terms are so commonly used in wireless networking that a CWNA would be ill-equipped to manage wireless networks without a full knowledge of them.

This chapter begins with basic RF mathematics and then moves on to specific applications of its use so that you can see the practical, real-world value in learning this information. It will help you pass the CWNA-108 exam, but it will also make you a better WLAN administrator.

Basic RF Math

You might be wondering why you have should back to high school days and study math to implement a wireless network. After all, you've been able to implement wired networks for years with very little math, other than counting the number of Ethernet ports needed for your users and making sure they supported the required speeds. Wireless is different. Because the wireless network uses an RF signal, you must understand the basics of RF math to determine if the output power of an RF transmitter is strong enough to get a detectable and usable signal to the RF receiver. You had to deal with similar issues with cabling, in that you could only use a CAT 5 cable of a maximum length, but you didn't really have to calculate anything most of the time. You simply knew you could not span a greater distance than that which was supported by the cabling type. The good news is that you don't really have to go back to high school. The bad news is that you might feel like it at times. I'll make this coverage of RF math as easy to follow as possible, but it will become a bit advanced by requirement.

You will need to know a few basic things to fully understand RF math. First, you'll need to understand the units of power that are measured in RF systems. Second, you'll need to understand how to measure power gains and losses. Third, you'll need to understand how to determine the output power you will need at a transmitter in order to get an acceptable signal to a receiver. If you are creating a point-to-point connection using wireless bridges, or if you are installing an access point in an access role, you will still need to understand these three basic concepts. In both wireless bridges and WLANs, a sufficient signal must reach the receiver listening on the other end of the connection.

Watt

The *watt* (W) is a basic unit of power equal to one joule per second. It is named after James Watt, an 18th-century Scottish inventor who also improved the steam engine, among other endeavors. This single watt is equal to one ampere of current flowing at one volt. Think of a water hose with a spray nozzle attached. You can adjust the spray nozzle to allow for different rates of flow. The flow rate is comparable to amperes in an electrical system. Now, the water hose also has a certain level of water pressure — regardless of the amount that is actually flowing through the nozzle. The pressure is like the voltage in an electrical system. If you apply more pressure or you allow more flow with the same pressure, either way, you will end up with more water flowing out of the nozzle. In the same way increased voltage or increased amperes will result in increased wattage, since the watt is the combination of the amperes and volts.

In wireless networks, outdoor links often use power levels measured in watts at the transmitter. In indoor wireless networks, the watt is too powerful, and most access points transmit at a much lower power level. For this, we use the milliwatt.

Milliwatt

WLANs do not need a tremendous amount of power to transmit a signal over an acceptable distance. You can see a 7-watt light bulb from more than 50 miles (83 kilometers) away on a clear night with line of sight. Remember, visible light is another portion of the same electromagnetic spectrum and so this gives you an idea of just how far away an electromagnetic signal can be detected. For this reason, many WLAN devices use a measurement of power that is 1/1000th of a watt. The unit of power is known as a *milliwatt*. 1 W, then, would be 1,000 milliwatts (mW).

Enterprise-class devices will often have output power levels of 1 mW to 100 mW while SOHO wireless devices may only offer up to 30 mW of output power (though some are marketed as high-power devices up to 1W). Some wireless devices may support up to 300 mW of output power, but these are the extreme exception to the rule. For example, the TP-Link EAP220 can transmit at up to 400 mW of output power in the 2.4 GHz band.

> When speaking of power levels at a transmitter, we speak of watts, milliwatts and dBm (discussed later in this chapter). At the receiver, the power levels are so low that using watts and milliwatts is not reasonable, so we use dBm only in most discussions.

For indoor use, it is generally recommended that you transmit at power levels of no more than 100 mW, and often, less is better. In most cases, the minimum gain that will be provided by any connected antennas is a 2 dBi gain, which you will read about later. This means that the output power would actually be approximately 160 mW in the propagation direction of this antenna. This usually provides sufficient coverage for indoor WLANs, and in many indoor WLANs today, less output power per access point is desired. In most cases, you want the output power of the access points to closely match the output power of your clients, for optimal link conditions. It is not uncommon for client devices to transmit at only 20-30 mW. Therefore, configuring access points to transmit at 100 mW can result in unfavorable link conditions.

However, outdoor WLAN links that are providing site-to-site links may use more power. The FCC limits the total output power from the antenna to 4 W for point-to-multipoint applications in the 2.4 GHz band and this, or the regulatory constraints in your area, must be considered when implementing WLAN solutions.

Decibel (dB)

The *decibel* is a comparative measurement value. It is a measurement of the difference between two power levels. For example, it is common to say that a certain power level is 6 dB stronger than another power level, or that it is 3 dB weaker. These statements mean that a 6-dB gain and a 3-dB loss has occurred, respectively.

A decibel is 1/10th of a bel. You could equally say that a bel is 10 decibels. The point is that the decibel is based on the bel, which was developed by Bell Laboratories in order to calculate the power losses in telephone communications as ratios. The definition of a bel is simple: 1 bel is a ratio of 10:1 between two power levels. Therefore, a power ratio of 200:20 is 1 bel (10:1) and 200:40 is .5 bel (5:1) and 200:10 is 2 bel (20:1). In the end, the decibel is a measurement of power that is used very frequently in RF mathematics.

You may have been asked the same question that I was asked as a child: Would you rather have $1,000,000 at the end of a month, or one cent doubled in value every day for a month? Of course, the latter option is worth more than $5,000,000 by the end of the month. This is the power of exponential growth. RF signals experience exponential decay, rather than growth, as they travel through space. This is also called logarithmic decay. The result is a quickly weakening signal. This power loss is measured with decibels.

The decibel is relative where the milliwatt is absolute. The decibel is logarithmic where the milliwatt is linear. To understand this, you'll need to understand the basics of a logarithm or you'll at least need a good tool to calculate logarithms for you, such as a spreadsheet application like Microsoft Excel.

> The decibel is used to measure differences in power levels and it is relative to an absolute value. Absolute values (watts and milliwatts) may be said to increase or decrease in decibels.

A logarithm is the exponent to which the base number must be raised to reach some given value. The most common base number evaluated is the number 10 and you will often see this referenced in formulas as log10. For example, the logarithm or log of 100 is 2 with a base of 10. This would be written:

log10(100) = ?

It is a fancy way of saying $10^2 = 100$, which is the shorthand way of saying 10 * 10 = 100. However, knowing the concept of a logarithm is very important in many RF-based math calculations, though you will not be tested on the complex formulas on the CWNA exam. You will, however, need to be able to calculate simple power level problems. How will you deal with these problems? Using the rules of 10s and 3s. This system will usually allow you to calculate RF signal power levels with close estimation without ever having to resort to logarithmic math. Here are the basic rules:

- A gain of 3 dB magnifies the output power by two.
- A loss of 3 dB equals one half of the output power.
- A gain of 10 dB magnifies the output power by ten.
- A loss of 10 dB equals one tenth of the output power.
- dB gains and losses are cumulative.

Now, let's evaluate what these five rules mean and the impact they have on your RF math calculations. First, 3 dB of gain doubles the output power. This means that 100 mW plus 3 dB of gain equals 200 mW of power, or 30 mW plus 3 dB of gain equals 60 mW of power. The power level is always doubled for each 3 dB of gain that is added. Rule five stated that these gains and losses are cumulative. This means that 6 dB of gain is the same as 3 dB of gain applied twice. Therefore, 100 mW of power plus 6 dB of gain equals 400 mW of power. The following examples illustrate this based on 9 dB of gain (3 dB added three times):

40 mW + 3dB + 3dB + 3dB = 320 mW

40 mW * 2 * 2 * 2 = 320 mW

Both formulas are saying the same thing. Now consider the impact of 3 dB of loss. 3 dB of loss halves the output power. Look at the impact on the following formula with 6 dB of gain and 3 dB of loss:

40 mW + 3 dB + 3 dB – 3 dB = 80 mW

40 mW * 2 * 2 / 2 = 80 mW

Again, both formulas are saying the same thing. You can see, from this last example, how the accumulation of gains and losses are calculated. Now, rules three and four say that a gain or loss of 10 results in a gain of 10 times or a loss of 10 times (divide by 10). Consider the following example, which illustrates rules 3, 4 and 5:

40 mW + 10 dB + 10 dB = 4000 mW (4 W)

40 mW * 10 * 10 = 4000 mW (4 W)

As you can see, adding 10 dB of gain twice causes a 40-mW signal to become a 4000-mW signal, which could also be stated as a 4 W signal. Losses would be subtracted in the same way as the 3 dB losses were; however, instead of dividing by 2 we would now divide by 10 such as in the following example with simply 10 dB of loss:

40 mW – 10 dB = 4 mW

40 mW / 10 = 4 mW

You should be beginning to understand the five rules of 10s and 3s. However, it is also important to know that the 10s and 3s can be used together to calculate the power levels after any integer gain or loss of dB. This is done with creative combinations of 10s and 3s. For example, imagine you want to know what the power level would be of a 12-mW signal with 16 dB of gain. Here is the math:

12 mW + 16 dB = 480 mW

But how was this calculated? The answer is very simple: add 10 dB and then add 3 dB twice. Here it is in long hand:

12 mW + 16 dB = 480 mW

12 mW + 10 dB + 3 dB + 3 dB = 480 mW

12 mW * 10 * 2 * 2 = 480 mW

Sometimes you are dealing with both gains and losses of unusual amounts. While the following numbers are completely fabricated, consider the assumed difficulty they present to calculating a final RF signal power level:

30 mW + 7 dB – 5 dB + 12 dB – 6 dB = power level

At first glance, this sequence of numbers may seem impossible to calculate with the rules of 10s and 3s; however, remember that the dB gains and losses are cumulative, and this includes both the positive gains and the negative losses. Let's take the first two gains and losses: 7 dB of gain and 5 dB of loss. You could write the first part of the previous formula like this:

30 mW + 7 dB + (-5 dB) = 30 mW + 2 dB

Why is this? Because +7 plus -5 equals +2. Carrying this out for the rest of our formula, we could say the following:

30 mW + 7 dB + (-5 dB) + 12 dB + (-6 dB) = 30 mW + 2 dB + 6 dB

or

30 mW + 8 dB = power level

The only question that is left is this: How do we calculate a gain of 8 dB? Remember, the rules of 10s and 3s. We have to find a combination of positive and negative 10s and 3s that add up to 8 dB. Here's a possibility:

+10 + 10 – 3 – 3 – 3 – 3 = 8

If we use these numbers to perform RF dB-based math, we come up with the following formula:

30 mW + 10 + 10 – 3 – 3 – 3 – 3 = 187.5 mW

30 mW * 10 * 10 / 2 / 2 / 2 / 2 = 187.5 mW

To help you visualize the math, consider the following step-by-step breakdown of the immediately preceding example:

30 mW * 10 = 300 mW

300 mW * 10 = 3000 mW

3000 mW / 2 = 1500 mW

1500 mW / 2 = 750 mW

750 mW / 2 = 375 mW

375 mW / 2 = 187.5 mW

In the end, nearly any integer dB-based power gain or loss sequence can be estimated using the rule of 10s and 3s. Table 3.1 provides a breakdown of dB gains from 1 to 9 with the expressions as 10s and 3s for your reference. From this table, you should be able to determine the combinations of 10s and 3s you would need to calculate the power gain or loss from any provided dB value. Always remember that, while plus 10 is actually times 10, plus 3 is only times 2. The same is true in reverse, in that minus 10 is actually divided by 10, and minus 3 is divided by 2.

Gain in dB	Expression in 10s and 3s
1	+ 10 – 3 – 3 – 3
2	+ 3 + 3 + 3 + 3 – 10
3	+ 3
4	+ 10 – 3 – 3
5	+ 3 + 3 + 3 + 3 + 3 – 10
6	+ 3 + 3
7	+ 10 – 3
8	+ 10 + 10 – 3 – 3 – 3 – 3
9	+ 3 + 3 + 3

Table 3.1: Expressions of 10s and 3s

When you add 3 dB, you double the absolute power. When you add -3 dB (or subtract 3 dB), you halve the absolute power. When you add 10 dB, you multiply the absolute power by 10. When you add -10 dB (or subtract 10 dB), you divide the absolute power by 10.

To test your ability to perform this math. Answer the following challenges:

1. If you start with 25 mW and introduce 17 dB of gain, what is the result?
2. If you start with 100 mW and introduce 13 dB of loss, what is the result?
3. If you begin with 30 mW and introduce 6 dB of gain, 2 dB of loss, 4 dB of gain, and 7 dB of loss, what is the result?

Challenge answers:

1. If you start with 25 mW and introduce 17 dB of gain, what is the result?

   25 mW + 17 dB = 1250 mW

   25 mW + 10 dB + 10 dB + (-3 dB) = 1250 mW

2. If you start with 100 mW and introduce 13 dB of loss, what is the result?

   100 mW + (-13 dB) = 5 mW

   100 mW + (-10 dB) + (-3 dB) = 5 mW

3. If you begin with 30 mW and introduce 6 dB of gain, 2 dB of loss, 4 dB of gain, and 7 dB of loss, what is the result?

   30 mW + 6 dB + (-2 dB) + 4 dB + (-7 dB) = 37.5 mW

   30 mW + 4 dB + (-3 dB) = 37.5 mW

   30 mW + 1 dB = 37.5 mW

   30 mW + 10 dB + (-3 dB) + (-3 dB) + (-3 dB) = 37.5 mW

As you can see by the challenge answers, creative usage of the rules of 10s and 3s can create a handy solution to power level calculations based on dB gains and losses.

dBm

dBm is an absolute measurement of power where the *m* stands for milliwatts. Effectively, *dBm* references decibels relative to 1 milliwatt or that 0 dBm equals 1 milliwatt. Once you establish that 0 dBm equals 1 milliwatt, you can reference any power strength in dBm.

Because a wireless receiver can detect and process very weak signals, it is easier to refer to the received signal strength in dBm rather than in mW. For example, a signal that is transmitted at 4 W of output power (4000 mW or 36 dBm) and experiences -63 dB of loss has a signal strength of .002 mW (-27 dBm). Rather than say that the signal strength is .002 mW, we say that the signal strength is -27 dBm.

The formula to get dBm from milliwatts is:

$$dBm = 10 * \log_{10}(Power\_mW)$$

For example, if the known milliwatt power is 30 mW, the following formula would be accurate:

$$10 * \log_{10}(30) = 14.77 \text{ dBm}$$

The result of this formula would often be rounded to 15 dBm for simplicity; however, you must be very cautious about rounding if you are calculating a link budget, because your end numbers can be drastically incorrect if you've performed multiple rounding calculations along the path. Table 3.2 provides a list of common milliwatt power levels and their dBm values.

One of the benefits of working with dBm values instead of milliwatts is the ability to easily add and subtract simple decibels instead of multiplying and dividing often huge or tiny numbers. For example, consider that 14.77 dBm is 30 mW as you can see in Table 3.2. Now, assume that you have a transmitter that transmits at that 14.77 dBm, and you are passing its signal through an amplifier that adds 6 dB of gain. You can quickly calculate that the 14.77 dBm of original output power becomes 20.77 dBm of power after passing through the amplifier. Now, remember that 14.77 dBm was 30 mW. With the 10s and 3s of RF math, which you learned about earlier, you can calculate that 30 mW plus 6 dB is equal to 120 mW. The interesting thing to note is that 20.77 dBm is equal to 119.4 mW. As you can see, the numbers are very close indeed. While I've been using a lot of more exact figures in this section, you'll find that rounded values are often used in vendor literature and documentation and they are sufficient for the CWNA-108 exam as well. Figure 3.1 shows a set of rounded power level charts that can be used for simple mW to dBm and dBm to mW conversion.

mW	dBm
1	0.00
10	10.00
20	13.01 (rounded to 13)
30	14.77 (rounded to 15)
40	16.02 (rounded to (16)
50	16.99 (rounded to 17)
100	20.00
1000	30.00
4000	36.02 (rounded to 36)

Table 3.2: mW to dBM Conversion Table (rounded to two precision levels)

Remember a few mW to dBm comparisons for the exam. Examples include 1 mW equals 0 dBm, 10 mW equals 10 dBM, 100 mW equals 20 dBm, and 1000 mW equals 30 dBm. Also remember that negative values are used to represent low milliwatt power levels. For example, -10 dBm is 0.1 mW and -20 dBm is 0.01 mW. Remembering a few key values and the rules of 10s and 3s should allow you to perform the conversions required for the CWNA-108 exam.

dBm	Watts	dBm	Watts	dBm	Watts
0	1.0 mW	16	40 mW	32	1.6 W
1	1.3 mW	17	50 mW	33	2.0 W
2	1.6 mW	18	63 mW	34	2.5 W
3	2.0 mW	19	79 mW	35	3 W
4	2.5 mW	20	100 mW	36	4 W
5	3.2 mW	21	126 mW	37	5 W
6	4 mW	22	158 mW	38	6 W
7	5 mW	23	200 mW	39	8 W
8	6 mW	24	250 mW	40	10 W
9	8 mW	25	316 mW	41	13 W
10	10 mW	26	398 mW	42	16 W
11	13 mW	27	500 mW	43	20 W
12	16 mW	28	630 mW	44	25 W
13	20 mW	29	800 mW	45	32 W
14	25 mW	30	1.0 W	46	40 W
15	32 mW	31	1.3 W	47	50 W

Figure 3.1: mW to dBm and dBm to W conversion table

dBi

dBi (the i stands for isotropic) is a measurement of power gain used for RF antennas. It is a comparison of the gain of the antenna and the output of a theoretical isotropic radiator. An isotropic radiator is an ideal antenna that we cannot create with any known technology. This is an antenna that radiates power equally in all directions. In order to do this, the power source would have to be at the center of the radiating element and be infinitesimally small. Since this technology does not exist, we call the isotropic radiator the ideal against which other antennas are measured. More details about dBi are provided in the later section titled "Isotropic Radiator." For now, just remember that dBi is a measurement of directional gain in power and is not an absolute power reference. The dBi value must be calculated against the input power provided to the antenna to determine the actual output power in the direction in which the antenna propagates RF signals. An antenna, for example, may indicate it offers 7 dBi of gain. The result would be that a 50-mW input power to the antenna (at the intentional radiator) results in 250 mW of output power in the designed direction of propagation (EIRP).

dBd

Antenna manufacturers have used both dBi, mentioned previously, and dBd to calculate the directional gain of antennas. Where dBi is a calculation of directional gain compared to an isotropic radiator, *dBd* is a calculation of directional gain compared to a dipole antenna. Therefore, the last d in dBd stands for dipole. Like dBi, dBd is a value calculated against the input power to determine the directional output power of the antenna.

What is the difference between dBi and dBd then? The difference is that a dBd value is compared with a dipole antenna, which itself has a gain of 2.14 over an isotropic radiator. Therefore, an antenna with a gain of 7 dBd has a gain of 9.14 dBi. Remember, to convert from dBd to dBi, just add 2.14. To convert from dBi to dBd, just subtract 2.14. To remember this, just remember the formula 0 dBd = 2.14 dBi.

> dBd is not seen as commonly as dBi today. This change is a good thing because it provides a more consistent standard for antenna gain measurement and specification. dBd is still often used in other wireless communication systems, but WLAN gear has moved toward dBi.

Implementation of RF Math

Now that you have learned the basics of RF mathematics, it's time to investigate some of the more advanced uses of RF math and implementation scenarios where it is required. This section will cover the following concepts:

- SNR and SINR

- RSSI
- Link Budgets
- System Operating Margins
- Fade Margins
- Intentional Radiators
- EIRP

SNR and SINR

Background RF noise, which can be caused by all the various systems and natural phenomenon that generate energy in the electromagnetic spectrum, is known as the noise floor. The power level of the RF signal relative to the power level of the noise floor is known as the signal-to-noise ratio or *SNR*. It is the difference between the signal strength and the noise floor, so don't let the term "ratio" confuse you. As SNR is used in WLANs, it is not really a ratio, but a dB value. Figure 3.2 illustrates the concept of SNR.

Think of it like this. Imagine you are in a large conference room. Further, imagine that hundreds of people are having conversations at normal conversation sound levels. Now, imagine that you want to say something so that everyone will hear you; therefore, you cup your hands around your mouth and yell. You could say that the conversations of everyone else in the conference room is a noise floor and that your yelling is the important signal or information. Furthermore, you could say that the loudness of your yelling relative to the loudness of all other discussions is the SNR for your communication.

In WLAN networks, the SNR becomes a very important measurement. If the noise floor power levels are too close to the received signal strength, the signal may be corrupted, or it may not even be detected. It's almost as if the received signal strength is weaker than it actually is when there is more electromagnetic noise in the environment. You may have noticed that when you yell in a room full of people yelling, your volume doesn't seem so great; however, if you yell in a room full of people whispering, your volume seems to be magnified. In fact, your volume is not greater, but the noise floor is less. RF signals are impacted in a similar way.

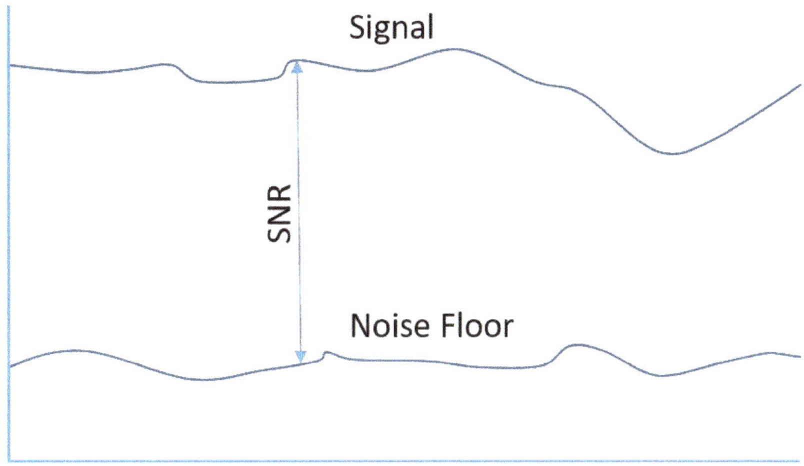

Figure 3.2: SNR Illustrated

Technically, SNR is defined as the difference between the noise floor and the signal in dB. The formula for calculating SNR is simple:

SNR = signal strength value in dBm – noise floor value in dBm

If the noise floor is rated at -95 dBm and the signal is detected at -70 dBm, the SNR is 25.

The SNR is extremely important in WLAN communications. The signal strength is not what actually determines the data rate you will achieve. Vendor specification sheets often indicate that you need a particular received signal strength in dBm to achieve a given data rate. However, there is some mystical assumed noise floor value behind these specifications and they rarely tell you what that value is. It is true that 2.4 GHz networks often have a noise floor of around -95 to -100 dBm and 5 GHz networks often see noise floor values of around -100 to -105 dBm, but these are just commonalities and not guarantees. You should always test for the noise floor in your environment. If your noise floor is -93 dBm for some reason, this will impact the data rates you can achieve with a signal strength of say -67 dBm. That is, your -67 dBm signal will be more like a -70 dBm signal to someone who has an environment with a noise floor of -96. This information may seem complicated and, if so, read through this paragraph again. The goal is to understand that the

SNR is what determines the modulation and coding you can use and, therefore, the data rate you can achieve. Modulation and coding are covered in detail in later sections of this book.

> Make sure you know how to calculate SNR. If given a noise floor rating value and a signal strength value, be prepared to calculate the SNR. Remember the simple formula of signal strength value – noise floor value = SNR. Know that the signal strength may be provided in mW and need converted to dBm, but the mW value will usually be a basic value such as -0.1 or -0.01.

In addition to the term SNR, the term SINR has become common in vendor literature and tech-speak for WLAN professionals. *SINR* is the signal to interference plus noise ratio. Like SNR, it is not a ratio, but a value in dB. The difference is that SINR is more momentary in nature than SNR. SNR looks at the noise floor at a given point in time and assumes it doesn't change drastically, which is usually a good assumption. However, sporadic interferers may generate RF energy for small bursts of time, during that time window, SINR is a better measurement of reality. Even if the SNR would allow for the reception of an RF signal at a given data rate, the SINR may not because of the temporary interference. Interferers are discussed more in the final chapter of this book on troubleshooting WLANs.

RSSI

The Received Signal Strength Indicator (*RSSI*) is an arbitrary measurement of received signal strength defined in the 802.11 standards. No absolute rule exists as to how this signal strength rating must be implemented in order to comply with the IEEE standard. Other than the fact that it is optional (though most vendors implement it), it should report the rating to the device driver, and it should use 1 byte for the rating, providing a potential range of 0 to 255.

The IEEE does specify that a RSSI_MAX parameter should exist, which would be 100 for some vendors and 60 for others (resulting in great variability as to what

RSSI actually means). The RSSI_MAX parameter allows software applications to determine the range implemented by the vendors, and then convert the rating value into a percentage. It would not be very beneficial if the client software reported the actual rating to the user. Because of the different ranges used by the different vendors, using the actual rating would result in unusual scenarios. By this, I mean that an RSSI rating of 75 in one client may be the same relative rating as an RSSI rating of 45 in another chipset (assuming they are using similar linear stepping algorithms internally). Therefore, many applications use percentages, if they even report true RSSI today.

If a client card reported an RSSI of 47 and used an RSSI_MAX of 60, the software application could process the following formula to determine the signal strength in percentage:

47 / 60 * 100 = 78.3% signal strength

How does the software know to use the maximum value of 60? From the RSSI_MAX parameter that is required by the IEEE standard. Most vendors have chosen to use an RSSI of 0 to represent a signal strength less than the receive sensitivity of the device and, therefore, a signal strength that is not usable. This variance is why most client software packages report the signal strength in percentage instead of RSSI. The formula to calculate percentages from RSSI values is:

Signal Strength Percentage = RSSI / RSSI_MAX * 100

Where RSSI is the rating specified for the specific vendor chipset and RSSI_MAX is the highest RSSI rating possible. The result is the signal strength percentage value that you see in so many WLAN client software packages.

Now, let's make this even more complex — just for fun. Earlier I said that a rating of 75 in one vendor may be the same as a rating of 45 in another vendor assuming they use the same linear stepping algorithm. By linear stepping algorithm, I'm talking about the connection between dBm and RSSI rating. For example, one might assume that a dBm of -12 gets an RSSI rating of 100 for a vendor that uses RSSI_MAX of 100 and that a dBm of -12 gets an RSSI rating of 60 for a vendor that uses RSSI_MAX of 60. It would make sense to assume that the RSSI_MAX parameter is equal to the same actual dBm signal strength with all vendors; however, since the IEEE leaves it up to the vendors to determine the details of RSSI implementation (mostly because it is an optional parameter anyway), the different vendors often use different dBm signal strengths for their RSSI_MAX parameter. What is the result of this complexity? You may show a 100% signal strength for one client device and show a lesser signal strength for another client device from the exact same location, but it is not always a factor of receive sensitivity when RSSI is used as the metric. Your assumption may be that the client device with the lesser signal strength is providing inferior performance, when in fact they are identical.

How can this be? Consider a situation where two vendors use a RSSI_MAX value of 100. However, one vendor (vendor A) equates the RSSI rating of 100 to -12 dBm and the other vendor (vendor B) equates the RSSI rating of 100 to -15 dBm. Now, assume that both vendors use a linear stepping scale for their ratings, where a decrease in dBm of .7 causes the RSSI rating to drop by 1. This means that, at -15 dBm, vendor B will report 100% signal strength, but vendor A will have dropped the RSSI rating four times to a value of 96 and report a 96% signal strength. You can see how one might assume that vendor B's client is performing better because it has a higher percentage signal strength when, in fact, the two clients simply use a different implementation of the RSSI feature.

Due to these incompatibility issues, RSSI values should only be compared with the values from other computers using the same vendor's devices. RSSI values should never be conceptualized as universal, or in any way determinant of the value in one vendor's adapter over another vendor's value. Apples must be compared with apples — or to avoid confusion - Ciscos with Ciscos and NETGEARs with NETGEARs.

The RSSI rating may also be arbitrarily used to determine when to re-associate (roam) and when to transmit. Vendors will decide what the lowest RSSI rating should be before attempting to re-associate to a BSS with a stronger beacon signal. Additionally, vendors must determine when to transmit. To do this, they must determine a clear channel threshold. This is an RSSI value at which it can be assumed that there is no arriving signal, and therefore, the device may transmit.

Ok, we made it through that information time for some advice. Use dBm for signal strength and ignore RSSI. Many tools today have an RSSI metric column or section, but they actually just report the signal strength in dBm. dBm is a much better, accurate, and universal metric for signal strength reporting.

Link Budget and System Operating Margin (SOM)

The term *budget* can be defined as a plan for controlling a resource. In a wireless network, the resource is RF energy and you must ensure that you have enough of it to meet your communication needs. This is done by calculating a link budget that results in a system operating margin (SOM). *Link budget* is an accounting of all components of power, gain, loss, receiver sensitivity and fade margin. This includes the cables and connectors leading up to the antenna and the antenna itself. It also includes the factor of free space path loss. The many concepts we've been talking about so far in this chapter are about to come together in a way that will help you make effective decisions when building wireless links. You will take the knowledge you've gained of RF propagation and free space path loss and the information related to RF math and use that to perform link budget calculations that result in a SOM.

When creating a financial budget, money management coaches often suggest to their clients that they should monitor how they are currently spending their money. Then they suggest that these individuals create a budget that documents this spending of money. The alternative would be to go ahead and create a financial budget without any consideration for what your expenses are. I'm sure you can see that the latter simply will not work. First, you have to know how much money you need to live, and then you design your budget around that knowledge.

Similarly, in WLAN links, you will need to first determine the signal strength that is required at the receiving device and then figure out how you will accomplish this with your link budget. The first calculation you should perform in your link budget is to determine the minimum signal strength needed at the receiver and this is called the receive sensitivity. The receive sensitivity is not a single dBm rating, but it is a series of dBm ratings required to communicate at varying data rates. For example, Table 3.3 shows a portion of the receive sensitivity scale for an Aruba Networks AP200 series AP.

There are actually two ways to think of the receive sensitivity: the absolute weakest signal the wireless radio can reliably receive, and the weakest signal the wireless radio can reliably receive at a specific data rate. The lowest number in dBm, which is -95 dBm in Table 3.3, is the weakest signal the radio can tolerate. This number is sometimes referenced as the receive sensitivity or the absolute receive sensitivity. In more accurate terminology, the receive sensitivity of a device is the complete series or system of sensitivity levels supported by the device.

The receive sensitivity ratings are determined by the vendors. They will place the radio in a specially constructed shielded room and transmit RF signals of decreasing strength. As the RF signal strength is decreasing, the bit error rate in the receiving radio is increasing. Once this bit error rate reaches a vendor-defined rate, the power level in dBm is noted, and the radio is configured to switch down to the next standard data rate. This process continues until the lowest standard data rate for that 802.11-based device (1 Mbps in 2.4 GHz or 6 Mbps in 5 GHz) can no longer be achieved, and this dBm becomes the lowest receive sensitivity rating. In the end, a lower receive sensitivity rating is better because it indicates that the device can process a weaker signal.

dBm Power Level	Data Rate
-95 dBm	1 Mbps (802.11b)
-88 dBm	11 Mbps (802.11b)
-92 dBm	6 Mbps (802.11g)
-74 dBm	54 Mbps (802.11g)
-91 dBm	7.2/14.4 Mbps (802.11n 20 MHz 2.4 GHz)
-71 dBm	72.2/144.4 Mbps (802.11n 20 MHz 2.4 GHz)
-88 dBm	15/30 Mbps (802.11n 40 MHz 2.4 GHz)
-68 dBm	150/300 Mbps (802.11n 40 MHz 2.4 GHz)
-93 dBm	6 Mbps (802.11a)
-75 dBm	54 Mbps (802.11a)

Table 3.3: Aruba Networks AP200 Series Receive Sensitivity Requirements

The reason you need to know the receive sensitivity rating is that it is the first of your link budget calculations. The SOM is the amount of received signal strength relative to the devices received sensitivity. If you have a device with a receive sensitivity of -95 dBm and the card is picking up the wireless signal at -65 dBm, the SOM is the difference between -95 dBm and -65 dBm. Therefore, you would use the following formula to calculate the system operating margin:

SOM = S – RS

Where S is the signal strength (the second link budget calculation used to determine the SOM) at the wireless device and RS is the receive sensitivity of the device. Plugging in our numbers looks like this:

SOM = (-65) – (-95)

The resulting SOM is 30 dBm. This means that the signal strength can weaken by 30 dBm, in theory, and a link can be maintained. There are many factors at play when RF signals are being transmitted and this number, 30 dBm, will act as a good estimate. You may be able to maintain the link with a loss of 32 dBm and you may lose the link with a loss of 25 dBm. The link budget is a good estimate and should not be taken as a guarantee for connectivity.

It is important to remember that our calculation of the SOM was based on the lowest data rate of 1 Mbps. In fact, you may want a higher data rate. If you want to have a minimum data rate of 54 Mbps based on Table 3.3 in 2.4 GHz, and the received signal strength is -65 dBm, you are left with a SOM of only 9 dB. That is not a lot of extra space, and it may be important to get more dB variance in the SOM of the link through antenna gain or output power settings.

> Think of the receive sensitivity rating of a WLAN adapter as its "emotional intelligence." The receive sensitivity determines how sensitive it is to the signals passing by it, much like a human's emotional intelligence level determines how sensitive he is to the signals put off by other humans (facial expressions, sighs, etc.).

It is rare to calculate the link budget or SOM for indoor connections. This is because most indoor connections are not direct line-of-sight type connections, but instead they reflect and scatter all throughout the indoor environment. In fact, someone can move a filing cabinet and cause your signal strength to change. It can really be that fickle.

Outdoor links, specifically bridge links, are the most common type of links where you will need to create a link budget and determine the SOM. A detailed link budget can be much more complex than that which has been discussed so far. For example, it may include consideration for earth bulge, the type of terrain and the local weather patterns. For this reason, some vendors provide link budget calculation utilities.

Let's consider an actual example of a link budget calculation. Figure 3.3 shows a site-to-site link being created across a distance of 200 meters with 802.11 bridges. Based on the output power of the bridge, the attenuation of the cables, the gain of the antennas, and the free space path loss, we can calculate the link budget since the receive sensitivity of both bridges is -94 dBm. The calculations are as follows (if we accept a minimum data rate as the baseline receive sensitivity):

Link Budget calculation 1: **100 mW = 20 dBm**

Link Budget calculation 2: **20 dBm – 3 dB + 7 dBi – 83 dB = -59 dBm**

Link Budget calculation 3: **(-59 dBm) – (-94 dBm) = 35 dBm**

SOM = 35 dBm

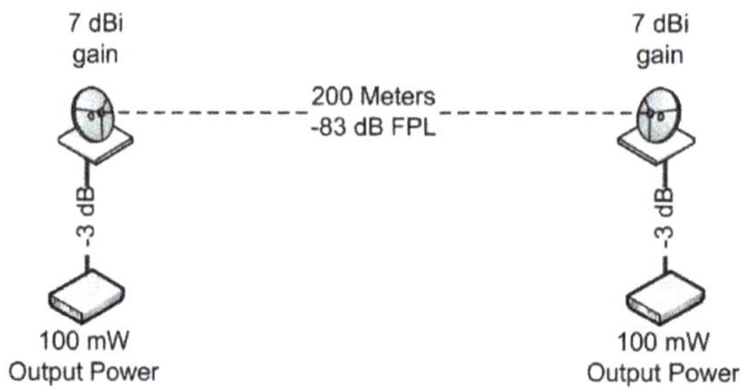

Figure 3.3: Link Budget Calculation

Fade Margin

Because of the variableness of wireless links, it is not uncommon to "pad the budget," much like a project manager may do for "risk factors" in a project. The padding of the budget is needed because the weather does change, trees grow, and buildings are built. These factors, and others, can cause the signal to degrade over time. By including a few extra dB of strength in the required link budget, you can provide a link that will hold up better under changing circumstances. The extra signal strength actually has a name and it is *fade margin*. You do not add to the link budget/SOM dBm value, but instead you take away from the receive sensitivity. For example, you may decide to work off an absolute receive sensitivity of -80 dBm instead of the -94 dBm supported by the devices in Figure 3.3. This would provide a fade margin of 14 dBm. It would also change our calculations, based on Figure 3.3, to a SOM of 21 dBm.

When you create outdoor bridge links, a fade margin is a practical requirement. Careful link budget calculations should be made to determine the SOM, and then you should pad that budget. Not drastically, but pad the budget. The fade margin will give you two things: a more consistent link and a longer lasting link. Without the fade margin, you may notice that the link drops periodically in certain seasons of the year, or the link simply fails to work after several months or years (due to changes in foliage or other environmental factors). Padding the budget with a fade margin helps in creating a more durable link.

For indoor communications, fade margins may not be required. Why? Because we rarely perform full link budget calculations for standard indoor WLANs. We depend on reflections and diffractions to get the signal to the proper end location within the environment. For indoor bridge links (connections to remote location in large buildings), you may want to calculate the SOM. For all other indoor WLANs, you'll likely just let the site survey do its job and ensure proper coverage and capacity in that way.

Intentional Radiator

The *intentional radiator*, in a WLAN transmission system, is the point at which the antenna is connected. The signal originates at a transmitter and may pass through connectors, amplifiers, attenuators, and cables before reaching the antenna. These components amplify or attenuate the signal resulting in the output power at the intentional radiator before entering the antenna. Regulatory agencies typically define rules regarding the power that can be delivered to the antenna and the power that can be radiated by the antenna. These are two different allowances, the first is the intentional radiator and the second is the antenna element. To understand this, you'll need to understand something called EIRP.

Equivalent Isotropically Radiated Power (EIRP)

The *Equivalent Isotropically Radiated Power* (*EIRP*) is the hypothetical power that is delivered by an intentional radiator to an imaginary isotropic antenna that would produce an even distribution of RF power with the same amplitude actually experienced in the preferred direction of the actual antenna. How's that for a technical definition? To make it simpler, it's the output power from the intentional radiator (output power from the transmitter plus any gains or losses leading up to the connection point of the antenna) plus the directional gain provided by the antenna. For the CWNA-108 exam, be sure to know the difference between the intentional radiator and EIRP.

Chapter Summary

In this chapter, you learned to perform RF math. First, you explore terms and concepts related to RF math, and then you learned the rules of 10s and 3s. Finally, you explored various ways in which this RF math is applied in practical WLAN implementations. In the next chapter, you will learn about antennas, antenna systems and accessories.

Points to Remember

Remember the following important points:

- Watt (W) is often used as a transmit power metric for outdoor links.
- Milliwatt (mW) is more often used as a transmit power metric for indoor links.
- dBm is decibels related to mW where 0 dB = 1 mW.
- Adding 3 dB of power doubles the original power level.
- Removing 3 dB of power halves the original power level.
- Adding 10 dB of power multiplies the original power level by 10.
- Removing 10 dB of power divides the original power level by 10.
- dB math with 10s and 3s is cumulative.
- The Fade Margin is extra dB of signal strength added to the link, usually by taking it from the receive sensitivity value.
- The SNR is the difference between the signal and the noise floor.
- SINR also considered interference sources above the noise floor, which is more realistic and is therefore growing in use as a term.
- dBi is decibels compared to an isotropic radiator and is used in antenna gain metrics.
- RSSI is an arbitrary signal measurement of little value when compared with dBm.

Review Questions

1. What metric is often used for indoor access point output power definition?
    a. W
    b. RSSI
    c. mW
    d. dBd

2. What signal metric has become very popular for both transmit and receive power levels?
    a. W
    b. RSSI
    c. dBm
    d. dBi

3. Why is the RSSI metric not typically a good signal metric for comparison across vendors?
    a. Each vendor specifies the RSSI percentage formula differently
    b. Each vendor specifies the RSSI_MAX value differently
    c. RSSI is no longer supported in WLAN devices
    d. RSSI is not a signal metric; it is a modulation method

4. You start with a power level of 50 mW and add 7 dB, what is the resulting power level?
    a. 500 mW
    b. 100 mW
    c. 200 mW
    d. 250 mW

5. An AP has an output power setting of 20 dBm. The antennas provide 5 dB of gain. What is the EIRP?
    a. 500 mW
    b. 320 mW
    c. 3200 mW
    d. 4 W

6. An AP has an output power setting of 13 dBm. The antenna is directly connected and has 3 dBi gain. What is the output power at the intentional radiator?
    a. 400 mW
    b. 40 mW
    c. 100 mW
    d. 20 mW

7. You are calculating an outdoor bridge link. You have determined the bridge output power levels, antenna gain, and losses incurred by all cables and connectors on both ends. What additional factor is missing?
   a. FSPL
   b. RSSI
   c. Fade Margin
   d. Wind resistance

8. How do you calculate the dBi of an antenna when the gain is provided in dBd?
   a. Subtract 2.14
   b. Add 2.14
   c. Multiple by 2.14
   d. Divide by 2.1

9. You are calculating power levels and have been given the following values: +3 dB, +5 dB, -21 dB, +11 dB, and +3 dB. How can you simplify the calculation?
   a. Just use the value +1 dB
   b. Ignore the negative values
   c. Ignore the positive values
   d. You cannot simplify it you must calculate all dB changes in sequence

10. What metric is preferred by engineers and is used to represent received signal strength?
   a. RSSI
   b. dB
   c. dBm
   d. Frequency

Review Answers

1. **C is correct.** mW (milliwatt) is more often used for indoor access points as output power levels rarely require above 100 mW.

2. **C is correct.** Decibels to milliwatts (dBm) is the most common metric used for both transmit and receive power levels today.

3. **B is correct.** The RSSI_MAX value can be up to 255 and vendors set it at varying values from 31 to 100 commonly. The end result is inconsistencies in what the metric means across vendors.

4. **D is correct.** 50 mW + 10 dB + (-3 dB) = 250 mW.

5. **B is correct.** The answer is 320 mW. First, 20 dBm is equal to 100 mW. Then, you can add 3 dB five times to get to 15 dB. That means you calculate 100 mW * 2 * 2 * 2 * 2 * 2 = 3200 mW. Now you have to remove 10 dB to get back to the 5 dBi gain of the antenna. 3200 / 10 = 320 mW.

6. **D is correct.** The output power of the AP is 13 dBm, which is equal to 20 mW. Given that the antenna is directly connected, the IR power is the same as the AP power, so it is 20 mW. The antenna gain only becomes a factor for EIRP and not for IR calculations.

7. **A is correct.** FSPL must be factored into outdoor bridge links.

8. **B is correct.** To convert from dBd to dBi, add 2.14 dB. To convert from dBi to dBd, subtract 2.14 dB.

9. **A is correct.** Because the values +3 dB, +5 dB, -21 dB, +11 dB, and +3 dB total +1 dB, you can simplify calculations to 1 dB of gain in this system.

10. **C is correct.** While RSSI is a common metric as well, engineers prefer dBm because it is an absolute power measurement, whereas RSSI is arbitrary and uniquely specified by WLAN vendors.

Chapter 4 — RF Antennas and Hardware

This chapter introduces the topic of antennas as they relate to WLANs. The good news is that in typical indoor office deployments today, most installations use APs with internal antennas, and you simply mount enough of them in the right locations and with the right settings to achieve the coverage and capacity you require. Okay, there is quite a bit of complexity behind determining those locations and settings, but that is CWDP-level information and will be left for that exam.

Here, we will cover basic antenna functionality, antenna charts and antenna types. We will also look at specific multi-antenna uses in modern devices, such as MIMO, MU-MIMO, STBC, MRC, and transmit beamforming, to name a few. So, raise your learning antennas, and let's get started.

Antenna Functionality

The antenna is the radiating element in an RF system. It is the device that actually causes RF waves to be propagated through space. It is also the device that receives the RF signals from other propagating antennas. Different antennas have different coverage capabilities and different characteristics, which you will learn about in this and the next section.

Visual (Physical) LOS

If you stand on top of a tall building, you can see for a very great distance. You may even be able to see for many miles on a very clear day. If you can physically see something, it is said to be in your visual line of sight (visual LOS) or just LOS, for simplicity. This is also called physical LOS. This *visual LOS* is actually the transmission path of the light waves from the object you are viewing (transmitter) to your eyes (receiver). Visual LOS is an apparently straight line from your perspective, but light waves are subject to similar behavior as RF waves, like refraction and reflection, and therefore the line may not actually be straight. Consider an object you are viewing in a mirror. The object is not directly in front of you and yet it appears to be, showing that visual LOS is not necessarily a straight line between two objects.

RF LOS

Because RF is part of the same electromagnetic phenomenon as visible light, behaviors similar to visual LOS exist. However, RF LOS is more sensitive than visual LOS to interference nearby the path between the transmitter and the receiver. You might say that more space is needed for the RF waves to be seen by each end of the connection. This extra space can actually be calculated and has a name: *the Fresnel Zone*. The Fresnel Zone is only important for bridge links, as indoor propagation patterns ensure the signals get through to the clients as long as the WLAN is implemented on proper design principles.

> When creating bridge links, start with the visual LOS, and then determine whether you have RF LOS. For long distance outdoor links, you will never have RF LOS if you don't have visual LOS; however, you may not be able to see as far as the link reaches, so you may have to rely on topographic maps of the area.

The Fresnel Zone

Before getting into specific antenna types, we need to explore a bit more math and some concepts related to outdoor links. The Fresnel zones (pronounced frah-nell) are named after the French physicist Augustin-Jean Fresnel and are a theoretically infinite number of ellipsoidal areas around the LOS in an RF link. Many WLAN administrators refer to the Fresnel zone when it is more proper to refer to the *first* Fresnel zone, according to the science of Physics. While it may be the intention of most WLAN administrators to reference the first Fresnel zone when they speak of *only* the Fresnel zone, it is important that you understand the difference. The first Fresnel zone is the zone with the greatest impact on a WLAN link, in most scenarios. The Fresnel zones have been referenced as an ellipsoid-shaped area, an American football-shaped area, and even a Zeppelin-shaped area. Figure 4.1 shows the intention of these analogies.

In this text, we will call Fresnel zone 1 1FZ from this point forward for simplification. Since 1FZ is an area surrounding the LOS, and this area cannot be largely blocked and still provide a functional link, it is important that you know how to calculate the size of 1FZ for your links. You'll also need to consider the impact of earth bulge on the link and 1FZ.

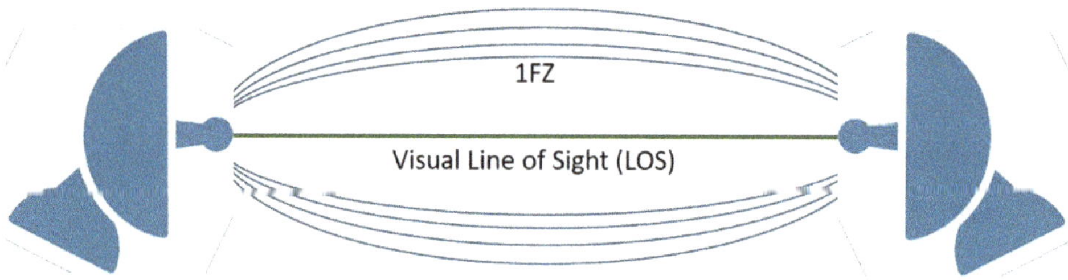

Figure 4.1: A First Fresnel Zone (1FZ) Representation

You will not need to memorize the 1FZ formulas provided here for the CWNA-108 exam; however, they will prove a useful reference for you when you need to create Point-to-Point (PtP) or Point-to-Multi-Point (PtMP) links in the future.

To calculate the radius of the 1FZ, use the following formula:

radius = 72.2 x √ (D / (4 x F))

Where D is the distance of the link in miles and F is the frequency used for transmission in GHz and radius is reported in feet (72.2 is the constant for feet and 17.32 would be the constant for meters). For example, if you are creating a link that will span 1.5 miles and you are using 2.4 GHz radios, the formula would be used as follows:

72.2 x √ (1.5 / (4 x 2.4)) = 28.54 feet

This formula provides you with the radius of the 1FZ, and doubling the result would give you the diameter, if you needed it to be calculated. However, it is important to realize that a blockage of the 1FZ of more than 40% can cause the link to become non-functional. To calculate the 60% radius, so that you can ensure it remains clear, use the following formula:

clearance radius = 43.3 x √ (D / (4 x F))

Where D is the distance of the link in miles and F is the frequency used for transmission in GHz and radius is reported in feet. Using the same example, we used to calculate the radius of the entire 1FZ, you will now see that the 60% clearance radius is only 17.12 feet. However, this leaves no room for error or change. For example, trees often grow into the 1FZ and cause greater blockage than they did at the time of link creation. For this reason, many WLAN engineers choose to use a 20% blockage or 80% clearance guideline and this is the recommended minimum clearance of the CWNP program as well. So how would you calculate this? Use the following formula:

$$\text{recommended radius} = 57.8 \times \sqrt{(D / (4 \times F))}$$

Once you've processed this formula, you will see that the recommended minimum of 80% clearance (recommended maximum of 20% blockage) results in a 1FZ radius of 22.8 feet.

As it is always better to be safe rather than sorry when creating WLAN links, you will probably want to make it a habit to round your Fresnel zone calculations upward. For example, we would round the recommended radius to 23 feet in our example.

> Remember that 80% clearance of the first Fresnel zone is recommended for 802.11 bridge links and 60% clearance is required. The additional 20% clearance allows for environmental changes without significantly impacting the link.

You might be wondering why we calculate the radius instead of the diameter. The reason is simple: we can determine where the visual LOS resides and then measure outward in all directions around that point to determine where the 1FZ actually resides. Remember, the 1FZ does not reside in a downward direction only. It might seem that way since we are usually dealing with trees and other objects protruding up from the ground as interference and blockage objects. However, it is entirely possible that something could be hanging down from a very high position — such as a bridge — and encroach on the 1FZ from above the visual LOS. Additionally, buildings and other objects can cause blockages from the sides. For example, if you are attempting to create a PtP link that has visual LOS between two buildings on either side of the link in a downtown area, the two buildings may encroach on the 1FZ resulting in insufficient signal strength for a consistent connection.

Another factor that should be considered in 1FZ blockage is the Earth itself. As you know, the Earth — it turns out — is round. When any two objects for that matter are farther apart, there will be a greater likelihood that the Earth is between them. This is demonstrated in Figure 4.2. Note the encroachment of the earth on the 1FZ over a significant distance.

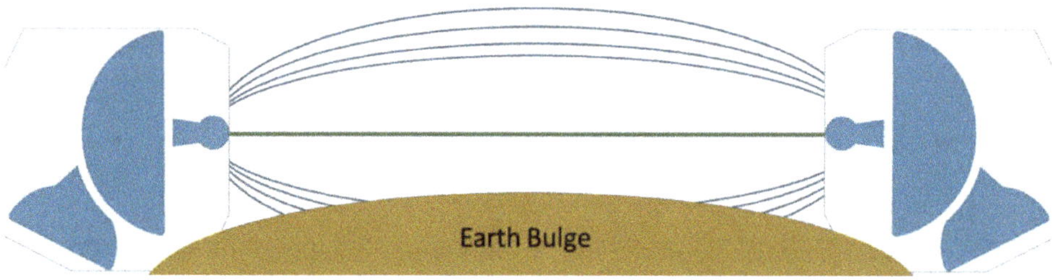

Figure 4.2: Earth Bulge Demonstrated

If you are creating wireless links over distances greater than 7 miles using WLAN technologies, you will need to account for earth bulge in your antenna positioning formulas. Again, you will not need to memorize the following formula for the CWNA examination, but you will need to know that earth bulge is a potential problem in outdoor wireless links over greater distances. The formula for calculating the extra height your antennas will need to compensate for earth bulge is:

$$\text{Height} = D^2 / 8$$

Where height is the height of earth bulge in feet, and D is the distance between antennas in miles. Therefore, if you are creating a 10-mile link, you would process the following formula:

$$100 / 8 = 12.5 \text{ feet}$$

Using our guideline of rounding up, we would raise the antenna height by 13 to 14 feet to accommodate for earth bulge.

It is important that you learn to deal with 1FZ obstructions. If the obstructions are coming up from the ground into the 1FZ and there are no obstructions anywhere above it, you can often solve the problem by simply raising the antennas involved in the communication link. For example, if there is a forest with maximum tree height of 23 feet that is between the two antennas, and there is a distance of 11 miles that must be spanned, we can calculate the needed height for the antennas including earth bulge with the following formula:

$$\text{minimum antenna height} = (57.8 \times \sqrt{(11 / (4 \times 2.4))}) + (121 / 8)$$

This might seem complex, at first, but it is a simple combination of the recommended 1FZ clearance formula and the earth bulge formula. The result is rounded up to 77 feet. This means you will need to install very high towers and you will also need to monitor the forest, though it is unlikely that the trees would grow that much more into the 1FZ in a few years. Additionally, you will likely be required to acquire permits for the towers in most regulatory domains.

If the obstructions are coming into the 1FZ from the sides — such as buildings intruding into the pathways, you will have to either calculate the 1FZ for a different frequency to see if you can get the clearance, or you will have to raise the antennas above the buildings. You may also be able to create a multi-hop link to "shoot" around the buildings if you can gain access rights to a third location that can be seen (RF LOS — including 1FZ) by both of your locations.

Notice that it was an option to calculate the 1FZ with a different frequency. Because the Fresnel zones are a factor of wavelengths and not a factor of antenna gain or beamwidth (covered later in this chapter), which is very important to differentiate, you can often implement a PtP link successfully using different frequencies. For example, the 77-foot antenna height to allow us to communicate over the top of the forest across 11 miles can be lowered to only 54 feet, if you are using 802.11 devices in the 5 GHz range. However, the trade-off is in distance. The 2.4 GHz signals are detected more easily than 5 GHz signals at a distance due to the receiving area of the antenna element and the length of the signal wave, but 5

GHz signals have a narrower 1FZ. The formula, when using the 5 GHz band changes to the following, assuming you use channel 149 for the link, which is centered on frequency 5.745 GHz:

$$\text{minimum antenna height} = (57.8 \times \sqrt{(11 / (4 \times 5.745))}) + (121 / 8)$$

An example of this is a link that travels only about a city block (0.1 miles). In the 2.4 GHz spectrum, the 1FZ radius would be approximately 6 feet. In the 5 GHz spectrum the 1FZ would only be about 4 feet. Remember, this means 6 feet or 4 feet out from the center point in all directions. Therefore, a 5 GHz link traveling between two buildings for 0.2 mile would require a space between the buildings of about 8-9 feet, while the 2.4 GHz link would need a space between the buildings of about 12-13 feet. These factors are important considerations.

Beamwidths

Different antennas have different *beamwidths* and this beamwidth is the measurement of how broad or narrow the focus of the RF energy is as it propagates from the antenna along the main lobe. The main lobe is the primary RF energy coming from the antenna. Beamwidth is measured both vertically and horizontally, so don't let the term width confuse you into thinking it is a one-dimensional measurement. Specifically, the beamwidth is a measurement taken from the center of the RF signal to the points on the vertical and horizontal axes where the signal decreases by 3 dB or half power. In the end, there is a vertical and horizontal beamwidth measurement that is stated in degrees. Figure 4.3 shows both the concept of the beamwidth and how it is measured, and Table 4.1 provides a table of common beamwidths for various antenna types (these antenna types are each covered in detail later in this chapter).

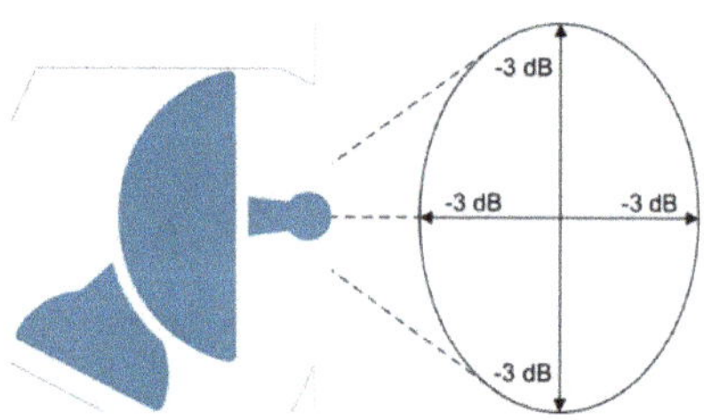

Figure 4.3: Beamwidth Concept and Measurement

Remember that the beamwidth is calculated where the signal reaches half power or -3 dB. It is calculated both horizontally and vertically and it indicates the focus of the beam in the antenna main lobe.

Some example antennas are listed in Table 4.2 with their horizontal and vertical beamwidths.

Antenna Type	Horizontal Beamwidth	Vertical Beamwidth
Omni-directional	360 degrees	7 to 80 degrees
Patch/panel	30 to 180 degrees	6 to 90 degrees
Yagi	30 to 78 degrees	14 to 64 degrees
Sector	60 to 180 degrees	7 to 17 degrees
Parabolic dish	4 to 25 degrees	4 to 21 degrees

Table 4.1: Common Beamwidths of Various Antenna Types

Antenna Model	Horizontal Beamwidth	Vertical Beamwidth
Cisco 9.5 dBi sector antenna	60 degrees	60 degrees
Cisco 2.2 dBi dipole antenna	360 degrees	55 degrees
Cisco Multi-band wall-mount (patch/panel) antenna	68 degrees	66 degrees
Hyperlink Technologies 2.4 GHz die cast grid antenna	8 degrees	8 degrees

Table 4.2: Specific Antenna Beamwidths

While beamwidth measurements give us an idea of the propagation pattern of an antenna, they are less than perfect at illustrating the actual areas that are covered by the antenna. For more useful visual representations, you will want to reference Azimuth and Elevation charts. However, when textual documentation of an antenna's characteristics is desired, the beamwidth is typically the best choice.

Azimuth & Elevation

Where the beamwidth calculations provide a measurement of an antenna's directional power, Azimuth and Elevation charts, which are typically presented together, provide a visualization of the antenna's propagation patterns. Figure 4.4 shows an example of an Azimuth chart and Figure 4.5 shows an example of an Elevation chart.

The difference between an Azimuth and Elevation chart is simple: The Azimuth chart shows a top-down view of the propagation path (to the left, in front, to the right and behind the antenna) and the Elevation chart shows a side view of the propagation path (above, in front, below and behind the antenna). Think of these charts in terms of a dipole antenna that is positioned vertically upright. If you are standing directly above it and looking down on it, you are seeing the perspective of an Azimuth chart. If you are beside looking at it from a horizontally level position, you are seeing the perspective of an Elevation chart.

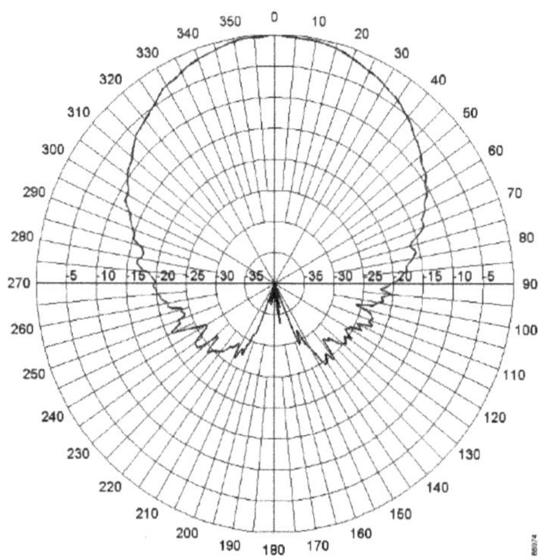

Figure 4.4: Antenna Azimuth Chart

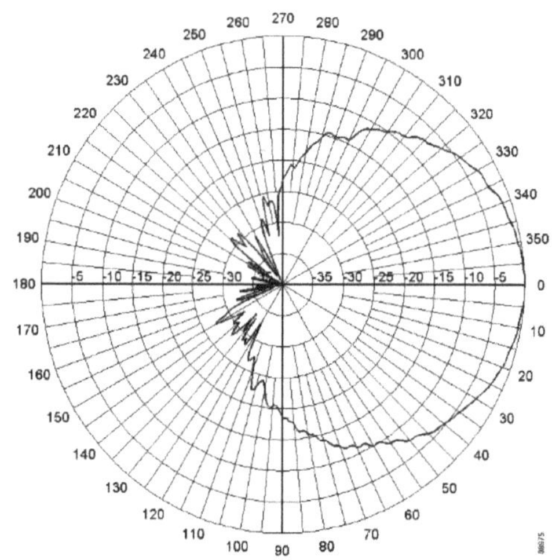

Figure 4.5: Antenna Elevation Chart

The Azimuth chart in Figure 4.4 is a chart of the Cisco 9.5 dBi sector antenna referenced in Table 2.2. As with most Azimuth charts, the direction of propagation is represented in the upward direction; however, the actual direction will depend on how you position the antenna — more on that in the later section titled

"Polarization." The chart is reporting the different signal strength you can expect at different degrees from the antenna. For example, at 90 and 270 degrees (to the immediate left and right of the antenna's intended propagation direction) you will see a loss of approximately 20 dB. Directly behind the antenna, at 180 degrees, you will see a loss of approximately 35 to 50 dB. This is a sector antenna and is intended to propagate its energy in one direction, but in a fairly wide path.

The Elevation chart in Figure 4.5 is for the same Cisco antenna. You will notice that the pattern of propagation is very similar to the Azimuth pattern. Like most Elevation charts, it is shown with the primary radiation direction to the right. Remember, this is intended to represent you looking at the antenna's propagation pattern from the side view. You can see that this antenna as very similar levels of loss along the same degree levels as the Azimuth chart.

> Azimuth charts show the propagation pattern from a top-down perspective and are also called horizontal charts. Elevation charts show the propagation pattern from a side perspective and are also called vertical charts.

Isotropic Radiator

The *isotropic radiator* is a fictional device or concept that cannot be developed using today's technology. Many say that it is not only impossible now, but because of the constraints of physics, will always be impossible. While the future may be debatable, we know that you cannot currently create an antenna that propagates RF energy equally in all directions. This truth is due to the fact that the antenna must have some length (it must exist) and it must receive power from some source (it must be connected to something). These two constraints alone make it impossible to create an isotropic radiator at this time.

Even though we cannot create such a device, it is a useful theoretical concept in that we can use it as a basis for measurements. In fact, dBi — as was stated earlier in the book — is a measurement of the gain of an antenna in a particular direction over the power level that would exist in that direction if the RF energy were propagated by an isotropic radiator. In other words, dBi is a measurement of the difference between the power levels at a point in space generated by a real antenna versus the theoretical isotropic radiator. Since we can all agree on the behavior of an isotropic radiator, we can all use it as a basis for such power level measurements. Figure 4.6 illustrates the concept of the theoretical isotropic radiator.

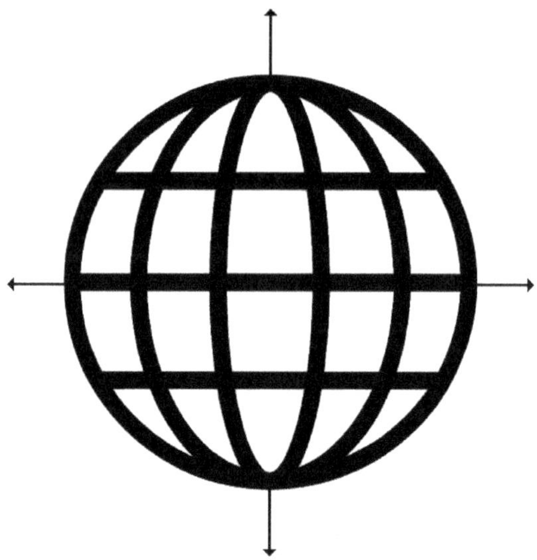

Figure 4.6: Theoretical Isotropic Radiator

The sun is often used as an analogy of an isotropic radiator. While this is an acceptable analogy, certain theories in physics — such as the hairy ball theorem, would even exclude the sun from being a true and complete isotropic radiator. However, it is one of the objects we've found to be closest to an isotropic radiator in that light certainly propagates from it in all directions. If we could analyze that light at the molecular level — or even the individual wave level, it is questionable as to whether the rays are truly radiated "equally" in all directions.

Polarization

A factor that greatly impacts the performance of RF antennas is the polarization of the antennas. Antenna polarization refers to the physical orientation of the antenna in a horizontal or vertical position — typically. Technically, it's about how the antenna is designed to be used, but in practical application, it's about how you position it.

You'll remember from previous discussions in this, and other chapters, that the electromagnetic wave is made up of electric and magnetic fields. The electric field forms what is known as the E-plane and the magnetic field forms what is known as the H-plane. The E-plane is parallel to the radiating antenna element and the H-plane is perpendicular to it. Simply stated, the E-plan runs alongside the antenna regardless of how you position it. If you put the dipole antenna upright, the E-plan is upright alongside it; however, if you tilt it, the E-plane is tilted as well.

The E-plane, or electric field, determines the polarization of the antenna since it is parallel to the antenna. Therefore, if the antenna is in a vertical position it is said to be vertically polarized. If the antenna is in a horizontal position, it is said to be horizontally polarized.

A vertically polarized omni-directional antenna propagates the signal horizontally, and a horizontally polarized omni-directional antenna propagates the signal vertically, which is not what you typically desire unless you are creating a bridge link between floors in a tall building. If you configure a link like this (horizontally between floors and along walls), a best practice would be to place the antennas approximately 2 feet out from the wall to prevent the wall from creating 1FZ interference. The link would likely work anyway, but communications can be improved with this consideration. The spacing of 2 feet should keep the 1FZ 80% clear for up to 60-70 feet.

The impact of polarization is seen when antennas are not polarized in the same way. For example, if you have an access point with the antennas positioned vertically (vertical polarization), and you have a USB client adapter with the antenna down (horizontal polarization), your connectivity will be less stable and, at greater distances, may even be lost. However, in most cases, due to indoor reflections, the polarization of antennas does not have as great an impact as it does with outdoor links. In outdoor links, the proper polarization of the antennas can make or break the connection.

Remember this: vertical polarization usually means that most of the signal is being propagated horizontally and horizontally polarization means that most of the signal is being propagated vertically, as previously stated. Therefore, the most popular polarization is vertical polarization.

Antenna Diversity

In preceding chapters, antenna diversity was briefly mentioned, here it is explained in more detail. *Antenna diversity* is a feature offered by many WLAN access points, routers and even some clients, that allows the device to receive signals using two antennas and one receiver. In a traditional antenna diversity implementation, only one antenna is used at a time, so this should not be confused with Multiple-Input/Multiple-Output (MIMO) configurations.

Remember that the wavelengths of RF signals are very short in the 2.4 and 5 GHz frequencies used by WLAN equipment. In fact, they are only about 5 inches long for 802.11 equipment operating at 2.4 GHz, and 2 inches long for 802.11 equipment operating at 5 GHz. This means that the antennas used in antenna diversity can actually be within just a few inches of each other and still receive very different signals. As you can imagine, based on what you've read in this book so far, RF signals are bouncing all around inside of the building that houses a WLAN. This activity means that a device can transmit a signal, and it may arrive at a receiving device from multiple angles with multiple signal strengths.

The device supporting antenna diversity will look at the signal that comes into each antenna during the frame preamble of a single WLAN frame and choose the signal that is best, on a frame-by-frame basis. If you're unfamiliar with frames, they will be covered in Chapter 7 in sufficient detail. The best frame preamble will determine which antenna is used to receive the rest of the current frame. Again, remember that there is only one receiver that has two connections and two antennas.

An additional type of diversity is Multiple-Input/Multiple-Output (MIMO) diversity. MIMO systems use more than one antenna in several different ways, but they can also support diversity. For example, a device may be a 2x3 device, which means that it can transmit on two antennas, but receive on three. Such a configuration would allow for diversity selection during frame reception providing MIMO receive diversity. Such a solution may also support maximal ration combining (or maximum ratio combining, depending on whose whitepaper you're reading) as discussed in the following section.

Advanced Antenna and RF Technologies

Wireless devices supporting the IEEE high throughput (HT) and very high throughput (VHT) physical layers, or 802.11n and 802.11ac respectively, can use several special techniques with the multiple antennas provided. These techniques include:

- Spatial Multiplexing – MIMO and MU-MIMO
- Transmit Beam Forming
- Maximal Ratio Combining
- Space Time Block Coding (STBC)

Spatial Multiplexing

Spatial Multiplexing (*SM*) uses advanced algorithms to create separate data streams for each transmitting antenna. It requires multiple radio chains, which are effectively radios linked to transmitting (Tx) and receiving (Rx) antennas. The 802.11n (HT) PHY can support up to four spatial streams with four radio chains and the 802.11ac (VHT) PHY can support up to eight spatial streams, though, at the time of writing, only 3x3:3 802.11n devices and 4x4:4 802.11ac devices were available. Such devices use MIMO chipsets.

The nomenclature 3x3:3 should be understood by the CWNA candidate. The interpretation is:

Tx chains x Rx chains: spatial streams

Therefore, a 3x3:3 device can transmit and receive on three radio chains and can send and receive three spatial streams for spatial multiplexing. A 3x3:2 device has three transmit antennas, three receive antennas, but can only process two spatial streams at a time.

If a spatial multiplexing link is to function at the highest possible data rate, the following factors must be true:

- The receiving device must have an equal number of radio chains to the transmitting device. For example, if the AP has four transmitting (Tx) and receiving (Rx) antennas, the client must also have four.
- The signal strength must be strong enough to avoid data rate switching to lower data rates.
- The link must be operating with spatial multiplexing enabled.

Complex spatial multiplexing allows for multi-user MIMO (MU-MIMO) in 802.11ac chipsets, as long as the devices support it. Early 802.11ac devices did not support MU-MIMO, but most new shipping devices do, at the time of writing. MU-MIMO allows the AP to transmit to multiple 802.11ac clients simultaneously. 802.11ac does not support uplink MU-MIMO communications. The forthcoming 802.11ax is expected to support uplink MU-MIMO.

For MU-MIMO to work, clients must be ideal locations to be grouped into MU groups and they must have data frames waiting for them in the AP at the same time. Because of these constraints, and others, MU-MIMO has not proven to be an exceptional performance booster in production 802.11ac WLANs. This is likely to change with 802.11ax due to more advanced modulation and channelization methods being included.

Transmit Beamforming

Transmit Beamforming (*TxBF*) is a special antenna technology that allows the signal to be focused on a specific destination. In order to use TxBF, the characteristics of the signal received at the remote wireless node must be known. Special communications called sounding frames occur between the AP and client to discover this information. TxBF uses multiple antennas and adjusts those antennas to simulate a sector array of antennas.

To better understand TxBF, you must first understand the phenomenon of multipath. Multipath occurs when the transmitted signal reflects, refracts, diffracts and scatters as it travels. The result is often that more than one copy of the signal arrives at the receiver. If two copies arrive at the receiver in phase with each other, upfade occurs and the signal strength is increased. If the signals arrive out of phase, the signal can be downfaded (resulting in a loss), corrupted, or canceled out completely.

TxBF devices work with the client, which must also be an 802.11n or 802.11ac client since 802.11b/g/a devices have no idea how to cooperate with TxBF to determine how to calibrate the transmissions so that multiple signals arrive in phase. Any time the client moves, the TxBF transmission must be recalculated. The result is that TxBF is best for nomadic roaming or in non-congested WLANs that will not be significantly impacted by extra frame transmissions for TxBF operations. With nomadic roaming, the clients move, but remain stationary most of the time.

Maximal Ratio Combining

Maximal Ration Combining (*MRC*) uses antenna diversity to increase the strength of the received signal through combination algorithms. Traditional antenna diversity uses only one antenna to receive a wireless frame, even if both antennas receive the signal fine because only one radio exists. With 802.11n and 802.11ac devices, MRC can combine the signals of the two antennas to increase signal strength at greater distances. The result is a reduction in dynamic rate switching activity, which lowers the data rate and the throughput of the wireless link.

Space Time Block Coding (STBC)

STBC is another diversity technique for improving an RF link's signal-to-noise ratio and may be applied when the number of transmitting antenna chains exceeds the number of receive antennas. STBC uses coding to transmit different, but known, copies of the data-stream from different antennas. As each of the different versions of the original signal reach the receiver, they are received and processed with specialized decoding techniques to provide signaling redundancy and optimized reception. Use of STBC allows the receiver to extract the original data with fewer errors than when a single transmit antenna is used, but both the transmitter and receiver must know the STBC coding scheme in order to take advantage of this mechanism.

One form of space time coding, Alamouti code, is used to spread one spatial stream over two space-time streams. This process takes a pair of data-stream bits and performs operations on them in consecutive time intervals. This is the basic

STBC method defined in the 802.11n PHY. When processing this sequence of two symbols from two space-time streams, the receiver is able to re-constitute the original data-stream, even in the presence of channel noise and distortion. STBC uses the time dimension, where consecutive symbol intervals contain the same pair of basic symbols but modified according to the code. A good choice of code, such as Alamouti, allows minimum complexity for the transmitter and receiver, but maximum improvement in SNR under normal channel impairments.

Antennas and Antenna Systems

Now that you understand several antenna system concepts, in this section, we will cover the basic types of antennas that are available to you, including their RF propagation patterns and their intended use.

Three primary categories of antennas are used today:

- Omni-directional
- Semi-directional
- Highly directional

In addition, variations on the implementation and management of these antenna types exist, which results in the sectorized and phased array antennas. These antenna types will also be addressed in this section. Finally, we'll review the MIMO (Multiple-Input/Multiple-Output) antenna systems that are used by the 802.11n and 802.11ac standard.

Omni-directional/Dipole Antennas

Omni-directional antennas, the most popular type being the dipole antenna in early days, and being internal device antennas today, are antennas with a 360-degree horizontal propagation pattern. In other words, they propagate most of their energy outward in a 360-degree pattern shaped much like a doughnut — though a very thick one, in low-gain omni antennas. The omni-directional antenna provides coverage at an angle upwards, downwards and directly out horizontally, as is shown in the Elevation chart in Figure 4.7.

Inspecting the Elevation chart in Figure 4.7 reveals that an omni-directional antenna propagates most of its energy to the right and left of the antenna (from a side view) and very little energy directly above the antenna. At the same time, the Azimuth chart shows a fairly even distribution around the antenna (from a top-down view). This is the common propagation characteristic of omni-directional antennas. Figure 4.8 shows a typical omni-directional antenna of the dipole design.

The omni-directional antenna is most commonly used indoors to provide coverage throughout an entire space; however, they have become more and more popular in outdoor usage for either hotspots or private access outdoor networks. Omni-directional antennas may be mounted on poles, masts, towers, ceilings or desktops and floors. They provide coverage on a horizontal plane with some coverage vertically and outward from the antenna. This means they may provide some coverage to floors above and below where they are mounted in some indoor installations. Most APs with internal antennas use omni-directional antenna patterns, though some use semi-directional patterns. Always consult the vendor specifications to determine the pattern implemented by a given AP.

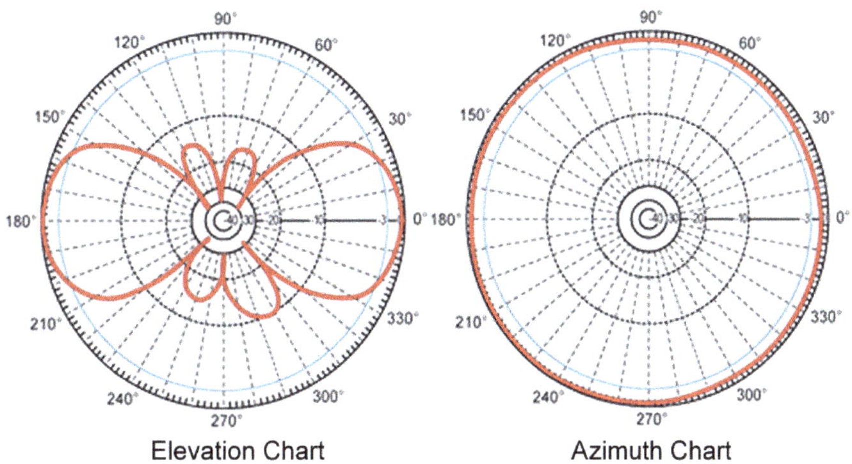

Figure 4.7: Omni-directional Elevation and Azimuth Charts

Figure 4.8: Omni-directional Antenna

Because all antennas use passive gain — they focus the RF energy — it is important to consider the impact of this passive gain on any antenna that you implement. In the case of omni-directional antennas, the result is that devices directly above or below the omni-directional antenna may have a very weak signal or even be unable to detect the signal. This is due to the primary signal being focused outwardly on a horizontal plane.

You can use antennas that have higher dBi gain such as 12 or 15 dBi omni antennas; however, you must keep the impact of these higher gain antennas in mind. As an example, consider the two Elevation charts side-by-side in Figure 4.9. The one on the left is from a 4 dBi omni antenna and the one on the right is from a 15 dBi omni antenna. You can clearly see the flattening of the signal. It is very plausible that a higher gain antenna, such the one on the right, could prevent users on the floors above and below the antenna to lose their connection. Ultimately, when using omni antennas, choosing between a higher gain and a lower gain is choosing between reaching people farther away horizontally (higher gain) or reaching people farther up or down vertically (lower gain). In most situations, you'll just place separate antennas (APs) on each floor of a multi-floor installation to get the coverage you need.

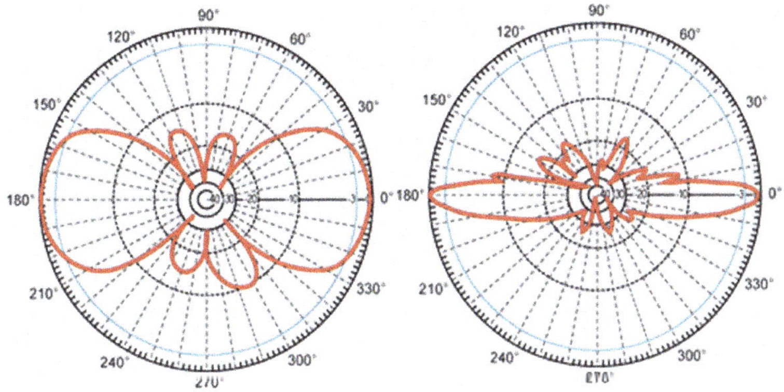

4 dBi 15 dBi

Figure 4.9: 4 dBi vs. 15 dBi Antennas

Semi-directional Antennas

Semi-directional antennas are antennas that focus most of their energy in a particular direction. Examples include patch, panel and Yagi antennas. (Yagi is pronounced yah-gee.) Patch and panel antennas come in flat enclosures and can be easily mounted on walls. Yagi antennas look a lot like TV antennas – a long rod with tines sticking out; however, the Yagi antennas are usually enclosed in a plastic casing that hides this appearance. Patch and panel antennas usually focus their energy in a horizontal arc of 180 degrees or less, where Yagi antennas usually have a 90 degree or less coverage pattern. Figure 4.10 shows examples of patch, panel and Yagi antennas.

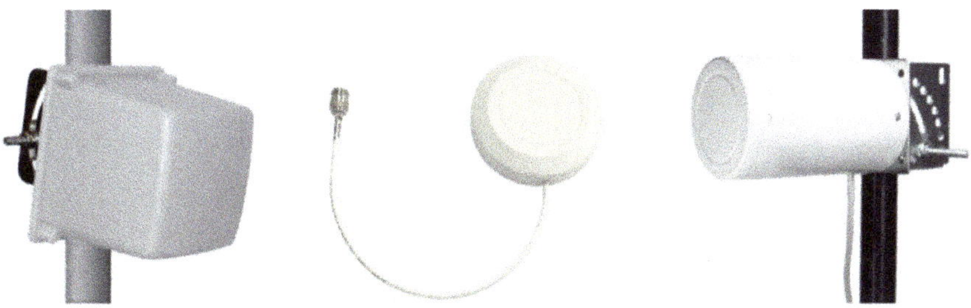

Panel Antenna Patch Antenna Yagi Antenna

Figure 4.10: Semi-Directional Antennas

The Azimuth and Elevation charts for Yagi antennas often look the same. They often have the same coverage pattern from the top-down view (horizontal coverage) as they do from the side view (vertical coverage). Figure 4.11 shows an example coverage pattern of a 9 dBi Yagi antenna. Panel antennas usually have a similar pattern to Yagi antennas except the "fish-like design" appears quite a bit fatter or thicker.

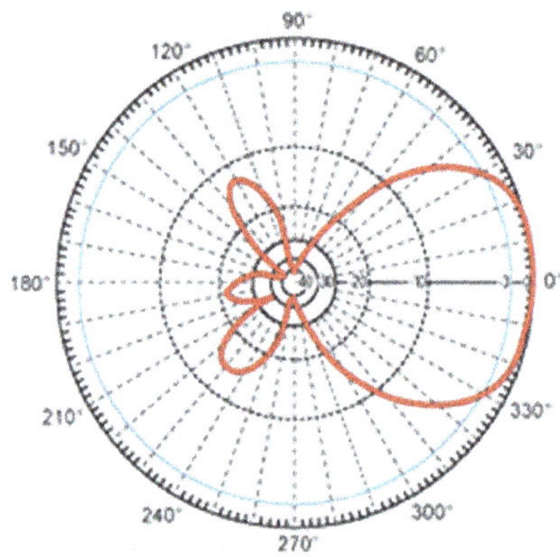

Figure 4.11: 9 dBi Yagi Antenna Pattern (Elevation Chart)

Semi-directional antennas are useful for providing RF coverage down long hallways or corridors when using Yagi-style antennas. They are also useful when providing RF coverage in "one" direction using patch or panel antennas. The patch and panel antennas will have some level of energy propagated behind their intended direction. This energy is known as the rear lobe. However, most of the energy will be directed inward. For this reason, patch and panel antennas are usually mounted on outside walls facing inward when they are intended to provide coverage inside an area only. Additionally, that can be used on the outside of a building to create an "external-only" hotspot that is open to the public or possibly less secure than the internal network.

Creatively using Yagi, patch and panel antennas can prevent the use of large numbers of omni antennas for many situations. For example, a single patch antenna placed on a wall facing inward may provide all the coverage needed, when two omni antennas would otherwise be needed. This is because the energy coming from the patch antenna is forced directionally inward instead of being force in all horizontal directions. The RF energy is going where it is needed instead of losing a third to half of it outside the walls of your facility. This, of course, assumes the AP can handle the capacity of the greater number of users. It is also assumed MIMO is not required when Yagi antennas are used. MIMO patch and panel antennas are available, but MIMO Yagi is not a typical installation.

A common misconception that enters at this point is the fear that using a Yagi, patch or panel antenna will get the signal to the client, but that it will not get the signal from the client to the access point (or the Yagi, patch or panel antenna). Stated another way, it is often assumed that you must use semi-directional antennas at the client if you use a semi-directional antenna at the access point; however, this is not the case.

I usually explain this by saying this, "When you place the megaphone over the antenna's mouth, it is smart enough to move it over its ear to listen." What I mean by this statement is simple: the very quality of the antenna that increases its gain in a particular direction, also allows it to "hear" better (have receive gain) from that same direction. Therefore, as Joseph Bardwell says, "If you can hear me, then I can hear you." For more information, see the website:

http://www.connect802.com/wcu/2005/newsletter_051001.htm

> Just accomplishing signal coverage is not sufficient in modern WLANs. Capacity is essential. For this reason, it is common to use MIMO antenna systems with more APs rather than attempting to cover large areas with SISO antenna systems.

Highly-Directional Antennas

Highly-directional antennas are antennas that transmit with a very narrow beam. These types of antennas often look like the satellite dish that is so popular with people who do not have access to wired cable television or do not desire to use it. They are generally called parabolic dish or grid antennas. The parabolic dish is the one that looks like a satellite dish and the grid antenna looks like an antenna with a curved grill grate behind it. Figure 4.12 shows examples of each antenna type.

Due to the high directionality of these antennas, they are mostly used for PtP or PtMP links. PtMP links will usually use an omni- or semi-directional antenna at the center and multiple highly or semi-directional antennas at the remote sites. They can transmit at distances of 35 miles or more and usually require detailed aiming procedures that include a lot of trial and error. By positioning one antenna according to visual LOS and then making small movements at the other antenna, accurate alignment can usually be achieved.

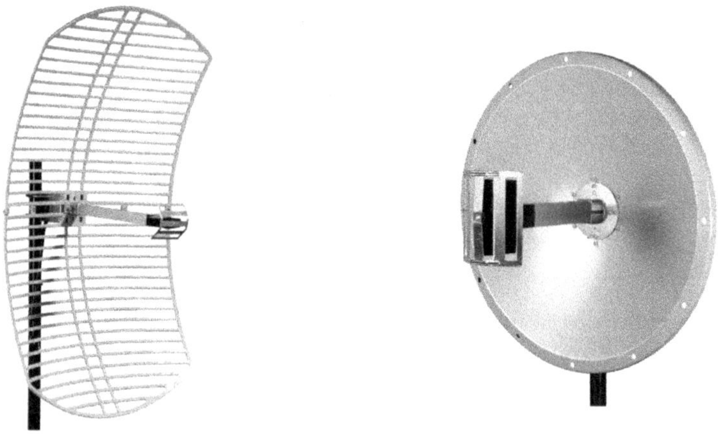

Figure 4.12: Parabolic Dish and Grid Antennas

The grid antenna provides the added benefit of allowing air to pass through the back panels so that the antenna does not shift as much as the parabolic dish in high wind load scenarios.

Sectorized and Phased Array Antennas

A *sectorized antenna* (or sector antenna) is a high-gain antenna that works back-to-back with other sectorized antennas. They are often mounted around a pole or mast and can provide coverage in indoor environments, such as warehouses, or outdoor environments, such as university campuses or hotspots. Figure 4.13 shows an example of sectorized antennas mounted on a pole.

A *phased array antenna* is a special antenna system that is actually comprised of multiple antennas connected to a single processor. The antennas are used to transmit different phases that result in a directed beam of RF energy aimed at client devices. Because phased array antennas are specialized and expensive, they are not commonly used in the WLAN market.

Multiple-Input/Multiple-Output (MIMO) Antenna Systems

The IEEE 802.11n amendment as ratified in September 2009, introduced MIMO (pronounced my-moe) technology. 802.11ac still supports MIMO and adds MU-MIMO support as well. MIMO is simply the use of more than one antenna at the same time. 802.11n or the HT PHY, and 802.11ac or the VHT PHY, support several MIMO solutions including spatial multiplexing, transmit beamforming and maximal ratio combining, as addressed earlier in this chapter. The solutions are different than traditional antenna diversity because multiple antennas are actually used at the same time to transmit and receive. With creative use of the antennas and changes in the underlying MAC layer of the 802.11 standard, bandwidth rates as high as 600 Mbps are optional for 802.11n and multiple Gbps for 802.11ac. Because the antennas used by MIMO devices are still standard antenna types connected to modern processing systems, MIMO will use the same antenna types discussed previously in this chapter. MIMO APs use internal omni-directional or semi-directional antennas and, when external antennas are supported, may use patch or panel MIMO antenna arrays.

Figure 4.13: Sectorized Antennas

Antenna Mounting

When mounting antennas, always abide by vendor recommendations. If the vendor provides a mounting kit, that's usually the best solution to use. However, if you have to create your own kit — because the vendor doesn't provide one, keep the following tips in mind:

- Be careful not to bend or twist antennas in an effort to "get them into" a specific location. Treat the antennas with love and care.
- Make sure the antennas are mounted firmly so that they are not continually moving and changing the resulting coverage patterns.
- Mount antennas according to their intended use. For example, antennas designed to be mounted on walls, of course, work best when mounted on walls.
- Pole or mast mount antennas usually come with a wrapping type of mounting kit. The wrappers may be simple zip ties or similar to them.
- Ceiling mount antennas are usually omni-directional and should be mounted in the center of the targeted coverage area.
- Wall mount antennas are usually semi-directional and should be mounted on the perimeter of the targeted coverage area so that they propagate inward.

As usual, if you have to climb a ladder or hang from a rafter to mount the antenna, please make sure you abide by safety best practices and regulations for your region. In many areas, this recommendation means that we abide by OSHA specifications. Whether OSHA has any influence on your area or not, please be careful and don't break a leg — in this case.

Antenna and RF Accessories

Several additional components should be understood by the WLAN administrator. These components include amplifiers, attenuators, lightning arrestors, grounding rods, mounting systems and RF cables and connectors.

Amplifiers

RF *amplifiers* are used to increase, or amplify, the RF signal strength. Amplifiers should be placed in-line between the AP or bridge and the antenna. In most cases, amplifiers are used to either make up for the signal loss in long cable runs, or they are used to increase the strength of the signal for longer bridge links. Amplifiers are not generally used for indoor installations because the output power of the typical AP is sufficient. Amplifiers may be used when the link budget calculations reveal that the signal strength will otherwise be too weak.

Remember, there are two types of gain: active and passive. Amplifiers create active gain because they increase the "amount" of RF energy being transmitted. Antennas create gain by focusing the energy in a specific direction so that the amount of energy going in that direction is higher, though the overall amount of energy is not increased. When active gain is employed in this way, it extends the signal's range and quality in all supported directions where a high gain antenna will only increase the signal's range and quality in the intended direction.

Amplifiers require a power supply since they are adding power to the RF signal. You will usually plug a DC converter into an AC outlet to power the amplifier or you will have to install a DC injector in-line with or before the amplifier. In a similar fashion to PoE, the DC injector places power on the RF cable that runs to the amplifier and the amplifier uses this power to amplify the signal. The power provided to the amplifier is sometimes called phantom voltage or phantom power.

There are two types of amplifiers: unidirectional and bidirectional. As the names imply, unidirectional amplifiers amplify the received or transmitted signal only and bidirectional amplifiers amplify both the sent and received signals. A bidirectional amplifier may be used with a stationary WLAN client that needs to increase both the received signal strength and the transmitted signal strength. A unidirectional amplifier could be used on each end of a bridge link. Both bridges could be configured with a unidirectional amplifier in the transmit path. Since they are both transmitting with stronger power, they should be able to hear each other well enough.

Finally, two variations are also available in amplifiers: fixed-gain and fixed-output. Fixed-gain amplifiers add a preconfigured amount of gain to the signal. For example, that may increase the strength by 6 dB. With this setup, whatever the input strength, it will be quadrupled by the amplifier — usually up to a certain maximum threshold. Fixed-output amplifiers are configured so that a certain range of input will always result in the same output power. For example, as long as the input power is from 5-50 mW, the output power will be 100 mW. This is an example of how one might function.

When purchasing an amplifier, you must be sure to match frequency response. In other words, if you are using a 2.4 GHz AP, you will need to use a 2.4 GHz amplifier. If you are using a 5 GH AP, you will need to use a 5 GHz amplifier and so forth. It is important to match frequency, or the system will not work.

Secondly, you will need to ensure that the amplifier matches your system in ohms and VSWR. Otherwise you will create an impedance mismatch and degrade the performance of the system, or even damage the equipment.

You should know the input power and gain you need based on the link budget calculations that you've made. If you need 400 mW of output power, for example, and you have a 100 mW output power AP, you will need a 6 dB gain amplifier. This is not considering the potential losses from cables and connectors.

When mounting the amplifier, it is usually best to mount it as close to the antenna as possible. This will prevent losses due to long cable runs before received frames enter the amplifier. When transmitting — particularly with fixed-output amplifiers — there should be as little component loss incurred as possible between the amplifier and the antenna. This means, again, installing the amplifier as close to the antenna as possible.

It is also important to note that some regulatory domains indicate that an entire RF system must be certified as a complete system. The system is inclusive of everything from the transmitting device to the antenna and everything in between. This means that, in some cases, if you purchase and install an amplifier, you will be running an illegal system.

Some vendors sell complete systems, including amplifiers, that are certified, and other vendors take devices sold by many vendors and get them certified with their amplifiers to increase their potential market base. Be sure to check your local regulatory constraints before using amplifiers with WLAN systems.

Attenuators

Attenuators do the opposite of amplifiers. They decrease the strength of the RF signal. RF attenuators may be fixed-loss or variable-loss. Fixed-loss attenuators reduce the signal by the same amount or to the same amount regardless of the input signal — assuming the input signal is strong enough to be reduced to that amount. Variable-loss attenuators can be set to varied levels of loss.

Attenuators are usually used when a system exceeds the allowed output power specified by the FCC. Variable-loss attenuators are useful because you may need to add or remove cable lengths over time and these different lengths of cable will results in different loss needs.

Like amplifiers, attenuators must be designed to work with the same frequency as the system being attenuated. Attenuators, unlike amplifiers, can be placed practically anywhere in-line between the AP or bridge and the antenna because strength of signal is less important than weakness of signal in this case.

Lightning Arrestors

Lightning arrestors are installed in order to redirect or shunt electric currents caused by close-proximity lightning strikes. They are not meant to protect against direct lightning strikes. Sadly, if your antenna is struck directly by lightning you will lose it and most likely the equipment to which it is attached.

Lightning arrestors are installed in series between the antenna and the AP or bridge (transceiver). Extra components installed between the lightning arrestor and the antenna, such as connectors or amplifiers, will not be protected by the lightning arrestor. For this reason, you should install the lightning arrestor closer to the antenna with nothing between it and the antenna — if possible. Lightning arrestors should be rated at <8 μs.

Among the features to look for in lightning arrestors are:

- Meets the IEEE standard of <8 μs
- Gas tube breakdown voltage
- Reusability
- Impedance
- Frequency response
- Connector types

The basic functionality of a lightning arrestor is as follows:

- Lightning strikes near the wireless antenna
- Transient currents are induced into the antenna or RF transmission line
- The lightning arrestor senses these currents and immediately causes a short in order to direct the current to earth ground

Grounding Rods/Wires

Grounding rods and wires make up a grounding system. The two main purposes of a grounding system are to provide a safe path for the current from lightning to travel to ground (or earth) and to ensure that all connected electrical systems share a common ground.

Lighting takes the path of least resistance when it strikes. If you intentionally create a path of least resistance, you can ensure that the power from the lightning does not damage your equipment (or at least reduce the likelihood that it will). In addition to the power of the lightning itself, the lighting traveling through high impedance systems generates a tremendous amount of heat. This heat, known as ohmic heating, can get so hot that it starts fires and melts metal. Since lighting prefers the path of least resistance (low impedance), an effective grounding system will ensure that the lighting's effects are limited.

You can ground your WLAN system by driving a copper rod, known as a grounding rod, into the earth. Next, connect your equipment to that rod using low impedance wires, known as grounding wires. This grounding rod should be eight feet deep in the earth for electrical installations, according to the National Electronics Council. Since this standard exists, it is usually used for grounding RF systems as well.

Since grounding can actually be a complicated issue and since it must be done right, I encourage you to hire a professional to ground your WLAN bridges and other devices (the entire system) when installing outdoors or on poles and cell towers.

Lightning arrestors do not protect against direct lightning strikes; however, you can install fiber between the WLAN bridge and the internal network to protect the internal network from the electrical energy generated from a lightning strike.

Mounting Systems

Much like the APs and bridges themselves, antennas can be mounted in various ways. These methods include pole or mast mount, ceiling mount and wall mount.

Pole or mast mount installations are usually performed outdoors. In these cases, public coverage areas can be created very easily by mounting an antenna on a pole or mast — usually with some type of U-brace — and connecting it to an AP or WLAN router. If the antenna installation will be sensitive to alignment variations, be sure to fasten the pole in a solid concrete base. This will keep the pole from moving by large enough amounts to impact the alignment in a short period of time.

Ceiling mounts are useful for omni-directional antennas. The antenna can be placed in such a way that it hangs down from the ceiling and provides coverage to the surrounding area. This is useful when you need to install an antenna in the center of a large open space and there are not poles or desks on which to place the antenna.

Wall mounts are used for both omni-directional and semi-directional antennas. With omni-directional antennas, the entire AP is usually mounted on the wall with the antennas attached directly to the AP. It is assumed that the RF signals will travel directly backward through the wall or will reflect and refract around it.

Patch and panel antennas are also often mounted on walls. They will then propagate their RF energy inward to the inner building coverage area.

Sector antennas are often mounted on poles or masts. In addition to mounting the sector antennas, you must be sure to provide the proper amount of downtilt. This will allow the sector antennas to cover the appropriate area.

Towers, Safety Equipment, and Concerns

Because OSHA standards and other requirements may prevent you from personally climbing a tower to install an antenna, you should consult a professional to have these antennas installed. In the rare case where the typical WLAN administrator can climb the tower, the proper safety equipment (harnesses, shoes, etc.) should be used in order to protect the climber. Check with your local safety enforcement agency (such as OSHA in the United States) to determine the guidelines for your area.

RF Cables and Connectors

RF cables are used to connect the transceiver to the antenna (and possibly other in series devices). Cables have different levels of loss and this should be considered when selecting the cabling for your system. Keep the following factors in mind when selecting RF cables for your implementation:

- Different cables have different levels of loss so not all cables are the same.
- Make sure the impedance of the cable matches the rest of your system.
- Be sure to select cable that is rated for the frequency you will be using.
- Check with the vendor to discover the loss incurred per foot or per 100 feet before selecting the cable.
- Higher frequencies mean greater loss in the same cable.
- Either master the art of building cables or hire a professional to cut the cables and install the connectors so you do not unnecessarily introduce extra loss.

RF connectors come in many shapes and sizes. The following types are common:

- N-TYPE
- SMA
- BNC
- TNC

In addition, there are common variations of these types, such as reverse polarity and reverse threading. These different types exist in an effort to comply with FCC and other regulations for components used in a wireless system. While dongles and pigtails exist, if they are used to convert from one type to another for the purpose of transmission, they may constitute a breach of regulatory agency regulations.

Figure 4.14: N-TYPE Connector

Figure 4.15: SMA Connectors

Figure 4.16: BNC Connector

Figure 4.17: TNC Connector

These connectors are found on the ends of cables, the back of APs and bridges and the ends of antennas (in the case of dipole or rubber ducky antennas).

RF splitters are installed in series between the transceiver and the antennas. The splitter receives a single input and has two or more outputs. They are not recommended for common use in WLANs but may be used with sectorized antennas. Other than these scenarios, RF splitters should be avoided in WLANs as they create insertion loss.

Chapter Summary

In this chapter, you learned about WLAN antennas and some calculations related to them. You learned about the different antenna types and antenna charts and specifications provided in relation to them. You also learned about various antenna cables and connectors. In the next chapter, you will begin to explore the details of the 802.11 standard.

Points to Remember

Remember the following important points:

- The first Fresnel zone should have a minimum of 60% clearance in bridge links and CWNP recommends 80% clearance.
- Earth bulge may be a factor in links over 7 miles in distance.
- Increasing antenna height can counter the impact of interference from trees and buildings and counter the impact of earth bulge.
- Omni-directional antennas radiate outward around the antenna with less signal directly above and beneath the antenna.
- APs usually use either omni-directional or semi-directional antennas when they use internal antennas.
- Semi-directional antennas radiate in a 180 degree or less range in the intended direction of propagation.
- Patch and panel antennas are examples of semi-directional antennas.
- Highly directional antennas include parabolic dish and grid antennas.
- The Azimuth chart shows the top-down view of the antenna and the Elevation chart shows the side view.
- Antenna beamwidth is calculated horizontally and vertically where the signal drops by 3 dB.
- RF cables, connectors and other components should be selected with the appropriate connectors for your system.

- Lightning arrestors help to protect systems from incidental energy generated by a nearby lightning strike. They do not protect from a direct lightning strike in most instances.
- When mounting on towers, it is best to hire trained and certified installation technicians, rather than doing it yourself.

Review Questions

1. When creating an outdoor bridge link, what is the recommended clearance for the first Fresnel zone?
    a. 60%
    b. 80%
    c. 20%
    d. 40%

2. You are creating a bridge link that spans 12 miles. You have inspected the area to identify trees and other interferers. What else should be considered?
    a. Weather in the area
    b. Daytime vs. nighttime propagation
    c. Earth bulge
    d. Other RF signals between the two endpoints

3. You are selecting an antenna for a MIMO AP that supports external antennas. You want to mount the antenna on the wall at the exterior of the target coverage area. What kind of antenna should you use?
    a. Yagi
    b. Omni
    c. Grid
    d. Patch

4. You are creating a very long distance bridge link in an area known for high winds. What kind of antenna would be useful for this scenario?
    a. Parabolic dish
    b. Omni
    c. Patch
    d. Grid

5. What kind of lightning arrestor is best for direct lightning strikes?
    a. Quick response arrestor
    b. Lead arrestor
    c. Water-based arrestor
    d. No lightning arrestor protects against direct strikes

6. At what signal loss level is the beamwidth of an antenna calculated?
    a. 6 dB
    b. 10 mW
    c. 3 dB
    d. 25 mW

7. Why are Yagi and semi-directional SISO antennas less popular for indoor WLAN deployments today?
    a. They are no longer supported by vendors
    b. They can't transmit OFDM signals
    c. They do not provide the capacity needed in modern WLANs
    d. They do not provide the signal quality of dipole antennas

8. What receive technology was introduced in 802.11n and allows the receiver to combine multiple copies of the received signal from multiple antennas to increase received signal strength?
    a. Spatial multiplexing
    b. MRC
    c. Transmit Beamforming
    d. 40 MHz channels

9. What is another term often used for an Elevation chart?
    a. Horizontal chart
    b. Vertical chart
    c. H-Plane chart
    d. Beamwidth chart

10. Which antenna specification defines the highest gain antenna?
    a. 7 dBi
    b. 6 dBd
    c. Standard dipole
    d. Isotropic radiator

Review Answers

1. **B is correct.** The required clearance is 60% and the recommended clearance is 80%.

2. **C is correct.** Because the link is over 7 miles, earth bulge must be considered.

3. **D is correct.** MIMO patch and panel antennas are available for such installations. They are semi-directional and should work well when installed on the exterior of the target coverage area.

4. **D is correct.** A grid antenna works well in high wind load areas. Parabolic dish and grid antennas both have high gain and are commonly used in very longs distance bridge links.

5. **D is correct.** Lightning arrestors do not protect against direct strikes. For protection of internal network equipment in such scenarios, fiber links should be created between the bridge and the internal network.

6. **C is correct.** The horizontal and vertical beamwidths are calculated at the point where the signal loses 3 dB of power.

7. **C is correct.** With 802.11n and 802.11ac APs, today, semi-directional and Yagi SISO antennas simply cannot implement the multiple spatial streams required for capacity on modern networks.

8. **B is correct.** Maximal Ratio Combining (MRC) allows the receiver to combine multiple received signals from multiple radio chains to achieve a boost in signal strength.

9. **B is correct.** An Elevation chart is also called a vertical chart and an Azimuth chart is also called a horizontal chart.

10. **B is correct.** 6 dBd is higher than 7 dBi, because it is equivalent to 8.14 dBi. An isotropic radiator would have no gain. A standard dipole is 2.14 dBi gain.

Chapter 5 — 802.11 PHYs and Network Types

Understanding how network technology works begins by understanding the language. This chapter begins by explaining the OSI Model and how it relates to 802.11 networks. Next, you will evaluate the many 802.11 PHYs. Then, you explore important functional concepts, like modulation, coding, channels, and more. Finally, you will take a tour of terminology related to service sets and WLAN bridging. Let's get under way.

OSI Model and 802.11

In order to help you understand how the various networking components work together to form a wireless network, I will first explain the Open System Interconnection (OSI) model. This model is not directly implemented in the TCP/IP networks that are most common today, but it is a valuable conceptual model that helps you to relate different technologies to one another and implement the right technology in the right way. When discussing wired and wireless networks, OSI model terminology creeps in on a regular basis. For example, you may read a whitepaper that covers "Layer 2 security solutions," or listen to a webinar where the presenter talks about "Layer 1 technologies." In both cases, the OSI model is the reference point.

According to document ISO/IEC 7498-1, which is the OSI Basic Reference Model standard document, the OSI model provides a *common basis for the coordination of standards development for the purpose of systems interconnection, while allowing existing standards to be placed into perspective within the overall reference model*. In other words, the model is useful for new standards as they are developed and for thinking about existing standards. For example, you can relate the TCP/IP protocol suite to the OSI model. Even though TCP/IP was developed before the OSI model, it can be *placed in perspective* in relation to the model.

The OSI model allows us to think about our network in chunks or layers. You can focus on securing each layer, optimizing each layer and troubleshooting each layer. This allows you to take a very complex communications process apart and evaluate its components. The OSI model is segmented into seven layers. The seven layers are (from top to bottom):

- Application

- Presentation
- Session
- Transport
- Network
- Data Link
- Physical

Each layer is defined as providing services and receiving services. For example, the Data Link layer provides a service to the Physical layer and receives a service from the Physical layer. How is this? In a simplified explanation, the Data Link layer converts packets into frames for the Physical layer and the Physical layer transmits these frames as bits on the chosen medium. The Physical layer reads bits off of the chosen medium and converts these into frames for the Data Link layer.

The layered model allows for abstraction. The higher layers do not necessarily have to know how the lower layers are doing their jobs. In addition, the lower layers do not necessarily have to know what the upper layers are actually doing with the results of the lower layers' labors. The abstraction gives you the ability to use the same web browser and HTTP protocol to communicate on the Internet whether the lower layer connection is a dial-up modem, a high-speed Internet connection, or somewhere in between. The resulting speed or performance will certainly vary, but the functionality will remain the same.

Figure 5.1 illustrates the concept of the OSI model. As you can see, data moves down through the layers, across the medium, and then back up through the layers on the receiving machine. Remember, most networking standards allow for the substitution of nearly any Data Link and Physical layer. While this example shows a wired Ethernet connection between the two machines, it could have just as easily been a wireless connection using the 802.11 standard for the descriptions of the Data Link and Physical layers. This example uses the 802.3 Ethernet standard and the 802.2 LLC standard (a layer within the Data Link layer) for the lower layers. The point is that the most popular upper layer protocol suite, TCP/IP, can work across most of the lower layer standards, such as 802.2 (Logical Link Control), 802.3 (Ethernet), 802.5 (Token Ring), 802.11 (WLANs) and 802.16 (WiMAX).

In order to fully understand the OSI model and be able to relate to it throughout the rest of this book, it is important that we evaluate each layer. You will need to understand the basic description of each layer and the services it provides to the networking process. I will define each layer and then give examples of its use starting with the topmost layer, which is the Application Layer, since this is the order in which they are documented in the standard.

It is important that you understand the basic operations that take place at each layer of the OSI model. It's also useful to know the primary components, such as switches, routers and hubs that function at each level. While not tested directly, indirect references to the OSI model will require this understanding.

Application Layer

The seven layers of the OSI model are defined in clause 7 of the document ISO/IEC 7498-1. The *Application layer* is defined in sub-clause 7.1 as the highest layer in the reference model and as the sole means of access to the OSIE (Open System Interconnection Environment). The Application layer is the layer that provides access to the other OSI layers for applications, and to applications for the other OSI layers. Do not confuse the Application layer with the general word "application," which is used to reference programs like Microsoft Excel, Corel WordPerfect and so on. The Application layer is the OSI layer that these applications communicate with when they need to send or receive data across the network. You could say that the Application layer exposes the higher-level protocols that an application needs to talk to. For example, Microsoft Outlook may need to talk to the SMTP protocol in order to transfer email messages.

Examples of Application layer protocols and functions include HTTP, FTP and SMTP. The Hypertext Transfer Protocol (HTTP) is used to transfer HTML, ASP, PHP and other types of documents from one machine to another. It is the most heavily used Application layer protocol on the Internet and, possibly, in the world. The File Transfer Protocol (FTP) is used to transfer binary and ASCII files between a server and a client. Both the HTTP and FTP protocols can transfer any file type.

The Simple Mail Transport Protocol (SMTP) is used to move email messages from one server to another and usually works in conjunction with other protocols for mail storage.

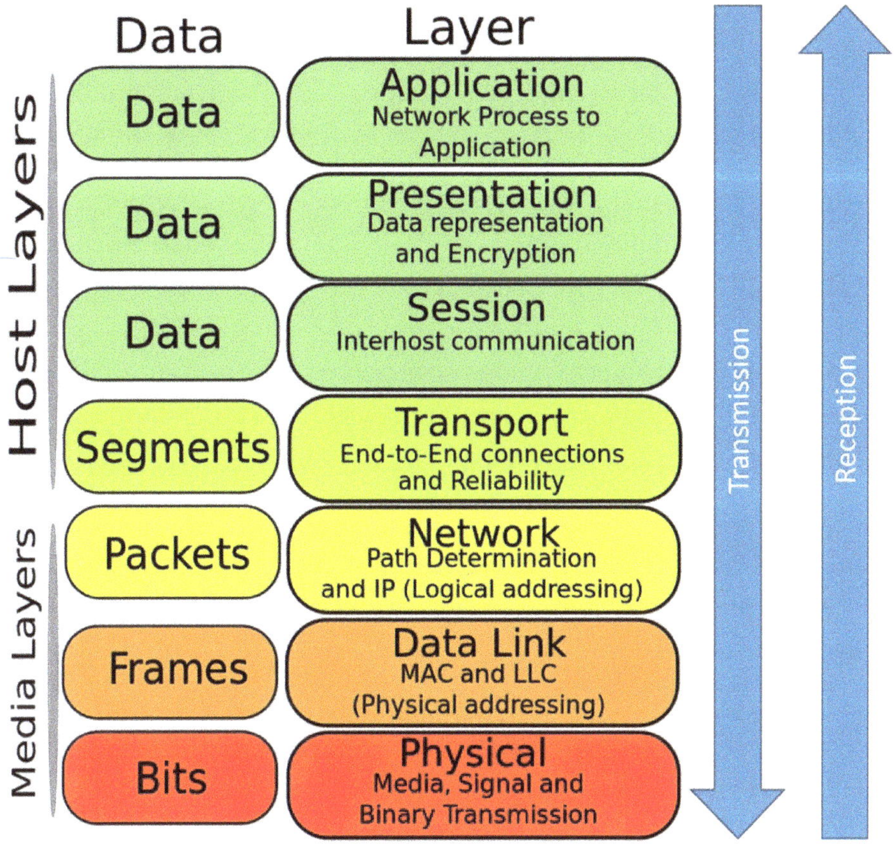

Figure 5.1: OSI Model

Application layer processes fall into two general categories: user applications and system applications. Email (SMTP), file transfer (FTP), and web browsing (HTTP) functions fall into the user application category as they provide direct results to applications used by users such as Outlook (email), WS_FTP (file transfer), and Firefox (web browsing). Notice that the applications or programs used by the user actually take advantage of the application services in the Application layer or

Layer 7. For example, Outlook takes advantage of SMTP. Outlook does not reside in Layer 7, but SMTP does. As examples of system applications, consider DHCP and DNS. The Dynamic Host Configuration Protocol (DHCP) provides for dynamic TCP/IP configuration and the Domain Name Service (DNS) protocol provides for name to IP address resolution. Both of these are considered system-level applications because they are not usually directly accessed by the user (though this is open for debate since administrators are users too and they use command-line tools or programs to directly access these services quite frequently).

The processes operating in the Application layer are known as application-entities. An application entity is defined in the standard as an active element embodying a set of capabilities which is pertinent to OSI and which is defined for the Application layer. Application entities are the services that run in Layer 7 and communicate with lower layers while exposing entry points to the OSI model for applications running on the local computing device. SMTP is an application entity as is HTTP and other Layer 7 protocols.

Presentation Layer

The *Presentation layer* is defined in sub-clause 7.2 of the standard as the sixth layer of the OSI model and it provides services to the Application layer above it and the Session layer below it. The Presentation layer, or Layer 6, provides for the representation of the information communicated by or referenced by application entities. The Presentation layer is not used in all network communications and it, as well as the Application layer and Session layer, is similar to the single Application layer of the TCP/IP model. The Presentation layer provides for syntax management and conversion, as well as encryption services. Syntax management refers to the process of ensuring that the sending and receiving hosts communicate with a shared syntax or language. When you realize this, you will realize why encryption is often handled at this layer. After all, encryption is really a modification of the data in such a way that must be reversed on the receiving end. Therefore, both the sender and receiver must understand the encryption algorithm in order to provide the proper data to the program that is sending or receiving on the network.

Don't be alarmed to discover that the ***TCP/IP model*** has its own Application layer that differs from the OSI model's Application layer. The TCP/IP protocol existed before the OSI model was released. For this reason, we relate the TCP/IP protocol suite to the OSI model, but we cannot say that it complies with the model directly. It's also useful to keep in mind the reality that the TCP/IP protocol is an implemented model and the OSI model is only a "reference" model.

Examples of Presentation layer protocols and functions include any number of data representation and encryption protocols. For example, if you choose to use HTTPS instead of HTTP, you are indicating that you want to use Secure Sockets Layer (SSL) encryption. SSL encryption is related to the Presentation layer or Layer 6 of the OSI model. SSL, the Netscape solution, and TLS, the IETF solution, both operate at Layer 6 of the OSI model.

Ultimately the Layer 6 is responsible, at least in part, for three major processes: data representation, data security and data compression. Data representation is the process of ensuring that data is presented to Layer 7 in a useful way and that it is passed to Layer 5 in a way that can be processed by the lower layers. Data security usually includes authentication, authorization and encryption. Authentication is used to verify the identity of the sender and receiver. With solid authentication, we gain a benefit known as non-repudiation. Non-repudiation simply means that the sender cannot deny the sending of data. This is often used for auditing and incident-handling purposes. Authorization ensures that only valid users can access the data and encryption ensures the privacy and integrity of the data as it is being transferred.

The processes running at Layer 6 are known as presentation entities in the OSI model documentation. Therefore, an application entity is said to depend on the services of a presentation entity and the presentation entity is said to serve the application entity.

Session Layer

The *Session layer* is defined in sub-clause 7.3 of the standard as providing the means necessary for cooperating presentation entities to organize and to synchronize their dialog and to manage their data exchange. This is accomplished by establishing a connection between two communicating presentation entities. The result is simple mechanisms for orderly data exchange and session termination.

A session includes the agreement to communicate and the rules by which the communications will transpire. Sessions are created, communications occur, and sessions are destroyed or ended. Layer 5 is responsible for establishing the session, managing the dialogs between the endpoints, and the proper closing of the session.

Examples of Session layer protocols and functions include the iSCSI protocol, RPC and NFS. iSCSI is a protocol that provides access to SCSI devices on remote computers or servers. The protocol allows SCSI commands to be sent to the remote device. The Remote Procedure Call (RPC) protocol allows subroutines to be executed on remote computers. A programmer can develop an application that calls the subroutine in the same way as a local subroutine. RPC abstracts the network layer, and allows the application running above Layer 7 to execute the subroutine without knowledge of the fact that it is running on a remote computer. The Network File System (NFS) protocol is used to provide access to files on remote computers as if they were on the local computer. NFS actually functions using an implementation of RPC known as Open Network Computing RPC (ONC RPC) that was developed by Sun Microsystems for use with NFS; however, ONC RPC has also been used by other systems since that time. Remember that these protocols are provided only as examples of the protocols available at Layer 5 (as were the other protocols mentioned for Layers 6 and 7). By learning the functionality of protocols that operate at each layer, you can better understand the intention of each layer.

The services and processes running in Layer 5 are known as session entities. Therefore, RPC and NFS would be session entities. These session entities will be served by the Transport layer.

Transport Layer

Layer 4, the *Transport Layer* is defined as providing transparent transfer of data between session entities and relieving them from any concern with the detailed way in which reliable and cost-effective transfer of data is achieved. This simply means that the Transport layer, as its name implies, is the layer where the data is segmented for effective transport in compliance with Quality of Service (QoS) requirements and shared medium access.

Examples of Transport layer protocols and functions include TCP and UDP. The Transmission Control Protocol (TCP) is the primary protocol used for the transmission of connection-oriented data in the TCP/IP suite. HTTP, SMTP, FTP and other important Layer 7 protocols depend on TCP for reliable delivery and receipt of data. The User Datagram Protocol (UDP) is used for connectionless data communications. For example, when speed of communications is more important than reliability, UDP is frequently used. Because voice data either has to arrive or not arrive (as opposed to arriving late), UDP is frequently used for the transfer of voice and video data.

TCP and UDP are examples of transport entities at Layer 4. These transport entities will be served by the Network layer. At the Transport layer, the data is broken into segments if necessary. If the data will fit in one segment, then the data becomes a single segment. Otherwise, the data is segmented into multiple segments for transmission.

Network Layer

The *Network Layer* is defined as providing the functional and procedural means for connectionless-mode (UDP) or connection-mode (TCP) transmission among transport entities and, therefore, provides to the transport entities independence of routing and relay considerations. In other words, the Network layer says to the Transport layer, "You just give me the segments you want to be transferred and tell me where you want them to go. I'll take care of the rest." This is why routers do not usually have to expand data beyond Layer 3 to route the data properly. For example, an IP router does not care if it's routing an email message or voice conversation. It only needs to know the IP address for which the packet is destined and any relevant QoS parameters in order to move the packet along.

Examples of Network layer protocols and functions include IP, ICMP and IPSec. The Internet Protocol (IP) is used for addressing and routing of data packets in order to allow them to reach their destination. That destination can be on the local network or a remote network. The local machine is never concerned with this, with the exception of the required knowledge of an exit point, or default gateway, from the local machine's network. The Internet Control Message Protocol (ICMP) is used for testing the TCP/IP communications and for error message handling within Layer 3. Finally, IP Security (IPSec) is a solution for securing IP communications using authentication and/or encryption for each IP packet. While security protocols such as SSL, TLS and SSH operate at Layers 4 through 7 of the OSI model, IPSec sits solidly at Layer 3. The benefit is that, since IPSec sits below Layer 4, any protocols running at or above Layer 4 can take advantage of this secure foundation. For this reason, IPSec has become more and more popular since it was first defined in 1995.

The services and processing operating in the Network layer are known as network entities. These network entities depend on the services provided by the Data Link layer. At the Network layer, Transport layer segments become packets. These packets will be processed by the Data Link layer.

Data-Link Layer

The *Data Link Layer* is defined as providing communications between connectionless-mode or connection-mode network entities. This may include the establishment, maintenance and release of connections for connection-mode network entities. The Data Link layer is also responsible for detecting errors that may occur in the Physical layer. Therefore, the Data Link layer provides services to Layer 3 and Layer 1. The Data Link layer, or Layer 2, may also correct errors detected in the Physical layer automatically.

Examples of Data Link layer protocols and functions include Wi-Fi (802.11), Ethernet, PPP and HDLC. Wi-Fi is the common name given to the 802.11 standard and is the primary topic of this book. Ethernet is the most widely used protocol for Local Area Networks (LANs) and will be the type of LAN you deal with when using most modern LAN technologies. Ethernet comes in many different implementations from 10 Mbps (megabits per second or million bits per second) to

1,000 Mbps in common implementation. Faster Ethernet technologies are being developed and implemented on a small scale today. The Point to Point Protocol (PPP) is commonly used for Wide Area Network (WAN) links across analog lines and other tunneling purposes across digital lines. The High-Level Data Link Control (HDLC) protocol is a solution created by the ISO for bit-oriented synchronous communications. It is a very popular protocol used for WAN links and is the default WAN link protocol for many Cisco routers.

The IEEE has divided the Data Link layer into two sublayers, the Logical Link Control (LLC) sublayer and the Medium Access Control (MAC) sublayer. The LLC sublayer is not actually used by many transport protocols, such as TCP. The varied IEEE standards identify the behavior of the MAC sublayer within the Data Link layer and the PHY layer as well.

The results of the processing in Layer 2 are that the packet becomes a frame that is ready to be transmitted by the Physical layer or Layer 1. So, the segments became packets in Layer 3, and now the packets have become frames. Remember, this is just the collection of terms that we use; the data is a collection of ones and zeros all the way down through the OSI layers. Each layer is simply manipulating or adding to these ones and zeros in order to perform that layer's service. Like the other layers before it, the services and processes within the Data Link layer are named after the layer and are called data-link entities.

Physical Layer

The *Physical Layer*, sometimes called the PHY, is responsible for providing the mechanical, electrical, functional or procedural means for establishing physical connections between data-link entities. The connections between all other layers are really logical connections, as the only real physical connection that results in true transfer of data is at Layer 1 — the Physical layer. For example, we say that the Layer 7 HTTP protocol on a client creates a connection with the Layer 7 HTTP protocol on a web server when a user browses an Internet website; in reality, this connection is logical, and the real connections happen at the Physical layer within a segment of the network.

It is really amazing to think that my computer — the one I'm using to type these words — is connected to a Wireless Access Point (AP) in my office, which is connected to my local network, that is in turn connected to the Internet. Through connections — possibly both wired and wireless — I can send signals (that's what happens at Layer 1) to a device on the other side of the globe. To think that there is a potential electrical-connection path between these devices and millions of others is really quite amazing.

It is Layer 1 that is responsible for taking the data frames from Layer 2 and transmitting them on the communications medium as binary bits (ones and zeros). This medium may be wired or wireless. It may use electrical signals or light pulses (both actually being electromagnetic in nature). Whatever you've chosen to use at Layer 1, the upper layers can communicate across it as long as the hardware and drivers abstract that layer so that it provides the services demanded of the upper layer protocols.

Examples of Physical layer protocols and functions include Ethernet, Wi-Fi and DSL. You probably noticed that Ethernet was mentioned as an example of a Data Link layer protocol. This is because Ethernet defines both the MAC sub-layer functionality within Layer 2 and the PHY for Layer 1. Wi-Fi technologies (802.11) are similar in that both the MAC and PHY are specified in the standard. Therefore, the Data Link and Physical layers are often defined in standards together. You could say that Layer 2 acts as an intermediary between Layers 3 through 7 so that you can run IPX/SPX (though hardly anyone uses this protocol today) or TCP/IP across a multitude of network types (network types being understood as different MAC and PHY specifications).

OSI Model and Wi-Fi

802.11 networks are related to the OSI Model in three ways:

- They carry upper layer (Layers 3-7) data across the RF medium.
- They define Layer 2.
- They define Layer 1.

Figure 5.2 illustrates the relationship between the OSI Model and 802.11 operations.

The terms MSDU, MPDU, PSDU and PPDU will be covered in Chapter 7, "802.11 MAC Operations."

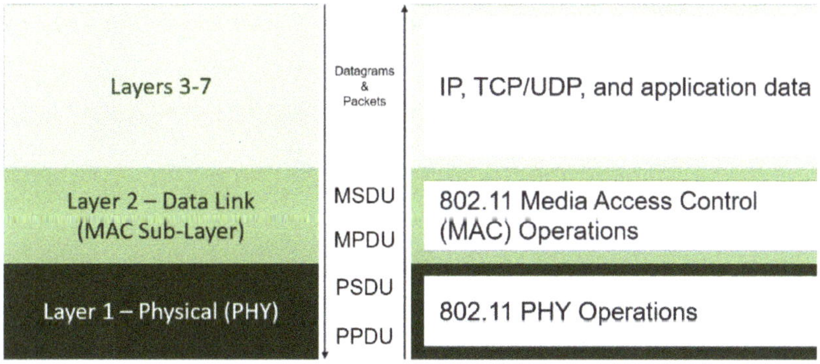

Figure 5.2: The OSI Model and 802.11

802.11 PHYs

The 802.11-2016 standard defines many different physical layers (PHYs) that provide for varying data rates, channel widths and operational frequency bands. Additionally, the 802.11ah amendment defines an added PHY ratified after 802.11-2016 was created and more PHYs will be added in the future, including 802.11ax, likely in 2019. The CWNA exam requires that you are aware of the features of the PHYs defined in 802.11-2016 and 802.11ah, including:

- Data rates (the speed of Wi-Fi frame transmissions supported by the PHY)
- Bands used (the frequency bands supported by the PHY)
- Supported technologies (for example, the devices that may use a particular PHY)

The channel width and modulation used have significant impact on the actual data rates available. Each PHY supports specific data rates based on the combination of channel width, modulation, coding and a few other features. Modulation and coding are addressed in detail later in this chapter. The data rates are not in some way arbitrarily variable (for example, going from 11 Mbps to 10.9 Mbps to 10.8 Mbps and so forth), but they are specific data rates supported based on combinations of these factors (for example, going from 11 Mbps to 5.5 Mbps to 2 Mbps and so forth).

The following subsections provide an overview of each PHY with all of the information required to select the appropriate hardware based on PHY support. This information is sufficient for both the CWNA-108 exam and the ability to implement and administer the technologies in a production WLAN.

DSSS

The oldest PHY still supported by modern 802.11 devices is *Direct Sequence Spread Spectrum (DSSS)*. DSSS uses a 22 MHz-wide channel and operates only in the 2.4 GHz band. Each channel is assigned based on a channel center frequency, such as 2.412 GHz for channel 1, and uses 11 MHz on either side of the center frequency. Therefore, channel 1 would use the range from 2.401 to 2.423 for the 22 MHz channel.

As with all PHYs introduced before 802.11n (HT), DSSS supports only one spatial stream. It is a SISO PHY. This statement simply means that the transceiver (transmitter/receiver) sends one stream of data and receives one stream of data at a time. 802.11n and 802.11ac, as well as future PHYs, will support multiple streams for transmission and reception, greatly increasing the available data rates.

Remembering the supported data rates for DSSS is simple. Only two data rates are supported: 1 Mbps or 2 Mbps. By today's standards, this PHY is very slow. However, the DSSS PHY is supported by all 802.11 devices that operate in the 2.4 GHz band, including the newest 802.11n devices. It is important to remember that he PHY preamble and headers will be transmitted at these low data rates regardless of the data rate at which actual Layer 2 frame data is transmitted — even in higher data rate capable PHYs.

In summary, the DSSS PHY supports data rates of 1 or 2 Mbps. It operates in the 2.4 GHz band and supports only a single spatial stream. All DSSS transmissions use a 22 MHz channel width. Figure 5.3 shows this summary information.

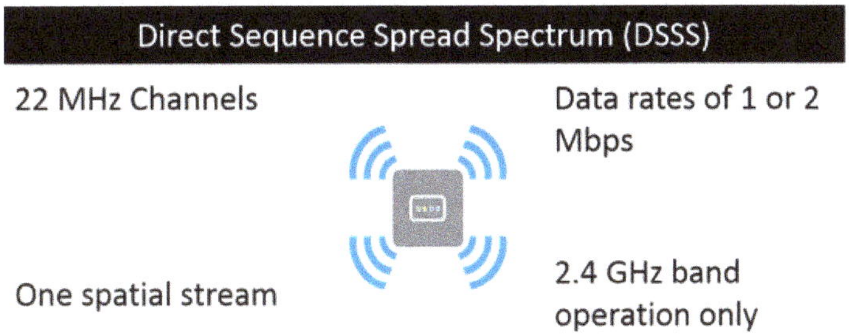

Figure 5.3: DSSS PHY Summary

HR/DSSS

The *High Rate/Direct Sequence Spread Spectrum (HR/DSSS)* PHY was released with the 802.11b amendment in 1999. It introduced more advanced modulation techniques allowing for data rates of 5.5 and 11 Mbps while still supporting the DSSS data rates of 1 and 2 Mbps. HR/DSSS uses the same 22 MHz-wide channels as DSSS and supports only a single spatial stream. Like DSSS, HR/DSSS operates only in the 2.4 GHz frequency band.

All newer PHYs operating in the 2.4 GHz frequency band are designed to be backward compatible with earlier PHYs in the same band. This statement means that an HR/DSSS device can communicate with a device that supports only DSSS. Additionally, though the details of the HT PHY have not yet been explored, an HT PHY device can communicate with a DSSS PHY device as long as they both operate in the same 2.4 GHz frequency band and the network configuration allows for it.

WLAN administrators may disable backward compatibility by disallowing lower data rates. However, this is a configuration constraint and not a radio or device constraint. With all data rates enabled, newer 2.4 GHz 802.11 devices can communicate with all older devices.

It is unlikely you will see actual 802.11b (HR/DSSS) or 802.11-Prime (DSSS) devices in production WLANs; however, this does not mean the PHY will not be used. If you allow the low data rates and a better AP is not available to which the client can roam (for example, an AP with a stronger signal), the client could fall back to these data rates. Instead, however, with proper configuration, the worst-case scenario for an HT (802.11n) device in 2.4 GHz should be MCS 0, assuming no 802.11b/g devices are in use, which operate at 6.5 Mbps (as a single-stream 20 MHz communication), 13 Mbps (as a two-stream 20 MHz communication), or 19.5 Mbps (as a three-stream 20 MHz communication) as the lowest data rates. These rates can be raised even higher when enabling the short Guard Interval (GI). You will learn more about this later in the chapter.

I will state this several times throughout this book: Even though 802.11n (HT) supports 40 MHz channels in 2.4 GHz, they should never be used. The available frequency space in 2.4 GHz is not sufficient for the use of 40 MHz channels.

Figure 5.4 provides a reference to this information about the HR/DSSS PHY.

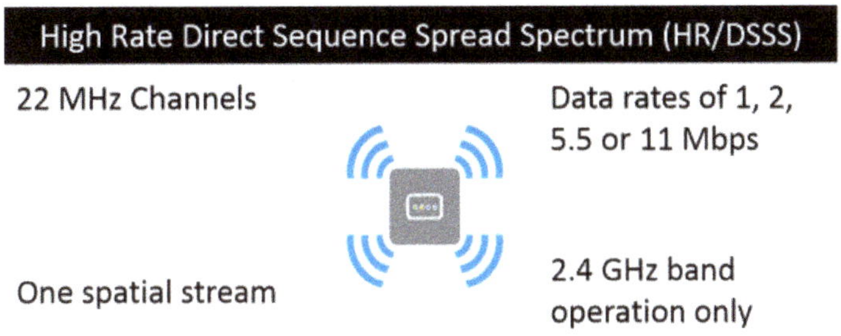

Figure 5.4: HR/DSSS PHY Summary

OFDM

The *Orthogonal Frequency Division Multiplexing (OFDM)* PHY was the first to support 5 GHz band operations. This PHY was made available through the 802.11a amendment in 1999. In addition to 5 GHz band support, OFDM was the first to use 20 MHz channels instead of 22 MHz channels. All modern PHYs that are based on OFDM use 20 MHz channels or some factor thereof. For example, they may use 20 MHz or 40 MHz channels (as well as 80 and 160 MHz). 802.11a was the first PHY amendment to use OFDM modulation and the PHY is named after the modulation. All PHYs introduced since 802.11a also use OFDM modulation but have a different PHY name to clearly differentiate them from 802.11a OFDM.

> The 802.11ax amendment will introduce a new modulation called OFDMA, but it has not been ratified at the time of writing. The PHY name currently proposed for 802.11ax is High Efficiency (HE), but this may change by the time of ratification. 802.11ax is not tested on the CWNA-108 exam.

OFDM still uses one spatial stream, but with enhanced modulation, it supports data rates of 6, 9, 12, 18, 24, 36, 48 and 54 Mbps. Notice that it does not support 1, 2, 5.5 or 11 Mbps. OFDM operates in 5 GHz and has no need to be backward compatible with DSSS or HR/DSSS. Figure 5.5 provides an overview of the OFDM PHY.

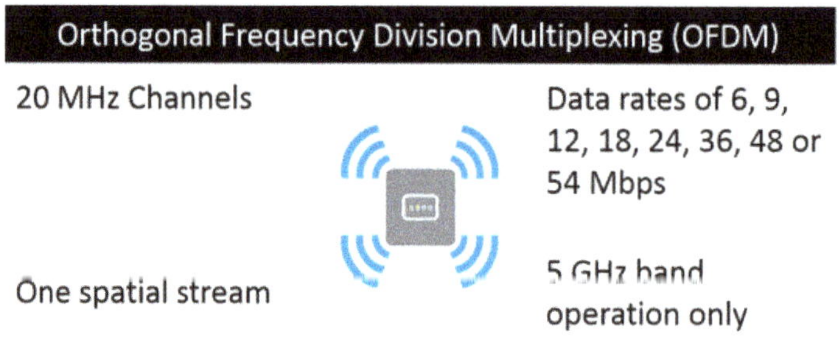

Figure 5.5: OFDM PHY Summary

ERP

The *Extended Rate PHY (ERP)* was introduced to bring the OFDM modulation down into the 2.4 GHz band. The implemented PHY features in 802.11 devices use the same OFDM modulation used in 802.11a 5 GHz devices and use 20 MHz channels. There are some slight differences in the way the PHY was implemented, but for the CWNA, it is sufficient to know that it provides the same basic functionality as OFDM provided in 5 GHz.

Operating in 2.4 GHz, all ERP devices (which are also called 802.11g devices based on the amendment that defined ERP) support backward compatibility with HR/DSSS and DSSS PHY devices. This fact is another differentiation when compared to the OFDM PHY in 5 GHz. OFDM required no backward compatibility. To accomplish backward compatibility, ERP (802.11g) devices still support the DSSS data rates of 1 and 2 Mbps and the HR/DSSS data rates of 5.5 and 11 Mbps. In addition, they support the same data rates as OFDM (802.11a) of 6, 9, 12, 18, 24, 36, 48 and 54 Mbps. To be clear, the ERP PHY supports only the data rates of 6, 9, 12, 18, 24, 36, 48 and 54 Mbps, but all devices implementing the ERP PHY also effectively implement the DSSS and HR/DSSS PHYs, so that the 1, 2, 5.5 and 11 Mbps data rates are also supported.

Figure 5.6 shows the summary information for the ERP PHY, and the important characteristics to know as a CWNA professional. Note that the data rates listed are those supported specifically by the ERP PHY but remember that ERP or 802.11g devices support backward compatibility by also implementing the DSSS and HR/DSSS PHYs, and the data rates they support.

Remember that ERP is the first OFDM-based PHY that operates in 2.4 GHz, and OFDM is the first OFDM-based PHY that operates in 5 GHz. HT and VHT, explained later in this section, are both OFDM-based as well.

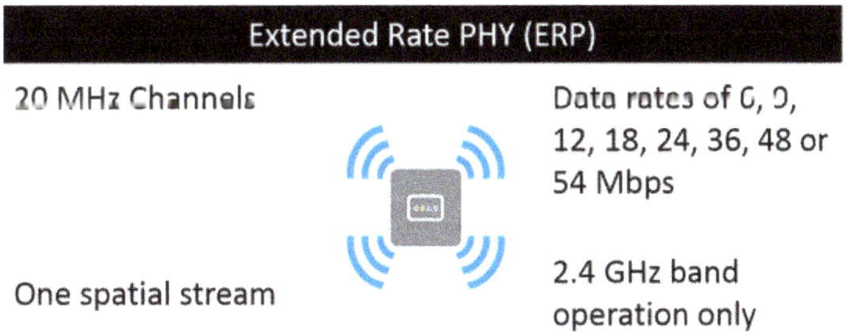

Figure 5.6: ERP PHY Summary

HT

The *High Throughput (HT)* PHY was introduced in the 802.11n amendment and offers several advantages over older PHYs. HT provides wider channels by combining two 20 MHz sections into a 40 MHz channel. So, HT provides either 20 MHz or 40 MHz channels. An AP offering a 40 MHz channel can still service 20 MHz clients on its primary channel. The second 20 MHz section is known as the secondary channel.

The primary channel will be one of the defined channel numbers, such as 1, 6, 11, 36 or 44. Then, the secondary channel, which provides the total of 40 MHz, will be the 20 MHz above or below the primary channel. When using the 20 MHz above the primary channel it is referenced as a +1 configuration. When using the 20 MHz below the primary channel it is referenced as a -1 configuration. When a device connects to an AP offering a 40 MHz channel and the connecting device supports only a 20 MHz channel, it will communicate with the AP using the primary channel. The 40 MHz client devices can use the entire 40 MHz channel.

Wider channels result in higher data rates even with no additional features. However, the HT PHY also introduces the capability to use multiple spatial streams through Multiple-Input/Multiple-Output (MIMO). MIMO takes advantage of RF propagation behaviors to send multiple concurrent streams of data from the transmitter to the receiver. The HT PHY supports up to four spatial streams; however, most devices support from one to three spatial streams when using the HT PHY today.

BEYOND THE EXAM: Chipsets and Features

The reality of the Wi-Fi market is that most devices use chipsets manufactured by organizations that specialize in chipset development, like QUALCOMM and others. When using these manufactured chipsets in clients or APs, the hardware vendor can only offer what the chipsets offer from a radio capability perspective. To get around this limitation, some vendors will actually create their own chipsets to incorporate advanced spectrum analysis capabilities and other features.

If the chipset used supports only three spatial streams, the device using that chipset can support no more than this. Given that 802.11n chipsets from these chipset manufacturers only support three spatial streams, devices are unable to support more. Additionally, the added benefit of a fourth spatial stream is not significant enough to warrant the research and development to make it happen. This is particularly true now that four stream 802.11ac chipsets are available in the 5 GHz band.

Another first for the HT PHY was the fact that it operates in either 2.4 GHz or 5 GHz. More channels are available in the 5 GHz band, so it is the preferred band, but many devices operate only in 2.4 GHz (even some of the newest devices being sold), so it must continue to be supported in nearly all implementations. It is important to know that HT devices may be 2.4 GHz-only, so as a WLAN administrator, you should select equipment with great care. It is best to select devices that support the 5 GHz bands whenever possible.

Finally, the HT PHY offers many more data rate possibilities than earlier PHYs. The actual data rates available will depend on the channel width (20 MHz versus 40 MHz), the number of spatial streams and the modulation and coding used. Some additional factors impact the available data rates, such as the GI. The maximum data rate achievable with the HT PHY, assuming a 40 MHz channel and the highest modulation and coding rate, is 600 Mbps. Most HT, or 802.11n, devices support maximum data rates of 150, 300 or 450 Mbps because the devices support from one to three spatial streams, but the standard allows for up to 600 Mbps.

It is beneficial to know that 2.4 GHz devices will support maximum data rates of 72.2, 144.4 and 216.7 Mbps (sometimes these numbers are rounded to 72, 144 and 217 Mbps) because they will only support 20 MHz channel widths in a proper implementation. While the 2.4 GHz devices could be configured to support the higher data rates offered by 40 MHz channels, they should not be. When using 40 MHz 2.4 GHz channels in a multi-AP deployment, the degradation in performance due to channel overlap is not worth the gains offered by 40 MHz channels. However, in 5 GHz standard deployments, 40 MHz channels can be beneficial, depending on the type of network being deployed.

Figure 5.7 shows the summary information for the HT PHY. The information in the figure is the primary information you should remember for the CWNA exam.

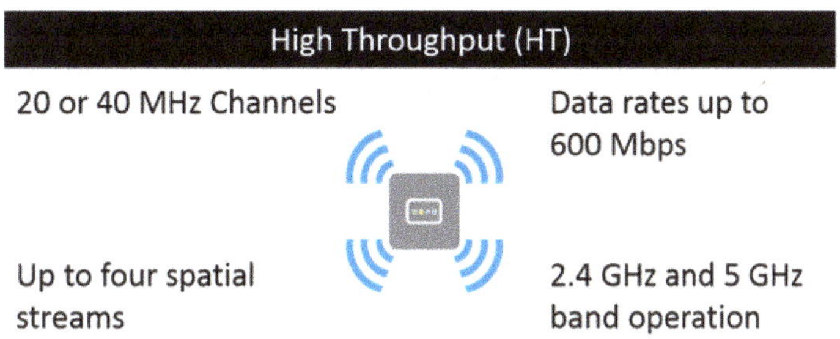

Figure 5.7: HT PHY Summary

VHT

The *Very High Throughput (VHT)* PHY moves 802.11 networks even further than the HT PHY. The VHT PHY now supports additional channel widths of 80 MHz and 160 MHz (though 160 MHz channels should not be used in enterprise deployments, and 80 MHz channels should rarely be used). The base channel width is still 20 MHz, but two, four or eight 20 MHz portions may be used to form the wider channels.

With the wider channels comes higher data rates, but VHT (802.11ac) also adds support for more spatial streams. A VHT device can use up to eight spatial streams. The first devices that were released supported three spatial streams, but devices are now on the market supporting four spatial streams. Whether we will see eight spatial streams is yet to be seen, simply because the general trend in client devices is to stay with fewer spatial streams, which reduces battery consumption and therefore extends battery life.

It is very important to know that the VHT PHY works only in the 5 GHz frequency band. There is no support for VHT in 2.4 GHz, unlike the HT PHY. The primary reason for this decision to limit VHT to the 5 GHz band was simply the lack of frequency space for wider channels in 2.4 GHz. Some vendors indicate that they have implemented 256-QAM (a new modulation introduced with VHT) in the 2.4 GHz band. They typically allow this because the chipset they use supports it. However, given that this is not according to the standard, you cannot assume that client devices will also support it. You should not plan on gaining the advantages of 256-QAM in 2.4 GHz even if the APs implemented allow for it.

Finally, VHT devices can achieve a maximum data rate of 6933.3 Mbps; however, this data rate would require eight spatial streams. Because 802.11ac devices implement no more than four spatial streams today, the real-world peak data rate is 3466.7 Mbps. To achieve this data rate of 3466.7 Mbps, the AP and client must both support four spatial streams and use a 160 MHz channel. Given the reality that few 802.11ac APs with be implemented with channels configured with a bandwidth of more than 40 MHz, it is more likely that you will see maximum data rates of 800 Mbps for four spatial streams on a 40 MHz channel.

Always remember that the data rate available for a link is constrained by the less capable device in the link. For example, if an AP is configured with a 40 MHz channel and supports four spatial streams, a four spatial stream client supporting a 40 MHz channel could potentially connect with a data rate of 3466.7 Mbps (according to the standard). However, a single stream 40 MHz client will connect to the same AP with a maximum data rate of 200 Mbps. As you can see, the real world is often very different from marketing literature, and even from the potential of the 802.11 standard.

Figure 5.8 provides a summary of this key information related to the VHT PHY.

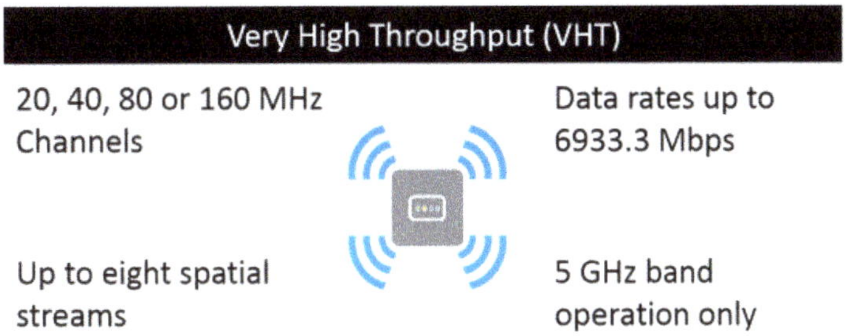

Figure 5.8: VHT PHY Summary

HE

802.11ax (High Efficiency (HE)) is the only PHY as of 2020 that will be fully supported in the 6 GHz band. OFDM is also supported but will not be the primary PHY used for data communications. Unlike 802.11ac, 802.11ax does allow operations in the 2.4 GHz band, but with a maximum optional channel width of 40 MHz, while 20 MHz channels are required in 2.4 GHz. 802.11ax also adds QAM-1024 (or 1024-QAM) modulation.

In addition, the HE PHY introduces Orthogonal Frequency Division Multiple Access (OFDMA) to 802.11 for the first time. This modulation scheme allows the channel to be divided into resource units that can be assigned to different stations, thereby implementing a form of Frequency Division Multiple Access (FDMA) to the 802.11 Wi-Fi channel.

In addition, the HE PHY introduces the following capabilities:

- BSS Color: The ability to tag the 802.11 frames identifying the BSS to which they belong. STAs performing Carrier Sense can ignore frames at different signal levels from other BSS networks than their own. The goal is to increase frequency efficiency and reuse.
- Target Wake Time (TWT): A new scheduled power management option that can also increase efficiency 802.11ax devices, particularly for power management on battery powered devices.
- Uplink MU-MIMO: The ability for multiple STAs to transmit to the AP at the same time and on the same frequencies. Like downlink MU-MIMO in 802.11ac, but the inverse.

Without question, the best new feature of 802.11ax will be support for the 6 GHz frequency band. With so many devices in 2.4 and 5 GHz using older radios, it will be several years before significant benefit occurs in those bands.

	Frequency Band	Channels	Bandwidth	Modulation	Data Rates
High Efficiency (HE) - required 802.11ax	2.4, 5 and 6 GHz	2.4, 5 and 6 GHz channels as previously defined.	20, 40 or 80 MHz (only 20 MHz channels are required in 2.4 GHz)	BPSK, QPSK, QAM-16 and QAM-64, QAM-256, QAM-1024	Based on Modulation and Coding Scheme (MCS) tables
High Efficiency (HE) - optional 802.11ax	2.4, 5 and 6 GHz	2.4, 5 and 6 GHz channels as previously defined.	20, 40, 80 or 160 MHz 160 MHz may be contiguous or 80+80 40 MHz optional in 2.4 GHz	BPSK, QPSK, QAM-16 and QAM-64, QAM-256, QAM-1024	Based on MCS tables

TVHT

The *Television High Throughput (TVHT)* PHY is not tested on the CWNA exam beyond awareness of the target use and frequencies used, due to the limited hardware available for deployment at this stage. This PHY is designed to take advanced of unused frequencies in the bands often used for television and other broadcasts. Because it is designed to use such spaces, it supports very narrow channel widths of 6, 7 or 8 MHz, depending on the regulatory domain in which it operates. Additionally, the channel widths of 6, 7 or 8 MHz (called Basic Channel Units, or BCUs) can operate as 1, 2 or 4 BCUs. Therefore, with two 7 MHz BCUs, the total frequency space available for the transmissions would be 14 MHz.

The maximum data rate supported, with the use of four 8 MHz BCUs (32 MHz of frequency space), is 568.9 Mbps. This is accomplished with four spatial streams. Like 802.11n devices, TVHT devices can use from one to four spatial streams.

Finally, TVHT operates in the frequency range from 50 MHz to 790 MHz and uses frequency space as allocated by the operating regulatory domain. Figure 5.9 provides a summary of this information related to TVHT.

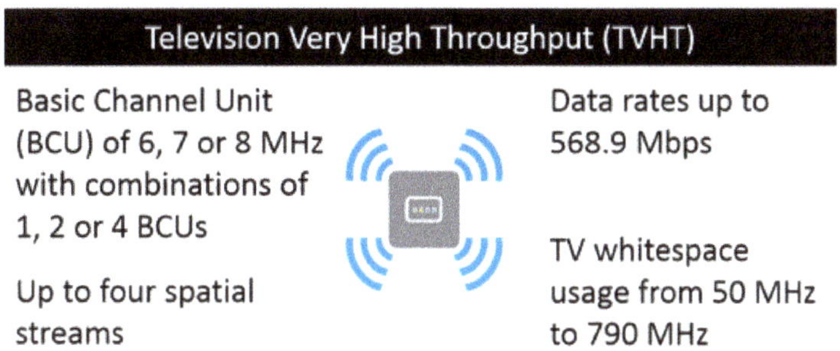

Figure 5.9: TVHT PHY Summary

S1G

The *Sub-1 GHz (S1G)* PHY was designed with long range, low data rate communications in mind and is defined in the 802.11ah amendment. It is ideal for Internet of Things (IoT) and industrial automation and monitoring networks. The S1G PHY operates on 1, 2, 4, 8 or 16 MHz channels and it appears likely that more devices will use the 1, 2 and 4 MHz channels as the likely use cases do not warrant the higher data rates.

The maximum data rate supported on the S1G PHY is 346.6667 Mbps. This rate is based on a 16 MHz channel and four spatial streams. Given the desire for extended battery life and little need for high data rates, as devices are released supporting the S1G PHY, we are likely to see many single stream devices. Such devices, operating on a likely maximum of 4 or 8 MHz channels, will achieve a maximum data rate of 8666.7 Kbps for a 2 MHz single stream device or 20,000 Kbps (20 Mbps) for a 4 MHz single stream device.

The actual frequencies used will vary greatly by regulatory domain but will all be less than 1 GHz. Some will operate in the 700 MHz range and others in the 900 MHz range and still others in between. Figure 5.10 provides a summary of this information related to S1G. For the CWNA, it is important to remember that all S1G PHY devices operate in the frequencies below 1 GHz.

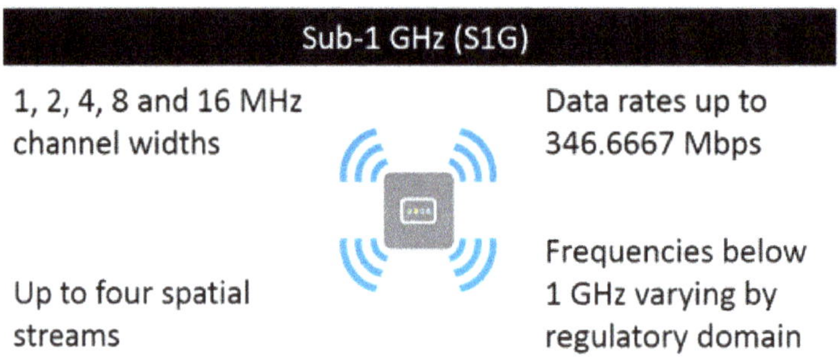

Figure 5.10: S1G PHY Summary

DMG

The *Directional Multi-Gigabit (DMG)* PHY is seeing some traction in consumer wireless routers and client devices. For example, a few vendors have released laptops supporting the DMG PHY, which was ratified in 802.11ad. This PHY operates in the 60 GHz frequency band and is a high data rate, low range specification. It specified three modulation methods:

- Control modulation
- Single Carrier (SC) modulation

- OFDM modulation (this mode is defined as obsolete and may be removed in a later version of the standard)

The DMG PHY supports the range from 57-64 GHz worldwide and from 57-66 GHz in Europe. The specific areas supported within these ranges varies by regulator domain. Figure 5.11 provides an overview of the frequency ranges supported.

The DMG PHY supports 1 to 4 channels depending on the region and channel 2 is supported by all regions and is therefore the default channel. The maximum data rate is 6756.75 Mbps, sufficient for high-definition video and other high throughput requirement transmissions. Remember, however, it works only over a very short range.

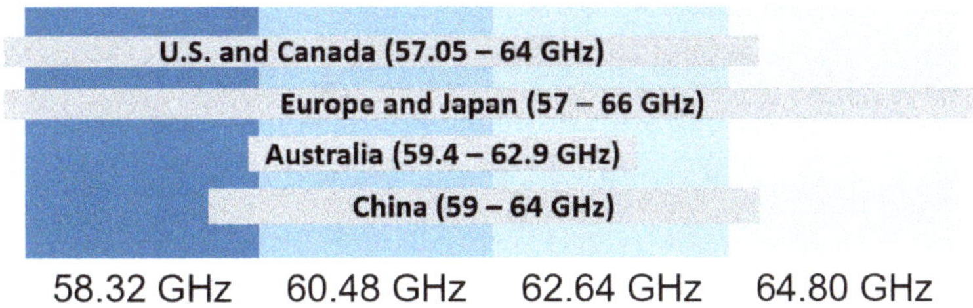

Figure 5.11: DMG Frequency Ranges by Region

802.11 PHYs Overview

Table 5.1 provides a summary of the basic details related to the various 802.11 PHYs.

PHY Layer	Max Data Rate	Max Spatial Streams	Band	Max Channel Width
DSSS	2 Mbps	1	2.4 GHz	22 MHz
HR/DSSS	11 Mbps	1	2.4 GHz	22 MHz
OFDM	54 Mbps	1	5 GHz	20 MHz
ERP	54 Mbps	1	2.4 GHz	20 MHz
HT	600 Mbps	4	2.4 and 5 GHz	40 MHz

VHT	6933.3 Mbps	8	5 GHz	160 MHz
HE	9602.8 Mbps	8	2.5, 5, and 6 GHz	160 MHz
TVHT	568.9 Mbps	4	54 to 790 MHz whitespaces only	32 MHz
S1G	346666.7 Kbps (346.6667 Mbps)	4	Sub-1 GHz – specifics vary by regulator domain	16 MHz
DMG	6756.75 Mbps	N/A	60 GHz	2160 MHz

Table 5.1: PHY Layers and Specifications

802.11 Functional Concepts

Several functional concepts should be understood to fully grasp the operations of 802.11 WLANs. These include modulation, coding, co-location interference, channel centers and widths, primary channels, adjacent overlapping and non-overlapping channels, throughput versus data rate, and available channels and constraining requirements, such as Dynamic Frequency Selection and Transmit Power Control.

Modulation and Coding

Modulation is a process by which some property of a carrier signal is modified to represent digital bits. Amplitude, frequency, and phase are the three basic elements of a waveform that can be varied.

Complex Modulation and Coding Schemes (MCS) are used in 802.11 WLANs. There are various types of modulation used in 802.11 WLANs including DBPSK, DQPSK, and QAM. There are various types of coding schemes as well, including Barker, CCK, PBCC, and FEC (convolutional).

PSK Modulation

Modulation constellations are a helpful visual tool for understanding modulation. These constellations show an in-phase and quadrature component of a carrier wave. Rotation around the center axis represents the phase component of a wave.

For example, in the BPSK constellation in Figure 5.12, you can see that there are only two phases here, both plotted on the in-phase axis. With phase shift keying modulation, the phase of the wave is shifted to reflect changes in data. The receiver is able to determine the original transmitted bit by identifying the phase of the wave it receives. If the phase matches the left side of the constellation, the radio signal is mapped to digital bit 0. Conversely, if the phase matches the right side of this constellation, the radio wave is signaling a 1.

You can see in the QPSK constellation that there are four possible data points. In terms of phase shifts, these could be 0, 90, 180, and 270-degree shifts. Because there are four possible data points, the QPSK modulation method can represent 2 bits of information. With BPSK, the receiver has a wider margin of error (180 degrees), whereas with QPSK, the margin for error is less (90 degrees) because there are more possible data points. However, QPSK is also twice as efficient as BPSK because the same carrier signal can represent 2 bits instead of 1.

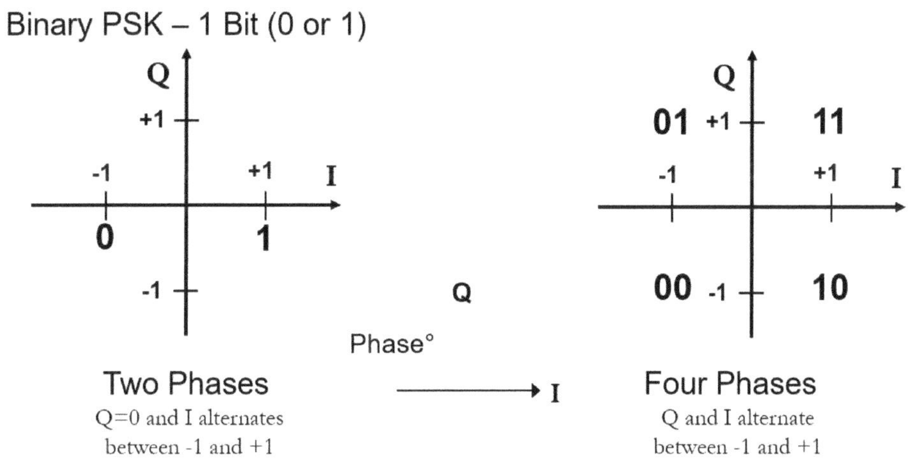

Figure 5.12: Binary PSK (BPSK) and Quadrature PSK (QPSK)

This text is an over-simplification of constellation diagrams, but it is important to understand that each modulation method is capable of varying the carrier wave in such a way that it represents some number of bits. Modulation methods with more data points can represent more bits, but they require better signal quality for the receiver to accurately discern what data was transmitted.

You will also learn that these modulation types can be "absolute" or "differential." In other words, with BPSK, the phase shifts are relative to an absolute phase value. The receiver must know the absolute phase value against which it is comparing received data, but that point is fixed. Conversely, differential BPSK (DBPSK) uses differential phase values. In DBPSK (or DQPSK), the phase shifts are compared with the phase setting of the previous signal measurement. In other words, if the phase is 0 on the first wave and the phase of the next wave is 180, then we know the phase shift was 180 degrees (180-degree phase differential). If the following wave is 180 degrees, we know that the phase shift was 0 degrees because the prior measurement was also 180. Then if the following wave is 270 degrees, we know that it shifted by 90 degrees from previous. In this way differential PSK modulation changes the encoded data only when the phase changes. DSSS and HR/DSSS utilize DBPSK and DQPSK, whereas OFDM (802.11a/g) utilizes BPSK and QPSK.

In IQ modulation, the *I* represents the in-phase component and the *Q* represents the quadrature component. It is a quadrature component as it represents 90 degree shifts. Given a 360-degree constellation, 90-degree shifts allow for four phases.

QAM Modulation

Quadrature Amplitude Modulation (QAM) modulation is a higher efficiency WLAN modulation type (compared with BPSK and QPSK) that is used with OFDM. QAM includes both a phase and an amplitude component, making it better suited to represent more data points. 16-QAM includes 16 data points, representing 4 bits (2^4 = 16), while 64-QAM includes 64 data points, representing 6 bits (2^6 = 64).

On the modulation constellations for BPSK or DQSK, we noted that each bit or bit sequence can be signified with a specific phase. With QAM modulation, both phase and amplitude are used in combination to indicate a specific bit sequence. Figure 5.13 shows QAM modulation constellations for 16-QAM and 64-QAM. 256-QAM is also available in 802.11ac. In the example of 16-QAM, each "quadrant" of the constellation contains four data points. By looking at the constellation shown here, 0010 and 0111 require the same phase adjustment, but the amplitude would change for each.

The phase on a constellation can be visualized with a circle around the center axis. Amplitude is visualized by the distance from the center axis, where higher amplitude is farther from center.

When compared with BPSK and DQSK, 16- and 64-QAM require better signal quality (signal strength compared to noise) at the receiver because the data points are "closer" together. 802.11ac introduces 256-QAM for even higher data rates than earlier PHYs.

OFDM Carriers

When used as part of 802.11-2016 clause 17 or 19 PHYs, Orthogonal Frequency Division Multiplexing (OFDM) subdivides a channel into 52 discrete subcarriers of 312.5 kHz each. Four subcarriers are used as "pilot" channels and not available for data transmissions. Forward Error Correction (convolutional) coding is used in the 802.11 OFDM PHY. Figure 5.14 shows this in illustration.

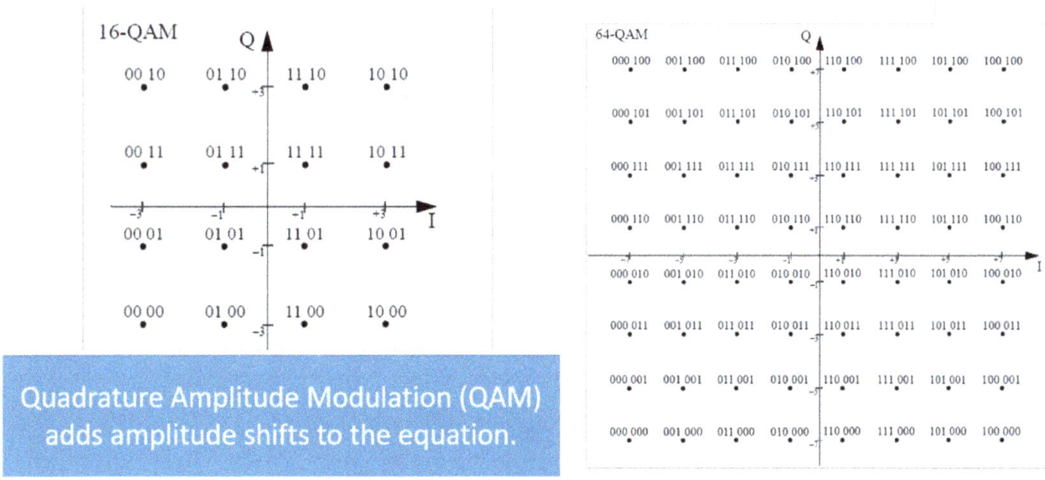

Figure 5.13: QAM Modulation

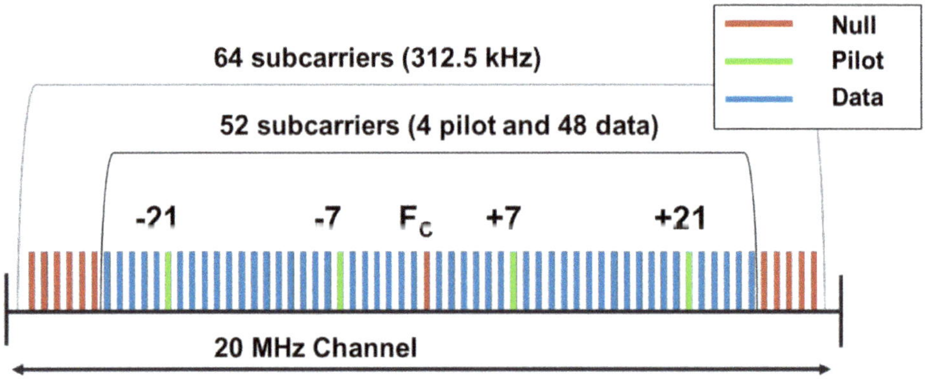

Figure 5.14: OFDM Subcarriers

Notice that the first six and last five subcarriers are null. The center is also a null subcarrier. Then, four of them are pilot subcarriers, which leaved 48 subcarriers to transmit actual data. This plays an important role in the next bit of information, because it is part of the overall calculation that results in a given data rate with OFDM communications.

This subcarrier information helps one to see why the data rates are what they are. For example (remember, a microsecond is 1/1,000,000 of a second):

- OFDM symbol duration = 4 microseconds
- Data-carrying subcarriers = 48
- Bits per subchannel = 6 (with the 64-QAM modulation)
- Bits per OFDM symbol = bits per subchannel X data-carrying subcarriers = 288
- Coding rate = ¾
- Useful bits per symbol = 288 X ¾ = 216 bits per symbol
- Bit rate per second = 216 / 4 microseconds X 1,000,000 = 54,000,000 bps or 54,000 Kbps or 54 Mbps

Now you can see why the data rate is stated as 54 Mbps instead of 72 Mbps. The coding rate uses bits for resiliency and these bits are not referenced in the data rates available but are assumed overhead to have an acceptable channel of communication.

Additionally, one should understand the basic concept of a symbol rate. The symbol rate is most simply defined as the number of symbols per second. It is also often called the baud rate (remember dial-up modems), or the modulation rate. It is technically the number of waveform changes or channel condition changes in each second and, remember, more than just a single bit is communicated in each symbol with the OFDM modulation. In fact, using 64-QAM, it can represent 6 bits and using 256-QAM it can represent 8 bits. So, with 48 data carriers, each sending a waveform during a symbol that represents 6 bits, you have the 288 total bits available for transmission during a symbol. But the symbol rate itself is the number of times per second that a symbol can be transmitted, and it is simply 1 second divided by the symbol duration. In this case, it is 250,000 (or 250k) with a 4-microsecond symbol.

> Many computer technicians studying data communications for the first time are tempted to use a 1024 factor for calculating bits per second at kilobit and megabit rates. For example, this would mean that 1,000,000 bps would be equal to 976.5625 Kbps. However, unlike computer memory and storage, with telecommunication data rates, the factor is 1000. Therefore, the math is easier and 1,000,000 bps is 1,000 Kbps or 1 Mbps.

Because you encounter so many references to small fragments of time when studying networking communications, the following list will be helpful:

- 1 millisecond (ms) = 1/1,000 of a second
- 1 microsecond (μs) = 1/1,000,000 of a second
- 1 nanosecond (ns) = 1/1,000,000,000 of a second

OFDMA

OFDMA uses resource units (RUs) to modulate symbols. An RU is a frequency range within the total channel width used to transmit to a specific STA. More RUs may be assigned to one STA than another. This allows an AP to transmit multiple frames to multiple STAs concurrently. Of course, wider channels have more RUs available. The actual modulation is the same as that used in 802.11ac, except for the added higher modulation rate with 1024-QAM (or QAM-1024).

Like OFDM, the subcarriers within an RU fall into multiple subcarrier types: Data, Pilot and Null (or Unused). Data subcarriers are modulated with data for transmission. Pilot subcarriers are used for phase information and tracking parameters. The Unused or Null subcarriers are used as guards between RUs and at the channel edges. They are also used as DC subcarriers.

When transmitting uplink OFDMA frames using RUs (from the clients to the APs), the AP will use trigger frames to inform the clients of the resource units they are to use for the uplink transmissions. This is known as triggered uplink access (TUA) in the 802.11ax amendment.

- OFDMA implements the *resource unit (RU)*
  - Modulation methods include BPSK, QPSK, 16-QAM, 64-QAM, 256-QAM, and 1024-QAM

Channel Width	Maximum RUs
20 MHz	9
40 MHz	18
80 MHz	37
160 MHz	74

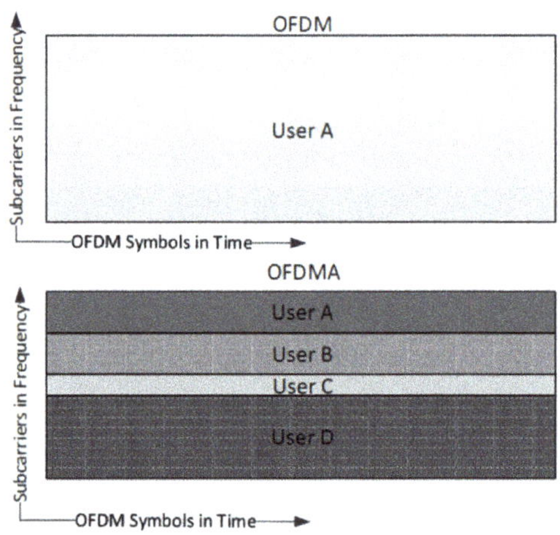

Coding Methods

Coding is the conversion of data bits into a series of symbol bits. It adds redundancy, which results in communication resiliency. Coding rates are a ratio of data bits to coded bits. For example, a coding rate of 2/3 indicates 2 data bits for every 3 coded bits. The earliest coding (DSSS) used a spreading sequence. OFDM-based PHYs use convolutional coding.

The spreading sequence used in 802.11 DSSS is comprised of 11 digits (chips) in the pattern — 10110111000 (transmitted from left to right). In DSSS, each data bit is multiplied, using Exclusive-OR (XOR) math, against a preset spreading sequence. The original 802.11 spreading code (Barker) is 11 digits in length, thus it codes each data bit into an 11-bit symbol. The symbol represents a single data bit, but by "spreading" the bit like this, it increases data recovery and resilience when transmitted across the wireless medium.

The Barker encoding process increases the bandwidth required to transmit a single bit by 11 times. Increasing the bandwidth allows a reduction in output power (without sacrificing reliability) when compared to narrow band transmission methods. With an 11-chip spreading code, the required channel bandwidth increases from 1 MHz to 11 MHz. However, the "occupied bandwidth" is wider, usually notated as 22 MHz, because of the 802.11 signal bandwidth requirements for DSSS signals.

802.11b HR/DSSS uses a more sophisticated spreading mechanism than Barker, known as Complementary Code Keying (CCK). Instead of using an 11-chip code for each bit, an 8-chip code is used. Further, the 8-chip code is not a fixed sequence like the Barker code. There are multiple different codes that can be combined with the data input, and the data input can be either 4- or 8-bit sequences (whereas Barker is 1 or 2 data bits).

Convolution coding encodes bits such that a series of input bits (data bits) are encoded to a series of code bits and use a convolution code ratio. They are represented as a ratio, such that for every x number of data bits, y number of code bits are used. Therefore, a 1/2 ratio tells us that one bit of data is encoded in every two bits of code. Likewise, with a ratio of 3/4, three bits of data are encoded for every four bits of code. The result is that higher data bits per encoding provide higher data rates.

In the end, the modulation and coding, partnered with channel bandwidth, the number of spatial streams, and the length of the GI, come together to result in the various data rates available in 802.11. This is even true for older PHYs, excepting the fact that they did not support multiple spatial streams (MIMO) and a short GI.

Modulation and Coding Schemes (MCS)

802.11n and 802.11ac use MCS tables instead of simply listing all possible data rates. Figure 5.15 shows an example MCS table from the VHT PHY for a 20 MHz single spatial stream configuration.

VHT-MCS Index	Modulation	R	N_{BPSCS}	N_{SD}	N_{SP}	N_{CBPS}	N_{DBPS}	N_{ES}	Data rate (Mb/s)	
									800 ns GI	400 ns GI (See NOTE)
0	BPSK	1/2	1	52	4	52	26	1	6.5	7.2
1	QPSK	1/2	2	52	4	104	52	1	13.0	14.4
2	QPSK	3/4	2	52	4	104	78	1	19.5	21.7
3	16-QAM	1/2	4	52	4	208	104	1	26.0	28.9
4	16-QAM	3/4	4	52	4	208	156	1	39.0	43.3
5	64-QAM	2/3	6	52	4	312	208	1	52.0	57.8
6	64-QAM	3/4	6	52	4	312	234	1	58.5	65.0
7	64-QAM	5/6	6	52	4	312	260	1	65.0	72.2
8	256-QAM	3/4	8	52	4	416	312	1	78.0	86.7
9	Not valid									
NOTE—Support of 400 ns GI is optional on transmit and receive.										

Figure 5.15: MCS Table for 802.11ac 20 MHz and 1 SS

The modulation and coding used, in relation to the channel bandwidth and number of spatial streams, are key determiners in the final data rate of the link. Given that a better signal quality is required for more complex modulation methods, devices fall back to less complex modulation methods as the signal quality degrades. The result is a lowering of the data rate. This process is called Dynamic Rate Switching (DRS). The 802.11 standard does not define the DRS algorithm from the perspective of signal quality, this definition is the responsibility of the chipset manufacturers. The 802.11 standard defines only that the STA must use a permitted data rate for transmission and this permitted list is found in Beacon frames and Probe Response frames from the associated AP.

Channel Centers and Widths

As you learned in the PHY section of this chapter, early 802.11 PHYs (DSSS and HR/DSSS) use 22 MHz-wide channels. All newer 2.4 and 5 GHz PHYs use 20 MHz channels or some factor thereof, such as 40, 80 or 160 MHz. 802.11n supports 20 and 40 MHz channels and 802.11ac supports 20, 40, 80 and 160 MHz channels. 802.11a/g supports only 20 MHz channels. This channel width is also called the channel bandwidth.

Figure 5.16 shows the channel plan for 2.4 GHz radios. Note that the center frequency for channel 1 is 2412 MHz or 2.412 GHz. All 2.4 GHz channel center frequencies are 5 MHz apart. Therefore, channel 2 is centered on 2.417 GHz, channel 4 is centered on 2.422 GHz and so forth. You need not memorize all the 2.4 GHz center frequencies. When you know the center frequency for channel 1, you know it for channels 1-11 too (based on increments of 5 MHz), which are the commonly implemented 2.4 GHz channels.

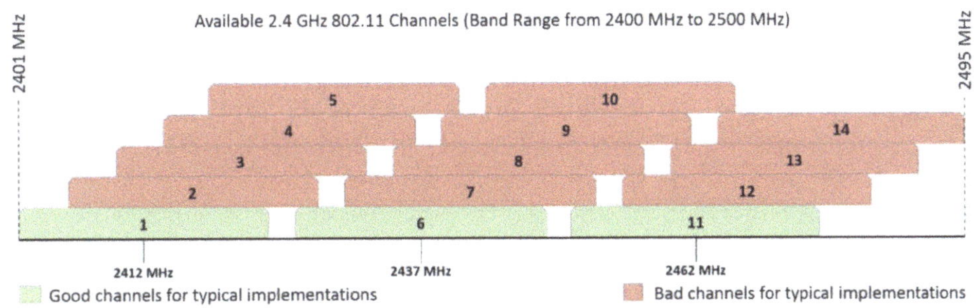

Figure 5.16: 2.4 GHz Channel Plan

The 2.4 GHz band ranges from 2.400 to 2.500 GHz; however, 802.11 devices use the range from 2.401 GHz to 2.495 GHz when all 14 channels are supported in the regulatory domain. North America supports only channels 1 through 11. Much of the rest of the world also supports channels 12-13. Channel 14 is supported in Japan for 802.11b operations. For this reason, most 2.4 GHz networks are configured to use only channels 1, 6, and 11. These channels do not overlap, and they are available in all regulatory domains so that no traveling client finds itself in a dead spot where only an AP on channel 13 or 14 is covering it.

It is important to note that 1, 2, 5.5 and 11 Mbps transmissions use 22 MHz channels, while all other data rates in 2.4 GHz use 20 MHz channels. This behavior is due to the use of older modulation methods when communicating with these lower data rates.

Notice, in Figure 5.16, that channel 2 overlaps with channel 1. This would be considered an adjacent overlapping channel. It is the next channel, but it is overlapping. As you will soon see, in the 5 GHz band, adjacent channels do not overlap. They are considered adjacent non-overlapping channels. If you start with channel 1, the first non-overlapping channel is channel 6. From there, the next non-overlapping channel is 11. Using adjacent overlapping channels or turning the output power up to high on an AP can result in ACI — adjacent channel interference — and significantly degrade the performance of the WLAN.

The term co-location interference is used in the 802.11 standard and should not be confused with ACI or Co-Channel Interference (CCI). *Co-location interference* references multiple 802.11 STAs in a single device that cause interference with each other.

When selecting channels in 2.4 GHz or 5 GHz, ensure that you use only channels allowed in your regulatory domain. All regulatory domains worldwide allow for the use of channels 1-11.

In the 5 GHz band, many more channels are available, and all 20 MHz channels are non-overlapping. Figure 5.17 shows the channel plan in 5 GHz. The channels in green can be considered safe all the time, from a bandwidth perspective. They are 20 MHz channels and should work in most installations. The exception is an installation that may require 40 MHz channels for bursts of higher throughput for some clients. The channels in yellow should only be used when required. If you do not require 40 MHz channels, it is often best to simply use 20 MHz channels. The channels in red should really make you stop and think before you implement them. They are 80 MHz channels and consume much of the frequency space, even in 5 GHz. Finally, the channels in black are simply no-go channels until the regulatory agencies give us enough frequency space to have at least 4-5 of them.

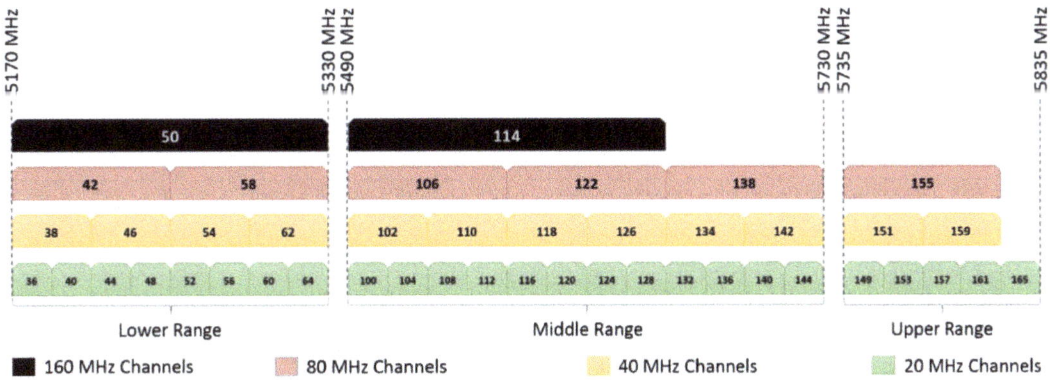

Figure 5.17: 5 GHz Channel Plan

It is also important to know that channels 52-144 require DFS in many regulatory domains. *Dynamic Frequency Selection (DFS)* is a radar detection and avoidance scheme defined for some regulatory domains on 5 GHz channels. On these same frequencies, military, weather, and other radars may be operating, thus it is desirable for WLANs to avoid interfering with them.

DFS is defined for the following reasons:

- Quieting the current channel for testing
- Testing for radar before using and while operating in a channel
- Discontinuing operations after detecting radar
- Detecting radar in current and other channels
- Requesting and reporting measurements in the current and other channels using Action frames and elements

- Association is based on supported channels (AP builds a table, which is used to determine the best future channels, if necessary)

The process of selecting and advertising a new channel to assist the migration of a BSS after radar is detected may not move all STAs successfully. The AP determines the new channel based on normal vendor proprietary channel selection algorithms, while also factoring in DFS measurement information and channel support information received by clients during association. Channel Switch Announcement frames (or CSA elements in Beacon/Probe frames) are used to announce the pending move of a BSS.

DFS may present a few common problems that should be considered when planning WLANs:

- Coverage holes (and thus, connectivity problems) will result if client devices do not support DFS channels. If some clients show signs of poor connectivity or complete coverage gaps, check for DFS support.

- Each vendor must follow the regulatory requirements for DFS. However, every vendor will implement its own proprietary channel-switch algorithms. For all vendors, a temporary service outage will occur if a DFS event occurs. At a minimum, the AP must send channel switch announcements, and then switch to a new channel. Some clients will inevitably miss the transition to a new channel, and some applications will suffer.

- To detect radar, spectrum analyzers may be used, but WLAN infrastructure devices will be more valuable. Because of the way that radar functions, it will often be received at fairly low power (relatively speaking) when compared with usual Wi-Fi interferers. It can be difficult to identify a radar in a spectrum analyzer.

- When DFS channels are used by the WLAN infrastructure and supported by clients, active scanning will be less efficient because the client must wait to hear a transmission before it can legally transmit. In the absence of devices currently operating on a channel, the client must wait a quiet time to measure for radar. This will impact scanning and roaming behavior.

Additionally, Transmit Power Control (TPC) is often required in regulatory domains. *TPC* was defined to serve two primary functions:

- *Power constraint* to prevent interference with satellite systems based upon regulatory transmit requirements
- *Dynamic power control* to improve client battery and overall cell performance

With TPC, the AP specifies power requirements for the client and uses the clients power capability information for operation:

- Specifies regulatory and local max Tx power for channel (Power Constraint element)
- Selection of Tx power for a transmission in a channel
- Adaptation of the transmit power
    - Can use any criteria to dynamically adapt Tx power for transmissions, but it is beyond the scope of the standard.

At the time of writing, some uncertainty still exists about the exact usage of the 6 GHz band throughout different regulatory domains. However, this slide presents the basic structure of the frequencies available.

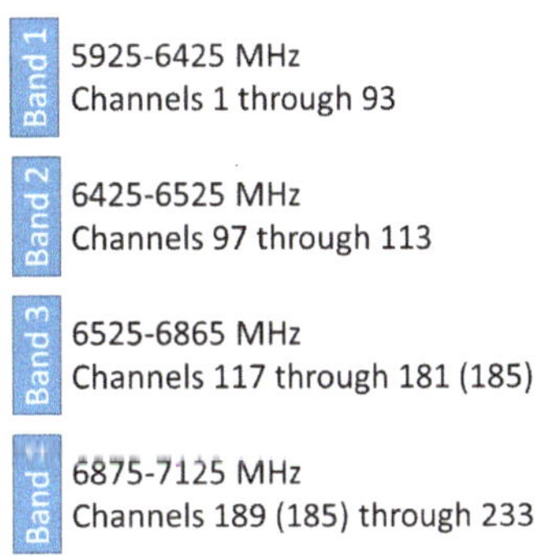

6 GHz Channels

- Band 1 is also called UNII-5
- Band 2 is also called UNII-6
- Band 3 is also called UNII-7
- Band 4 is also called UNII-8

It is expected that Europe will use at least Band1 or portions thereof and the rest of the world will follow either Europe or the United States, which allows use with some restrictions of the entire 6 GHz range described.

There is a 40 MHz gap between the top of what is called the 5 GHz band and the bottom of the 6 GHz band.

6 GHz is only supported by 802.11ax devices and the majority of 802.11 activity will remain on the 2.4 and 5 GHz bands through 2023 simply due to the time it takes to adopt the new technology.

Throughput vs. Data Rate

Data rate and throughput are not the same. Figure 5.18 illustrates this point. The data rate range in the image represents real-world 802.11ac networks with four spatial streams and a maximum of an 80 MHz channel. The throughput range, on the right of the image, is a rough estimate of potential values for a single STA. The throughput rate will vary greatly, depending on the 802.11 PHYs in use, the number of STAs in the BSS, the number of detectable STAs in other BSS networks on the same channel, and the signal quality at the receivers.

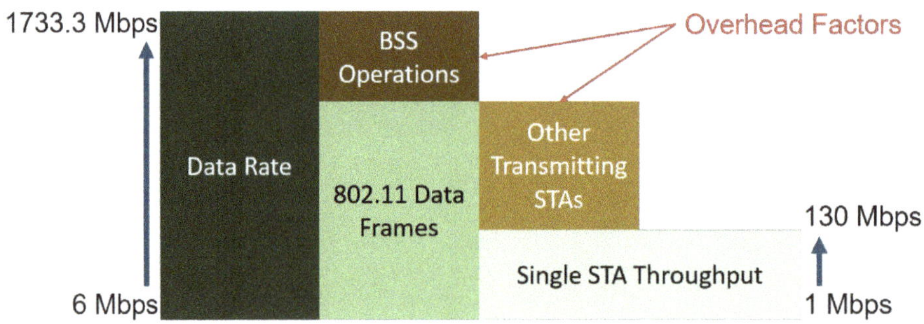

Figure 5.18: Data Rate vs. Throughput

802.11 Service Set Components

The concept of a service set, to which previous chapters allude, has not been fully defined up to this point. The service set is the basic logical entity that exists in IEEE 802.11 WLANs. Service sets and stations are both defined and explained in this section. Additionally, the distribution system is explored in order to understand how a WLAN is spread throughout a service area using both wired and wireless technologies. Finally, this section provides an overview of roaming within IEEE 802.11 WLANs.

Stations, BSSs and BSAs

The 802.11 standard defines an entity known as a station and uses the term station (STA) to refer to this entity. The STA is defined as any device that has an IEEE 802.11-compliant MAC and PHY interface to the wireless medium (WM). This means that the following devices, among others, would all be considered valid STAs assuming they use IEEE 802.11-compliant radios and drivers:

- Access Points (AP)
- Laptops, desktops and servers with wireless NICs
- Tablets and mobile phones with 802.11 radios
- Residential gateways (mostly known as wireless routers)
- Wireless print servers
- Wireless presentation devices
- Wireless bridges
- Wireless gaming adapters (mostly just wireless bridges)
- Wireless VoIP (Voice over IP) handsets
- 802.11 door locks
- 802.11 video cameras

This list is just a partial list. Many devices use 802.11-compliant radios and drivers and are capable of acting as an 802.11 STA. However, it is also possible to use

802.11 devices in ways that are not defined within the IEEE standards and in ways that are not compatible with IEEE-based WLANs. These implementations usually have alternate MAC or PHY layers that change the functionality of the 802.11-compliant device in such a way that it is no longer considered a valid 802.11 STA. In other words, the device is capable of operating in an 802.11-compliant manner, but the drivers or software cause it to function in a non-compliant manner. This can be done because an 802.11 device uses a radio that has the capability of behaving in ways that are not in compliance with the standard. Keep in mind that the standards are used to ensure interoperability between devices and non-standard devices should be avoided in common WLAN implementations.

It might seem odd to you to think of an AP as a STA if you are used to wired networks where you use the term "client station" to mean an end-user's computer; however, even in wired networks, any computer or device using the 802.3 Ethernet standard is technically a node on the network. Where node is the generic term for any Ethernet device, STA is the generic term for any 802.11 device. If the device can communicate on an 802.11-conformant WLAN, it has a set of STA services and is able to participate in a Basic Service Set.

This brings us to the phrase Basic Service Set. The *Basic Service Set (BSS)* is defined as a set of stations that have successfully synchronized after one station has executed the START primitive. In later chapters, you will learn about the Distributed Coordination Function (DCF), which is the way in which devices contend for access to the wireless network. The stations that are all cooperating together in the same DCF group form a BSS. There is more than one type of BSS. There is a BSS that is more dynamic (independent) and another that is more static (infrastructure). You can also combine more than one BSS together to form a logical group through which a STA may pass without network interruption, called an ESS. The following sections describe the components of a BSS, the BSS types, and the process of starting and joining a BSS.

A *primitive* is defined as a basic operation or capability supported by a system. The term primitive is used throughout the 802.11 standard to define various logical components that may be implemented in wireless STAs.

The BSA is the basic service area. This is the conceptual area within which BSS members may communicate. Another way of saying this is to say that the BSA is the physical space within which the STAs that are participating in the BSS may communicate with each other. In some BSS implementations, the client STAs communicate directly with each other, and in other BSS implementations, the client STAs communicate with each other through the AP STA. In either case, the BSA is the physical space boundary within which these STAs may communicate with each other. The BSA will vary or change slightly over time in most environments and can vary greatly in some environments. This is due to changes in the physical space such as atmospheric changes, physical item placement, the number of people present, and so forth. This is why a particular signal strength can be seen at a location at one time, and a very different signal strength can be seen at the exact same location at another time. Though the AP has not moved, environmental conditions change, and this results in a varying BSA. Figure 5.19 illustrates the BSA boundary in green.

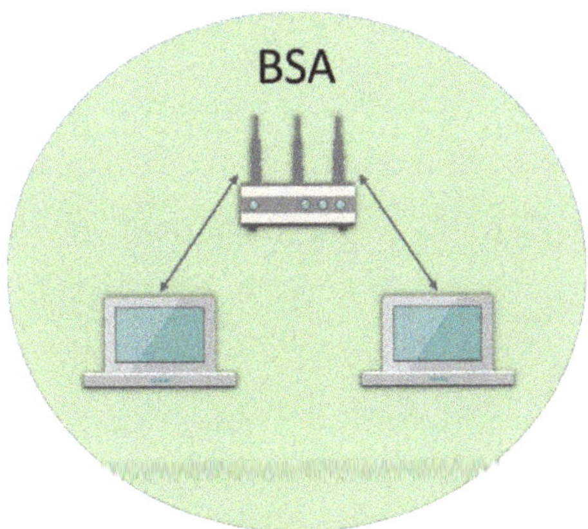

Figure 5.19: BSA Boundary

Ad Hoc Mode and IBSS

The dynamic topology offered by the 802.11 standard is the independent BSS (IBSS). This is sometimes called an ad-hoc wireless network. An *IBSS* is a collection of STAs that are communicating with each other directly, without the use of an AP. In order for a STA to be able to communicate with another STA, they must be within RF range of each other. There is no relaying of signals from one STA to another through the various STAs in the IBSS. For this reason, an IBSS is constrained to a fairly small BSA.

As an example, consider Figure 5.20. In this figure, the STAs on the right and the left can communicate with each other at an acceptable rate. All the STAs in between the right and left STA can also communicate with STAs A through E effectively. However, STA F is represented in the lower portion of Figure 5.20 and its signals cannot reach the other STAs in the IBSS and the other STAs' signals cannot reach it. STA F is outside the BSA of the IBSS.

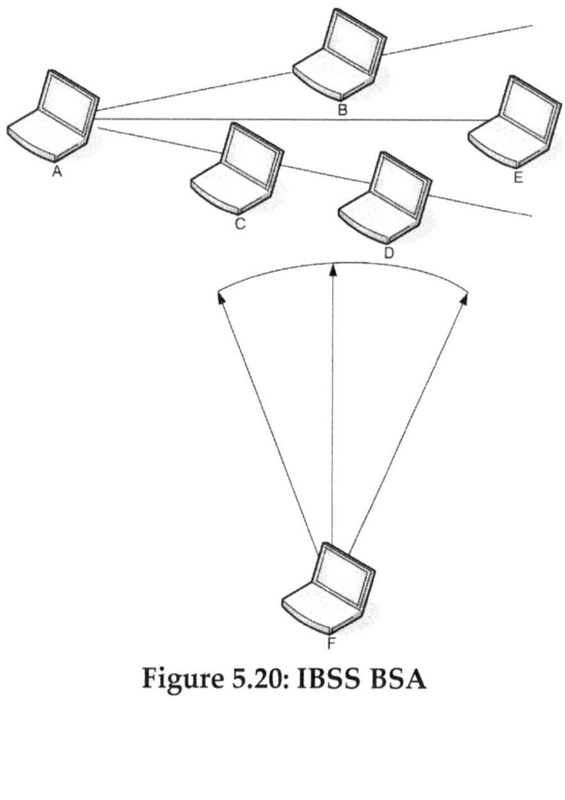

Figure 5.20: IBSS BSA

Infrastructure Mode and ESS

When a wireless AP station is used, an infrastructure BSS (simply called a BSS) is implemented. At the same time an extended service set (ESS) is made possible. An *ESS*, as defined in 802.11-2016 is a collection of one or more BSSs typically sharing the same SSID (defined in the next section). The following is the definition of an ESS pulled directly from the standard:

A set of one or more interconnected basic service sets (BSSs) that appears as a single BSS to the logical link control (LLC) layer at any station associated with one of those BSSs.

The definition in the standard reveals that a single BSS can indeed form an ESS. The phrase, "a set of one… BSSs" reveals such. This is also seen in the definition of the *Extended Service Area (ESA)*, which is the ESS equivalent of the BSA. The ESA is defined as being *larger than or equal to a BSA*. You may be wondering why I am taking the time to explain the technical reality that an ESS exists, as soon as an infrastructure BSS exists. Allow me to explain.

The vast majority of WLAN literature indicates that you create an ESS when you group more than one BSS together using the same SSID. This is an incorrect understanding, and it leads to a misunderstanding of what an ESS really is. An ESS is more like a Windows Workgroup than a Windows Domain, for those familiar with Windows environments. There is no real "central" controller of the ESS or requirement for more than one BSS to have an ESS. The ESS exists logically the moment the first AP comes online and forms a BSS — just like a Windows Workgroup exists as soon as a computer comes online using the workgroup name. The first computer cannot prevent other computers from using the same Workgroup name or control their access to the Workgroup. An ESS is the same in that the first AP that comes online starts the ESS and additional APs can join the ESS by using the same SSID (they may even become a virtual enlarged ESS even though they have different SSIDs) and this is not controlled by the initial AP. Though some WLAN switches or controllers may be able to control entry to an ESS, this is not part of the 802.11 standard.

Another angle of approach would be to say that the SSID is actually the ID of the ESS and not the BSS. In fact, as stated earlier, the BSS has an ID and it is the BSSID. The ESS has an ID too and it is usually the SSID. This understanding clears up the confusion that results in common questions like:

- What is the difference between the SSID and the BSSID?
- Where is the ESSID set?
- How do you "create" an ESS?

The answer to the first is that every service set has a machine friendly name (the BSSID) and a people-friendly name (the SSID). The answer to the second question is nowhere. There is no such thing as an ESSID, and where it is referenced in the earlier 802.11 documents, it is used only as a shorthand reference in logical process representations. The answer to the third question is that you create an ESS when the first AP comes online and executes a START primitive as an Infrastructure BSS. The IEEE has not determined how to make multiple ESSs into one ESS, but they have suggested that this is something that needs to be addressed in the future.

Ultimately, you might consider that two perspectives on the creation of an ESS exist. The first is that the ESS is logically created when the first AP comes online, as it is now available for other APs to join. By this, I mean that there is only one BSS, and most people think of an ESS as "more than one BSS," but the ESS exists the moment the first AP comes online. The second is that the ESS is created physically, or in reality, when the second AP is brought online and configured to use the same SSID or uses some other method of joining the ESS on the shared distribution system.

Though an ESS is often said to have an ESSID, in fact, the 802.11 standard does not identify such an element as an ESSID. The ID of an ESS is effectively the SSID that is shared among the infrastructure BSSs that are participating in the ESS. The BSSID of each infrastructure BSS is used by the APs in order to track which STA is associated in which BSS, so that data can be transferred from one BSS to another within the ESS. If you had a hard time following all those acronyms, make sure you read this note more than once.

Effectively, an ESS exists when the first AP comes online and defines the SSID. As each new AP, which is connected to the same distribution system, comes online with the same SSID as the first, it joins the existing ESS. This reveals that the ESS is really using the SSID to determine which APs should participate in the ESS. This is the default behavior that is most frequently implemented. The exceptions to this basic functionality would include when roaming specifications are implemented and configured to use RADIUS to control the APs that are allowed in the ESS, or when some other proprietary protocol is used to constrain the APs that can participate in the ESS.

It is also important to note that there is no requirement that STAs be able to roam between BSSs within an ESS without losing their upper layer connections, such as IP addresses and MAC layer connections. BSSs that form an ESS may overlap to allow for such roaming, or they may not overlap at all and allow for nomadic type connections.

BSSID and SSID

The *SSID or Service Set Identifier* is used to indicate the identity of an ESS or IBSS, depending on the implemented topology. The SSID can be from 2 to 32 characters in length and is normally sent in the Beacon frames. A STA seeking to join a WLAN may send probe request frames including the SSID of the desired WLAN. If an AP "hears" the probe request frame and it uses the same SSID, it will respond with a probe response frame. The STA that transmitted the original probe request frame may now authenticate and, if successful, associate with the BSS.

The *Basic Service Set Identifier (BSSID)* should not be confused with the SSID. The BSSID is a 48-bit identifier that is used to uniquely identify each service set. The BSSID is usually the MAC address of the STA within the AP in an infrastructure BSS.

In an IBSS, the BSSID will be an ID generated based on the rules for locally administered addresses according to Clause 5.2 of the IEEE 802-1990 standard. This means that the first bit (the individual/group bit) will be 0 to indicate an individual, and the second bit (the universal/local bit) will be 1 to indicate that the address is a locally administered address. The remaining 46 bits of the BSSID will be generated using an algorithm that minimizes the likelihood of other STAs generating the same BSSID.

Where the SSID identifies the service set, which may extend across multiple BSSs, the BSSID is unique to each BSS in an ESS, or to each independent BSS.

Distribution System (DS)

The *Distribution System (DS)* is defined as a system used to interconnect a set of BSSs and an integrated LAN to form an ESS. Additionally, the DS is used for the transfer of communications between the APs in the ESS. The communication that occurs between APs may be proprietary to the AP vendor, it may be according to a non-IEEE specification, or it may be in accordance with the 802.11r Fast Transition (FT) standard. Every AP has a DS within it, regardless of whether it is connected to other APs across some other shared system such as Ethernet. The DS is composed of two parts: the Distribution System Media and the Distribution System Services.

Distribution System Medium (DSM)

The *Distribution System Medium (DSM)* is the medium or set of media used for communications among APs in the ESS. The most popular medium in use today is certainly Ethernet, but the IEEE standard allows for the use of other media such as Token Ring (though highly unlikely today) or even another form of wireless.

Distribution System Services (DSS)

The *Distribution System Services (DSS)* are composed of the services that provide the delivery of frame payloads between stations that are in communication with each other over a shared instance of wireless medium (WM) and in the same infrastructure BSS. In other words, the DSS provide communications between stations in the same BSS. At this point, the 802.11 standard does not specify the full delivery path from a STA in a BSS to a station in another BSS, or from a STA in a BSS to a network node outside the BSS. Usually, each AP drops the frames out the connected portal (usually Ethernet), and hopes the Ethernet infrastructure (routers, switches, etc.) knows how to reach the destination.

Starting and Joining a BSS

The process of starting a BSS differs depending on whether it is an IBSS or an ESS (infrastructure BSS). In the case of an IBSS, the first station coming online starts the

IBSS. In the case of an ESS, the AP starts the BSS when it comes online. The following sections provide more details on the startup of an IBSS or an ESS. A client STA joins a BSS after locating it through the 802.11 authentication and association process covered in later chapters.

Starting an IBSS

An IBSS is started when the first station comes online. Specifically, the station processes an MLME-START.Request primitive with the parameter BSSType set to independent. This station sets the SSID to use in the IBSS, and all other stations that wish to join the same IBSS must use the same SSID. Additionally, this first station will set the BSSID according to the guidelines specified in the 802.11 standard. A station may scan before attempting to start the IBSS, or the station may start the IBSS without performing a scan first.

Starting a ESS

An infrastructure BSS (ESS) is started when the AP is started. The AP will process an MLME-START.Request primitive with the parameter BSSType set to infrastructure. The AP sets the SSID to use in the ESS. The BSSID will likely be the MAC address of the AP. At this starting point, the AP will specify the parameters to be used within the ESS. These parameters are presented in Table 5.2. Other parameters are also configured, and Table 5.2 is intended only to provide an understanding of the basic startup process for an ESS.

BSS Parameter	Description
SSID	The SSID to use for the ESS.
PHY Parameter Set	The parameter set used by the PHY that is being implemented. For example, OFDM, DSSS, etc.
Beacon Period	The Beacon Period to be used in the ESS.
Data Rates	Information such as supported data rates.

Table 5.2: ESS Startup Parameters

Layer 1, Layer 2, and Layer 3 Roaming

When a station associates with an AP in a BSS, it is joining a potentially larger network (the ESS). If the station moves out of the range of the initial AP, it may disassociate and re-associate with another AP that is participating in the same ESS. This process of reassociation is known as *roaming*. Roaming provides mobility, but there are different types of mobility. This section will present the different types of mobility and then covers the basics of roaming. Details about secure roaming are covered in "Chapter 11 — Security Solutions for WLANs."

Mobility Types

There are three basic types of mobility that can occur in an IEEE 802.11 WLAN.

- No-Transition — static or local movement.
- BSS-Transition — moving around to different BSSs within an ESS.
- ESS-Transition — Moving from a BSS in one ESS to a BSS in another. The IEEE states that upper layer connections are not guaranteed and are likely to be lost.

The first, no-transition, indicates that the station will not transition from one BSS to another, while attempting to maintain upper layer connections. In other words, it stays in range of its BSS. Figure 5.21 illustrates a no-transition type of mobility.

The second type is what is most commonly referenced as roaming. The BSS-transition mobility model is one that does allow for the maintenance of upper layer connections, while moving from one BSS to another within the same ESS. This is also called seamless roaming and is represented in Figure 5.22.

The third type occurs when a station moves from a BSS in one ESS to a BSS in a different ESS. Since an ESS can be thought of as a "virtual" LAN even though it may spread across massive areas, it is logical that you can maintain upper layer connections while roaming within an ESS (BSS-transition). However, separate ESSs can be thought of as separate "virtual" LANs and it is also logical that you will lose upper layer connections while roaming from one ESS to another (ESS-transition). There are technologies which allow for roaming between ESSs while still maintaining upper layer connections, but they are not part of the 802.11 standard. IEEE 802.11 does not specify such a technology, but proprietary solutions do exist.

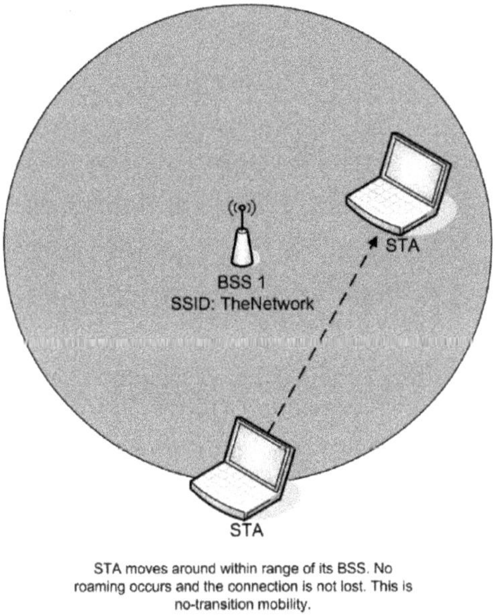

Figure 5.21: No-Transition Mobility

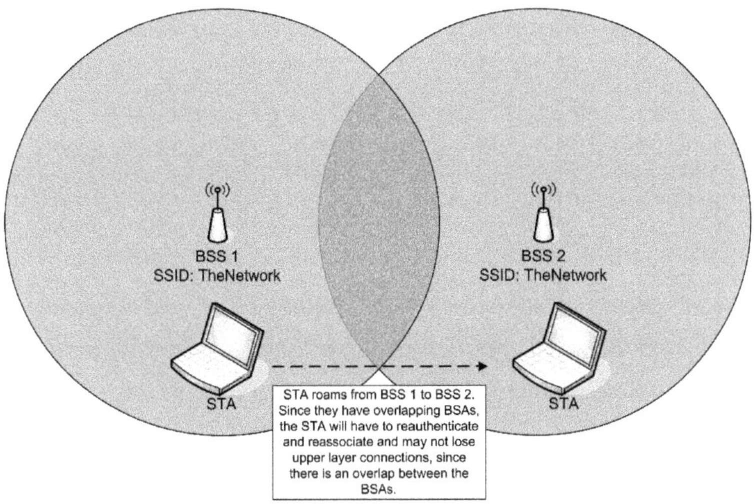

Figure 5.22: Seamless Roaming (BSS-Transition Mobility)

General Roaming Overview

In general, you could say that there are two types of roaming: seamless and reconnecting. *Seamless roaming* would be that roaming which allows a station to move its association from one BSS to another without losing upper layer connections. Think of it like being able to start a large FTP download while associated with one BSS, and then walking to another area where you are re-associated with another BSS within the ESS. Seamless roaming allows the FTP download to continue and not fail. Seamless roaming is usually an implementation of BSS-transition mobility.

Reconnecting roaming would require a new connection to the FTP server and, unless the server supported failure resume, a restart of the download. BSS-transition mobility may fall into this category if there is no association hand-off operation that can be performed between the two BSSs, even though they are in the same ESS.

Because the 802.11 standard did not specify the details of how roaming should occur early on, it is possible to implement a WLAN using APs from different vendors (or even different model APs from the same vendor) that cannot communicate with each other and will not allow for seamless roaming. If you want to purchase differing hardware and still allow for seamless roaming, you will have to either purchase and test the hardware to ensure that the APs can interoperate, or you should ensure that both vendors provide support for the same roaming procedures, such as Opportunistic Key Caching (OKC) or Fast BSS Transition (FT) based on the 802.11 standard.

IEEE 802.11r was ratified in 2008. More detail of 802.11r will be covered in Chapter 11 from both a functional and security perspective. You will be tested on the capabilities of 802.11r (FT) on the CWNA exam, but not on the functionality.

Wireless Bridging

Early on, there was little difference between WLAN bridges and WLAN APs other than their intended use. Today, WLAN bridges often provide capabilities not provided by APs in bridge mode and APs often provide capabilities beyond what a WLAN bridge placed in AP mode can provide. This is because there is a limited amount of RAM and processing power in these devices, and features or capabilities must be sacrificed to provide the best support for the intended use.

It is important to note that the 802.11 standards do not specify a bridging mode or describe a WLAN bridge device. However, WLAN bridging usually employs some proprietary modifications to the WLAN MAC layer, and also incorporates much of the 802.11 standard for the MAC layer. In some cases, the bridges will implement both a standard 802.11 MAC and the proprietary bridging capabilities at the exact same time. While it may be possible to create a bridge link with two WLAN bridges from two different vendors, it is not recommended. This is because they may be incompatible with each other, and you'll only lose valuable installation time trying to get them to work together. Because of proprietary bridging methods that may be implemented, the devices may not work together. However, if using a standard AP in bridge mode, you may indeed be able to create a bridge link using devices from different vendors.

In a wired network, a bridge is a device that connects two otherwise disconnected networks, and quite often converts between one network type and another. WLAN bridges may perform this type of function, where two WLAN bridges associate with each other and bridge two wired LANs across the 802.11 link. They may also work to bridge between a WLAN and a wired LAN, such as a Wireless Workgroup Bridge.

There are two fundamental modes of operation for wireless bridges: root and non-root. There are multiple usage scenarios that include various mixtures and configurations of these two modes. Generally speaking, only one wireless bridge can be in root mode, and any number of wireless bridges can be in non-root mode; however, modern bridge devices also allow for creative combinations of non-root mode and standard AP functionality.

When one bridge is in root mode and only one other bridge is associated with it, that other bridge must be in non-root mode. This type of link is a point-to-point (PtP) link and is common between buildings. When one bridge is in root mode and multiple other bridges associate with it, those other bridges must be in non-root mode. This type of link configuration is a point-to-multipoint (PtMP) configuration. Figures 5.23 and 5.24 show PtP and PtMP links respectively.

Many bridges from various vendors can perform in the following modes:

- Root bridge, which is the same as previously described
- Non-root bridge without clients, which is the same as previously described
- Non-root bridge with clients, which provides a mix of acting as an AP and a bridge at the same time (this is similar to a repeater, only it connects to a root bridge instead of a standard AP)

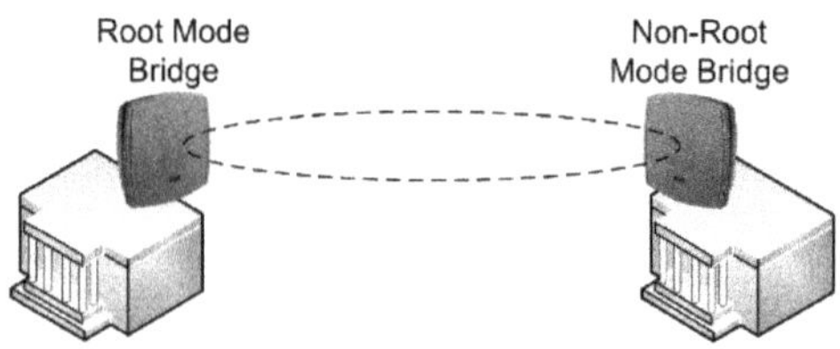

Figure 5.23: PtP Bridging

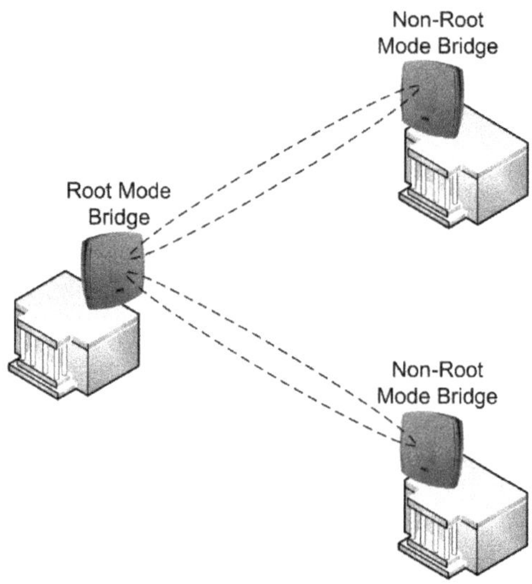

Figure 5.24: PtMP Bridging

In addition to these modes of operation, many bridges can act as a root mode AP (a standard AP), a repeater AP, and even a site survey client. Other bridges can only operate in the root and non-root bridge modes.

It is absolutely essential that you remember the rules for root and non-root bridge modes. Only one bridge in a link or set of links can act as the root bridge at a time. If it is a PtP link, one is the root and one is the non-root bridge. If it is a PtMP link set, one is the root and the others are non-root bridges.

Bridge Alignment

APs usually operate in an environment where the signal goes out from the AP's antenna in all directions, or in a wide swath. This is because they are usually serving clients that are within a few feet or yards. Wireless bridges are different. They are used to create connections that span a few hundred feet to a few miles. For this reason, alignment is crucial. As an illustration of why alignment is more important at greater distances, consider Figure 5.25. Notice how the bottom set of circles shows that the line of sight completely misses the arrow in the left of the two circles from the farther distance. The top set of circles shows that, even though the alignment is off by the same angle, the line of sight still connects with the arrow in the left circle, when the two circles are closer together.

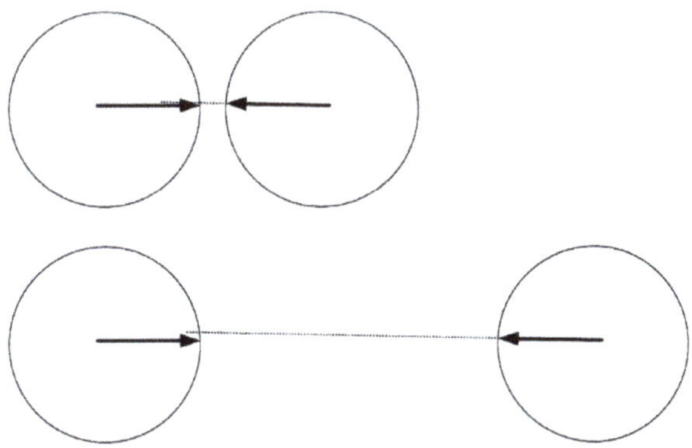

Figure 5.25: Misalignment at Distance

In much the same way, two bridges that are farther apart are more sensitive to alignment issues. You can also demonstrate this sensitivity with a flashlight that can focus the light beam. Shine the beam on a wall near you and move it as slightly as you can. You will likely notice a small change in its location on the wall. Now shine it on a wall trace as far from you and move it in the same way. The movement will be much greater because of the distance between you and the wall.

You can align bridges using many methods, including:

- Monitoring of RSSI or signal strength while slightly adjusting the bridges.
- Through LEDs that signify alignments. A few bridges have such a feature.
- Using a voltmeter that attaches to a connector on the bridge. The higher voltage reading indicates better alignment.
- For close bridge links, less than 200 or 300 yards within a campus, you may be able to perform data throughput tests to verify acceptable alignment. This may be helpful with less expensive bridges, or when using APs in bridge mode, since they are less likely to have alignment features.

When you implement very long-distance links (more than 2-5 kilometers) you may choose to use GPS equipment to assist in the location of your bridges. Due to terrain issues and other factors, such as weather, that have greater impact over longer distances, it is recommended that you attend training courses that focus specifically on building long distance wireless links. The knowledge given here should be sufficient for a WLAN administrator, but you will need more in-depth knowledge if you plan to implement what could be called WAN links using standard 802.11 hardware.

Chapter Summary

In this chapter, you learned about 802.11 PHYs and network types in detail. You explored the OSI Model as it related to WLANs, and then evaluated each significant PHY in the 802.11 standard. Next, you learned about 802.11 functional concepts, like channels, data rates, throughput and more. Finally, you explore 802.11 network service set concepts and WLAN bridging. In the next chapter, you will explore device types used on 802.11 networks.

Points to Remember

Remember the following important points:

- The 802.11 standard specified operations of wireless devices at Layer 1 and Layer 2 of the OSI Model.
- The OSI Model includes seven layers from top to bottom: Application, Presentation, Session, Transport, Network, Data-Link, and Physical.
- The ERP and OFDM PHYs both support maximum data rates of 54 Mbps, but the ERP PHY operates in 2.4 GHz and the OFDM PHY operates in 5 GHz.
- HT is the only PHY that operates in both 2.4 GHz and 5 GHz.
- VHT is a 5 GHz-only PHY
- HT introduced MIMO and VHT introduced MU-MIMO.
- HT supports a maximum of 4 spatial streams and VHT supports a maximum of 8 spatial streams.
- All PHYs use 20 MHz channels or some factor thereof, except for the DSSS and HR/DSSS PHYs, which use 22 MHz channels.
- HT supports up to 40 MHz channel widths and VHT supports up to 160 MHz channel widths.
- The BSA is the area in which clients can connect to and pass frames through the BSS.

- An infrastructure BSS (or simply BSS) uses an AP, and an independent BSS (IBSS) does not.
- An ESS is one or more BSSs sharing a distribution system medium (such as an Ethernet network).
- The SSID is the name of the BSS, the BSSID is the underlying identifier that differentiates one BSS from another, when they use the same SSID.
- When using wireless bridges, the longer the link, the more important the alignment.
- In a wireless bridging configuration, one bridge will act as the root bridge, and the other(s) will act as non-root bridges.

Review Questions

1. Which layers of the OSI Model are most impacted by the 802.11 standard?
    a. Layer 1 and Layer 2
    b. Layer 2 and Layer 3
    c. Layer 3 and Layer 4
    d. Layers 3-7

2. What PHY first introduced MIMO?
    a. VHT
    b. OFDM
    c. HT
    d. HR/DSSS

3. What PHY uses a 22 MHz-wide channel?
    a. S1G
    b. DMG
    c. HR/DSSS
    d. OFDM

4. What PHY operates in 60 GHz?
    a. S1G
    b. DMG
    c. TVHT
    d. ERP

5. How many channels are available in 2.4 GHz worldwide?
    a. 14
    b. 13
    c. 11
    d. 3

6. What is the maximum channel width that should ever be used in 2.4 GHz?
    a. 20 MHz
    b. 22 MHz
    c. 40 MHz
    d. None of these

7. What is the maximum channel width supported by 802.11ac (VHT)?
    a. 40 MHz
    b. 80 MHz
    c. 160 MHz
    d. 2160 MHz

8. What defines a BSS as opposed to an IBSS?
    a. A BSS does not use an AP and an IBSS does
    b. A BSS uses an AP and an IBSS does not
    c. A BSS supports multiple STAs and an IBSS supports only one
    d. An IBSS supports roaming and a BSS does not

9. What identifies one BSS from another when multiple BSSs use the same SSID?
    a. ESSID
    b. BSSID
    c. Switch MAC address
    d. VLAN

10. How many bridges can be in root mode in a PtMP configuration?
    a. 4
    b. 3
    c. 2
    d. 1

Review Answers

1. **A is correct.** Layers 1 and 2 of the OSI Model are most impacted by the 802.11 standard. The 802.11 standard defines MAC operations, which is a sublayer of the Data-Link layer (Layer 2) and it defines PHY operations for the Physical layer (Layer 1).

2. **C is correct.** The HT PHY (802.11n) first introduces MIMO to the 802.11 standard and the VHT PHY (802.11ac) first introduces MU-MIMO.

3. **C is correct.** Of those listed, only HR/DSSS uses a 22 MHz-wide channel. DSSS also uses a 22 MHz-wide channel.

4. **B is correct.** The DMG PHY (802.11ad) operates in the 60 GHz band.

5. **C is correct.** While 14 channels are specified for 2.4 GHz, only 11 of them are available in all regulatory domains.

6. **B is correct.** If using ERP or HT, only 20 MHz channels should be used. However, because HR/DSSS and DSSS are still supported in 2.4 GHz, the maximum channel width that should be used is 22 MHz. 40 MHz HT channels should be avoided in 2.4 GHz due to insufficient frequency space.

7. **C is correct.** An 802.11ac device can support 20, 40, 80, or 160 MHz channels, though 160 MHz channels should not be used in most enterprise deployments today, due to insufficient frequency space.

8. **B is correct.** A BSS uses an AP STA and an IBSS does not. A BSS is an infrastructure network and an IBSS is an ad-hoc network.

9. **B is correct.** The BSSID, which is formatted like a MAC address, uniquely identifies a BSS.

10. **D is correct.** In either a PtP or PtMP configuration, only one bridge can be in root mode. All other bridges will be in non-root mode.

Chapter 6 — 802.11 Network Devices

In this chapter, you will learn about WLAN APs and the features and capabilities they provide. In the process, you will also explore WLAN controllers, cloud management solutions and distributed AP solutions, as well. Next, you will shift gears to how these APs are commonly powered today, which is with Power over Ethernet (PoE). Finally, you will explore the various client devices used on WLANs, and how they are typically configured.

WLAN Infrastructure Devices

Access Points (APs) are the most frequently installed infrastructure (non-client) devices. APs provide access to the WLAN and may bridge to a wired LAN. APs provide a point of access to the WLAN and get their name from this functionality. Each BSS has one and only one AP. When multiple APs work together to form a larger network throughout which clients may roam, they form an ESS. While each BSS has only one AP, a single AP may provide more than one BSS.

In most cases, an AP will provide connectivity to a wired LAN or WAN for wireless client stations (STAs); however, this does not have to be the case. APs are often used in core mode at construction sites to form controlled and secure networks that are entirely wireless (with the exception of the power cords connected to the APs), as just one example of the use of APs where access to wired networks is not the intent.

Autonomous Access Points are APs that contain the software for complete management of the WLAN processes within themselves. Autonomous APs were the only kind of APs in early WLANs, until the lightweight AP was later developed. *Lightweight Access Points* are APs that contain limited software and depend on centralized WLAN switches or controllers to provide the remaining functionality. No standard for implementing lightweight versus autonomous APs exists, and the way in which they are implemented varies from vendor to vendor. Autonomous APs are sometimes called fat or thick APs, and lightweight APs are also called thin APs. Figure 6.1 shows a network implementation using autonomous APs and Figure 6.2 shows the use of lightweight APs. As you can see in these two images, the implementation will not look any different in the physical world, but in the internal software world of these devices, things are very different.

In the lightweight APs, much less of the work is happening at the AP and much more of the work is happening at the controller or switch.

Some APs can act as either an autonomous or lightweight AP depending on the configuration determined by the WLAN administrator. When used as an autonomous AP, all the AP software features are enabled. When used as a lightweight AP, many of the AP software features are disabled, or simply controlled by the centralized WLAN switch or controller. In most cases, the AP must be loaded with a different firmware version to act as an autonomous AP than that used to act as a lightweight AP.

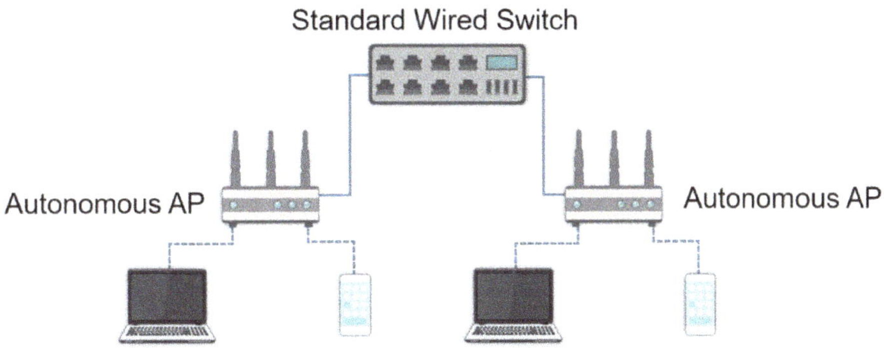

Figure 6.1: Autonomous AP Implementation

When lightweight APs are brought online (powered up and connected to the WLAN controller through their Ethernet port), they are automatically configured by the WLAN controller or switch. The automatic configuration may include the installation or update of firmware (internal software used to run and manage the AP). In the past, some vendors have shipped their lightweight APs with no firmware loaded, and the firmware was installed when it first connected to the WLAN controller. Symbol (now owned by Motorola) did this with their 5100 series WLAN switches and access ports (another term for a lightweight AP).

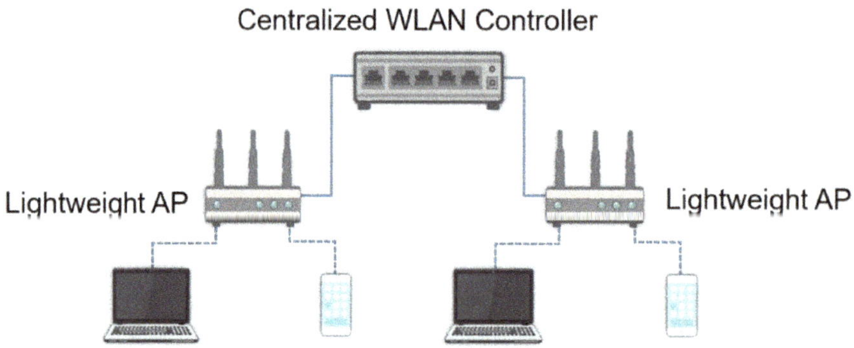

Figure 6.2: Lightweight AP Implementation

When an AP is converted to become a lightweight AP, features may include:

- Automatic updates of firmware files
- Support for multiple ESSs and BSSs with BSSIDs in a single AP
- Support for multiple VLANs
- Centralized management of all APs
- Automatic management of QoS features
- More encryption types than those supported by the AP internals

Autonomous APs that are converted to lightweight APs may also lose capabilities, such as access through the serial port, support for wireless bridging and repeater operational modes, and other vendor-specific features. Generally speaking, you gain centralized management and you may lose unique features of the autonomous AP; however, since conversion of autonomous APs to lightweight APs is usually only supported when the same vendor APs are used as the WLAN controller being implemented, few features are available in the AP itself, when used as an AP, that are not in the WLAN controller's software.

Two additional models have been offered in the WLAN space in recent years. These include cloud-based APs and distributed APs. Figure 6.3 and 6.4 show these models respectively

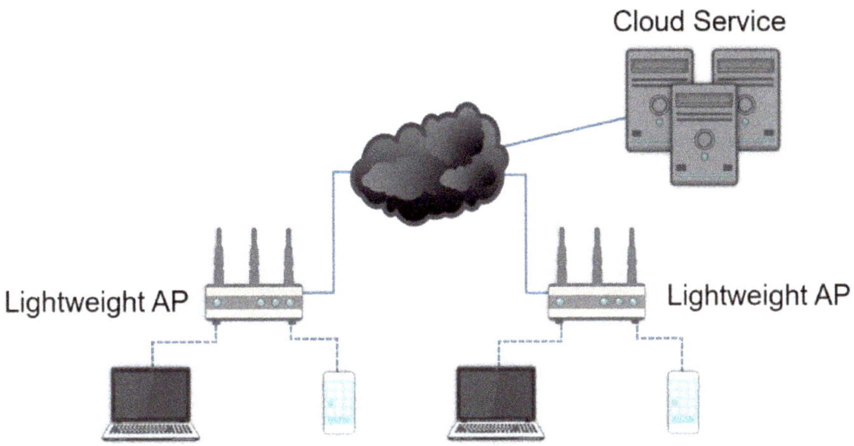

Figure 6.3: Cloud-Based APs

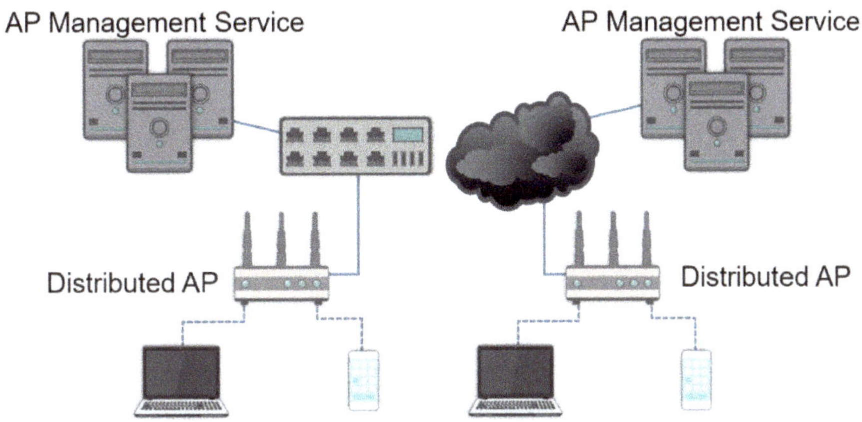

Figure 6.4: Distributed APs

Cloud-based APs are basically lightweight APs that use a cloud service for management and control, instead of a local on-network controller. Cloud-based APs are less likely to use centralized data forwarding than controller-based APs, due to the constraint of network bandwidth. However, the data traffic may be logged and reported to the cloud for statistical tracking and performance analysis.

Cloud APs are generally configured by first provisioning them in the cloud service (entering some form of AP ID in the cloud account) and then connecting the AP to the local network, which provides DHCP connectivity and Internet access. The AP will automatically locate the cloud service and join your subscription. Then, you can push firmware and configuration settings from the cloud service to the AP.

Depending on the cloud service provider/AP vendor, cloud-based APs may be distributed in function or they may function more like controller-based APs.

Distributed APs may be managed by an on-network management service or by a cloud-based management service. Over the years, the lines between a distributed AP and cloud-based AP have blurred somewhat in that cloud-based APs can often be managed in the same way as distributed APs were managed.

The hallmark of distributed APs is that they are completely self-sufficient, once configured, and the management service need not be available for them to function. Some cloud-based APs cease to function or lose some functionality (of the WLAN) if the cloud service is not available.

For example, cloud-based APs may use a cloud-based authentication service for 802.1X/EAP authentication. Distributed APs will nearly always use an on-network RADIUS server for 802.1X/EAP authentication. For this reason, even with cloud-based architectures, it is recommended that you use on-network RADIUS for authentication.

An additional concept is that of a Wireless Network Management System (WNMS). Initially, a WNMS was used as a way to manage autonomous APs. Today, many, if not most, WNMS solutions are used to manage controllers that manage APs, as represented in Figure 6.5. The WNMS may itself be a controller that manages the other controllers, or it may be a server with management software installed.

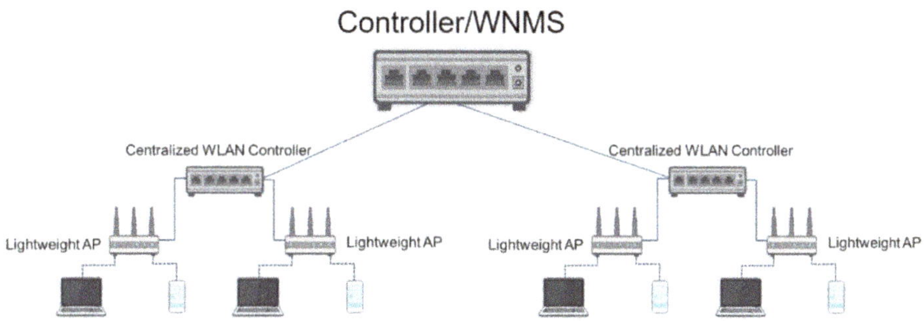

Figure 6.5: A WNMS Solution

In recent years, a few vendors have also come into the wireless space that offer WNMS for multiple AP vendors, through custom firmware loaded onto the devices. For example, you can acquire seven different APs from seven different vendors and load a special firmware on them that allows them to be managed by the WNMS. This is just one of the many recent innovations in WLAN management.

APs Defined

An AP is basically a small computer that includes one or more radios and usually one Ethernet port. Inside the AP is a processor and memory. In fact, one of the big differences between enterprise-class APs and those designed for SOHO (small office/home office) implementations is the processing power and the amount of memory available in the AP. Many WLAN administrators are surprised when they first learn that many APs either run a flavor of Linux or can run Linux through flash updates. It is important to remember that you may lose support from the device vendor if you flash the device with an operating system that is not supplied by the vendor. For example, firmware is available for many SOHO WLAN routers that turn them into more enterprise-like devices with advanced features usually only provided in WLAN switch/AP combination installs. These features include VPN endpoint support for client connections, more powerful filtering, and centralized management and control. Again, if a WLAN administrator chooses to install such a firmware, she will likely lose all support from the hardware vendor.

APs, both autonomous and lightweight, come in many shapes and sizes. Some have internal antennas and others use external, detachable antennas. They come in round encasements, rectangular and other shapes. Some are designed for mounting on walls or ceilings and some are designed to be placed on desktops or shelves. Figure 6.6 shows multiple APs from various vendors.

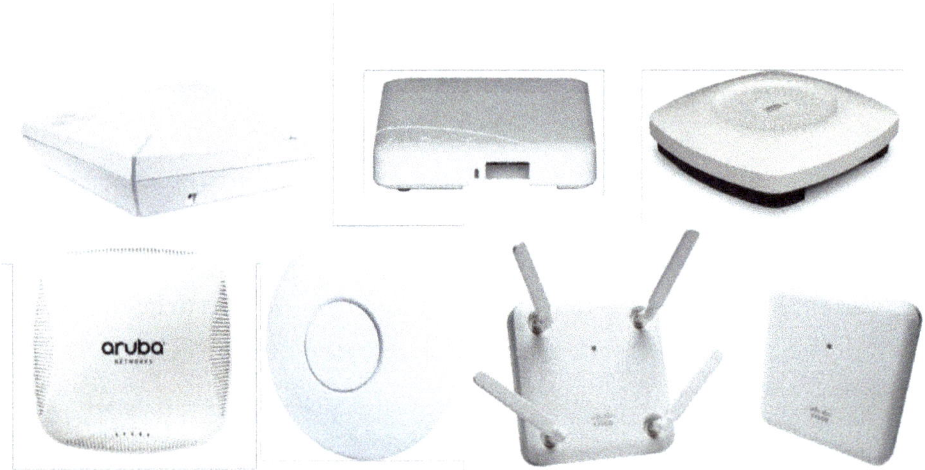

Figure 6.6: Multiple APs

APs come with common features and require various configuration processes. The following sections document each of these important factors. First, the common features will be covered, and it is important to note that, while these features are common, they are not available in all APs. Second, I will walk you through the basic installation and configuration of an AP.

By common features, I mean features that are commonly seen in APs, and not necessarily features that are common to all APs. Some APs will have all of the features listed here and more, while others may lack one or more of the listed features. Features that will be covered include:

- Operational Modes
- IEEE Standards Support

- Internal or External Antennas
- Filtering
- Removable and Replaceable Modules
- Variable Output Power
- Ethernet and Other Wired Connectivity
- Power over Ethernet Support
- Security Capabilities
- Management Capabilities
- Mounting Options

Operational Modes

The 802.11 standard defines an AP only as a STA that provides access to the distribution services via the wireless medium for associated STAs. It does not define the three common operational modes that are found in APs. These modes (root, bridge and repeater) are specific implementations of a WLAN STA for varied purposes and, in some cases, they may be proprietary rather than matching an IEEE standard. For example, in bridge mode, an AP is implementing a network functionality that is not directly stipulated in the 802.11 standard. Root mode is the closest to the 802.11 standard and many APs meet the 802.11 standard exactly when running in root mode.

The first and default mode offered by most APs is root mode, also called access mode. An AP operating in root/access mode is providing wireless clients with access to the WLAN, and usually a wired network. Root/access mode is the default mode of operation for all WLAN devices sold as APs. Some WLAN bridges are really APs that come with the operating mode set to bridge mode, and they are nothing more than a standard AP operating in bridge mode. Full-function WLAN bridges will implement a complete 802.1D bridging feature set. When APs operate in root/access mode, they may still communicate with each other, but the communications are not related to bridging. In root mode, inter-AP communications are usually related to the coordination of STA roaming. Figure 6.7 shows a typical installation of an AP in root mode.

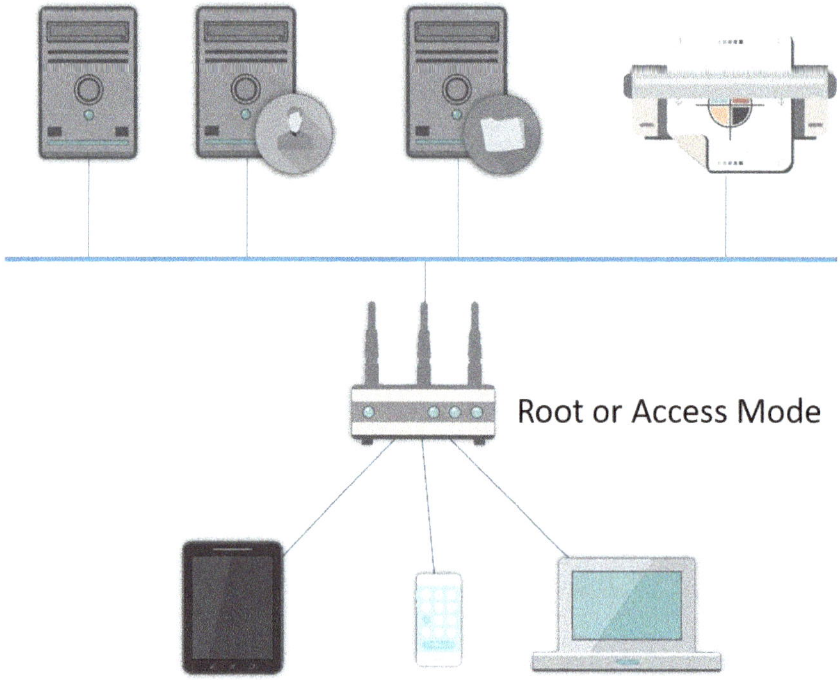

Figure 6.7: AP Implemented in Root Mode

Bridge mode is used to create a link between two access points. When only two APs are used, a point-to-point link is created. When more than two APs are involved, a set of point-to-multipoint links are created. In a bridge mode implementation, the APs involved usually associate only with each other and do not accept client STA associations. Exceptions to this exist, but it is not the normal implementation, since it would reduce the throughput available for the bridge link connection.

The final mode, repeater mode, is used to extend the range of a WLAN beyond its normal usable boundaries. The repeater AP acts as the AP for clients that would otherwise be out of range of the distant AP operating in root mode. Where a root AP is the connection point for many clients and is a client to no other APs, the AP in repeater mode is a client to the AP in root mode, while also accepting connections from client stations itself.

Repeater mode in a WLAN AP should not be confused with the functionality of an Ethernet repeater. Ethernet repeaters regenerate the received signal in order to allow it to travel farther than it would otherwise travel. They do not decapsulate and encapsulate data as a WLAN repeater will. The AP running in repeater mode will decapsulate the data frames received from the clients and encapsulate them for transmission to the root mode AP. In other words, the WLAN AP in repeater mode will receive data from the WLAN clients associated with it and then retransmit that data to the root mode AP with which it is associated. Figure 6.8 shows an AP operating in repeater mode to provide access to remote clients.

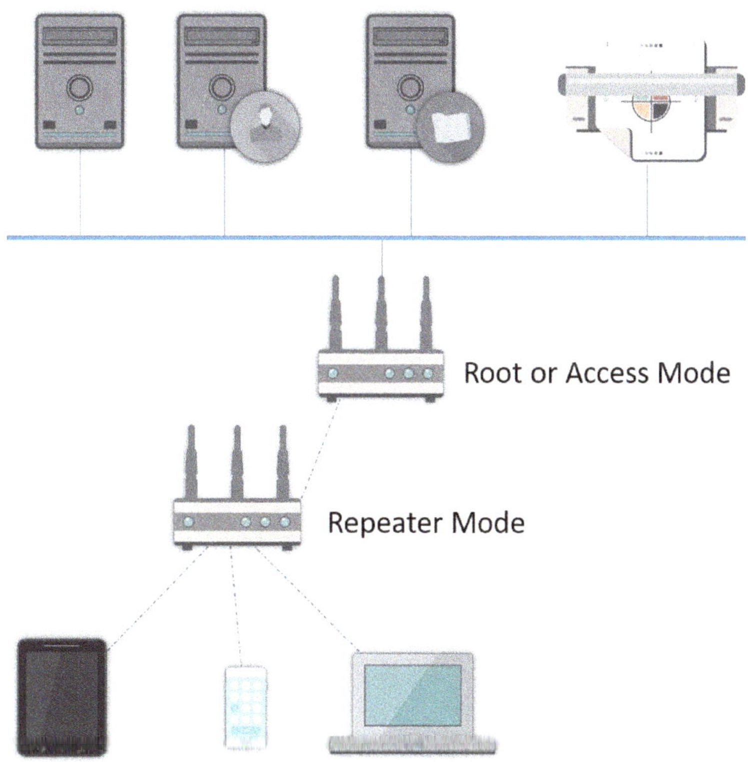

Figure 6.8: AP in Repeater Mode

Keep in mind that an AP operating in repeater mode must be able to communicate with the clients associated with it, as well as the root mode AP with which it is associated. Because of this, the repeater mode AP will usually have to implement a Basic Service Area (BSA) that overlaps with the BSA of the root mode AP by at least 50 percent. This reduces the overall coverage area that may be provided if each AP were operating in root mode and forming an ESS and using separate channels; however, Ethernet connectivity is not always available to provide for the preferred implementation, and repeater mode may be used in these scenarios.

Today, mesh APs are a better choice over repeater mode APs. Mesh APs can form an 802.11-standard or proprietary mesh network and use the different radios in the APs for different purposes. For example, they can use the 2.4 GHz radios to form client STA connections and the 5 GHz radios to build the mesh network.

IEEE Standards Support

APs on the market today support a wide range of 802.11 amendments and PHYs, but it is difficult to find hardware that supports some of the older PHYs, such as FHSS, and in reality, you probably won't want them. Most equipment supports HT, ERP, HR/DSSS, DSSS or OFDM. Remember that a device that implements the 802.11g amendment (ERP) will almost always support backward compatibility, which means it must support DSSS and HR/DSSS. Additionally, a device that implements the 802.11n amendment (HT) in the 5 GHz band will almost always support backward compatibility with 802.11a (OFDM). Of course, HT devices in the 2.4 GHz band will usually support backward compatibility with ERP, which results in backward compatibility with HR/DSSS and DSSS. The good news for networks containing mostly newer clients is that most APs allow you to disable backward compatibility.

The vendors usually report the standards support as 802.11ac, 802.11n, 802.11g, 802.11b, 802.11 or 802.11a. Many devices are said to be 802.11b/g/n devices. This simply means that the devices implement the HT PHY, which is capable of communicating with ERP, HR/DSSS, and DSSS PHY devices as well. If a device is said to be 802.11a/b/g/n/ac compatible, it means it has support for 802.11n (2.4 GHz) and 802.11ac (5 GHz) with backward compatibility in both the 2.4 GHz and 5 GHz unlicensed bands.

In addition to the PHYs that are supported, you should consider the standards-based security features that you may require. The vast majority of devices on the market today support 802.11i security. Some still support only WPA security, but thankfully these devices are becoming harder to locate. Most modern APs will support both WPA and WPA2 with pre-shared keys (PSK) at a minimum, and the vast majority will support WPA and WPA2 Enterprise, which utilizes a RADIUS authentication server.

Another standards-based feature to consider is Quality of Service (QoS). If you need support for QoS extensions, you should ensure that the AP has support for 802.11e or the Wireless Multimedia (WMM) certification by the Wi-Fi Alliance. These QoS features will be very important if you intend to support Voice over WLAN or video conference over the WLAN.

Newer APs tend to support the newer IEEE standards while also supporting older standards. One of the benefits of a newer HT or VHT PHY-based device is that it can communicate at the high data rates with other HT or VHT PHY devices, and it can also communicate at the lower data rates when communicating with older devices. Of course, the protection mechanism kicks in whenever an older device is associated with the AP. This protection mechanism means that the AP will transmit a frame that can be understood by the older machine(s) before transmitting the frame that can only be understood by the newer machine(s). The first frame is used to cause a backoff timer to kick in on the older machines, so they will not interfere during the newer PHY frame transmission. Protection mechanisms reduce overall throughput, but they are better than downgrading all STAs to the lowest common denominator.

In addition to the benefit of backward compatibility with the older PHYs, HT and VHT PHY devices are able to support more data rates than older devices. So, as the data rate changes it does not necessarily drop by half at a single step, like an older HR/DSSS device does when it goes from 11 Mbps to 5.5 Mbps in one step.

Finally, APs may not support utilization in every regulatory domain. You should be sure to verify that the APs you are purchasing are authorized for use within your regulatory domain. IEEE 802.11h specified support for European nations and

802.11j specified support for the regulatory domain of Japan. For more specific information regarding your regulatory domain, check with the regulation management organization in your country.

Internal or External Antennas

Many enterprise-class APs support external, detachable antennas. Some SOHO APs or home routers may also support detachable antennas. Detachable antennas are becoming less common with the release of 802.11n devices and 802.11ac devices for enterprise office deployments. That is not to say that they do not exist, they are just less common since the antennas must be configured appropriately for the MIMO technology to function properly. Detachable antennas are beneficial from at least two perspectives: the physical location of the antenna and the selection of a different antenna type.

The ability to move the physical location of the antenna to a different location than that of the AP is a valuable one. You can use RF cabling to move the antenna to a location that is more practical for the transmission and reception of RF signals and locate the AP itself closer to power outlets. This can be advantageous when you do not have power outlets close to the RF signal transmission and reception location. But with PoE, today, this is not as big of an issue. The bigger issue today is likely aesthetics. The customer does not want to see that big ugly AP box. So, you can place a smaller, less invasive, antenna where it needs to be, and hide the AP in the plenum or elsewhere.

The second benefit is that of replacing the antenna with a different antenna type. You may want to provide coverage down long narrow corridors (patch or panel antennas), or you may want to provide coverage in an area horizontally with as little RF energy propagating upward and downward as possible (higher gain omni antennas). Whatever the motivation, a detachable antenna provides you with the capability to better control how the RF energy is radiated from the antenna, and therefore how the AP provides coverage in the BSA. Figure 6.9 is an example of a MIMO-compatible patch antenna, supporting an AP with three spatial streams in both 2.4 GHz and 5 GHz.

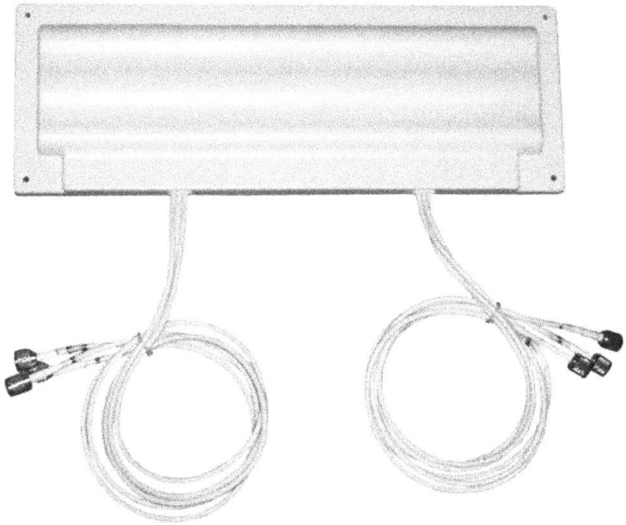

Figure 6.9: A MIMO External Patch Antenna

Filtering

Most APs offer two kinds of filtering at a minimum. The first kind is MAC address filtering and the second is protocol filtering. Filtering functionality provides the WLAN administrator with the capability to limit which STA frames can pass through the AP, based on the hardware configuration of the STA (MAC address) or the protocol being used, such as HTTP.

MAC filtering has often been referenced as a security solution, but it should not be thought of as such. It may be useful from the perspective of making it harder to accidentally associate with the wrong AP, but MAC filtering should not be considered a security solution in WLANs. This is because MAC spoofing (stealing a MAC address from a valid STA) is easy to do and step-by-step instructions are readily available on the Internet. The only common value seen from MAC filtering today is its use in specific association limitation scenarios. For example, a training center near my home office uses laptop computers in the training rooms. They do not want the laptop computers to be moved from room-to-room, but instead want them to stay in designated rooms. The simple solution was to use MAC filtering in the AP in each room. Each room's AP contains the MAC addresses of the laptops that are supposed to be in that room. The AP's output power is throttled back to reduce the coverage area provided. Now, if someone takes a laptop from the designated room to another room, the laptop will have to associate with an AP with a very weak signal in the remote room. Throughput suffers, and in most cases, the laptops cannot connect in such scenarios because the rooms are far enough apart. Again, if this were being done as a security solution, it would be a very bad idea. Any moderately skilled cracker can spoof a MAC address very quickly. So, I cannot emphasize enough that MAC filtering should not be considered a security solution.

Protocol filtering can be used to disallow specific protocols, or only allow specific protocols. This feature usually allows for filtering of both the frames arriving through the radio and the Ethernet port. You may also filter only the radio-side (wireless) frames, or only the wired frames, depending on the AP and vendor. Some APs can filter out frames based on the actual file extensions the user or machine is trying to access on the Internet. For example, if the user attempts to access a WMV file and the WLAN administrator has chosen not to allow access to such streaming media for performance reasons, the AP can disallow such requests. Most APs can blindly block all HTTP requests or FTP requests, and other such Internet protocols.

An additional kind of filtering is that of wireless STA to wireless STA filtering. It is also called direct access blocking, peer-to-peer blocking, and peer blocking. Some APs will allow you to create Virtual APs (VAPs) within one physical AP. You can then determine if wireless STAs associated with one VAP can communicate with wireless STAs associated with another VAP (inter-VAP filtering). You can also determine if wireless STAs can communicate with other wireless STAs associated with the same AP (intra-VAP filtering). Finally, you can disallow all client-to-client communications and only allow the STAs to use the AP for access to the wired medium. This type of filtering can be useful when you want one physical AP to service public and private clients. The public clients may have limited access to the network, and therefore, to the private clients. The private clients may have normal access to the network. In this way, one AP effectively provides access to both internal users, and public guests.

Removable, Configurable and Replaceable Radio Modules

Some APs are designed to support one PHY only, while others are designed to allow for multiple radios, and therefore multiple PHYs. These multiple radio APs are usually called dual-radio APs, or dual-band APs, because one radio is needed for the HT PHY (2.4 GHz), and another is needed for the VHT PHY (5 GHz). Additionally, some devices are multi-band devices. They support 2.4 GHz, 5 GHz and 60 GHz. At the time of writing, these are only client devices, but such multi-band options are likely to be seen in home wireless routers at the very least, in the future.

Many APs support upgradable modules so that you can install the latest 802.11ac radios without replacing the APs. Other APs are software-configurable, so that one radio is fixed to 5 GHz, and the other can be configured to either 2.4 GHz or 5 GHz, according to your needs.

Variable Output Power

Variable output power provides the WLAN administrator with the capability of sizing cells more accurately. Remember, this should not be considered a security solution by itself because a remote client with a powerful WLAN card and the right antenna can often still pick up the signal of the WLAN, and also transmit data to the WLAN. However, as an RF management philosophy, cell sizing makes a lot of sense.

As an example, consider a facility with the need for four different WLANs (for security reasons or otherwise), that must coexist in a fairly small space. Throughput is not a paramount concern, since the users of the WLAN perform minimal data transfers, though these data transfers happen several times per hour. Figure 6.10 shows a simplified floor plan of this facility. In order to implement the four distinct WLAN BSAs (cells), APs can be installed in areas A and D that use antennas (MIMO patch antennas, for example) that direct the majority of the RF energy outward. These antennas could be mounted on the walls near areas B and C and facing away from them. In areas B and C, APs could be installed centrally to the areas and using standard omni-directional antennas. These APs could have the output power settings lowered to ensure that there is minimal overlap into areas that are not intended for coverage by these APs.

Of course, a scenario like this can be implemented to provide unique configuration parameters for each BSA; however, you must remember that this type of cell size reduction does not of itself equal security, but it would help in RF spectrum management in small areas that need different types of WLAN access, such as that depicted here.

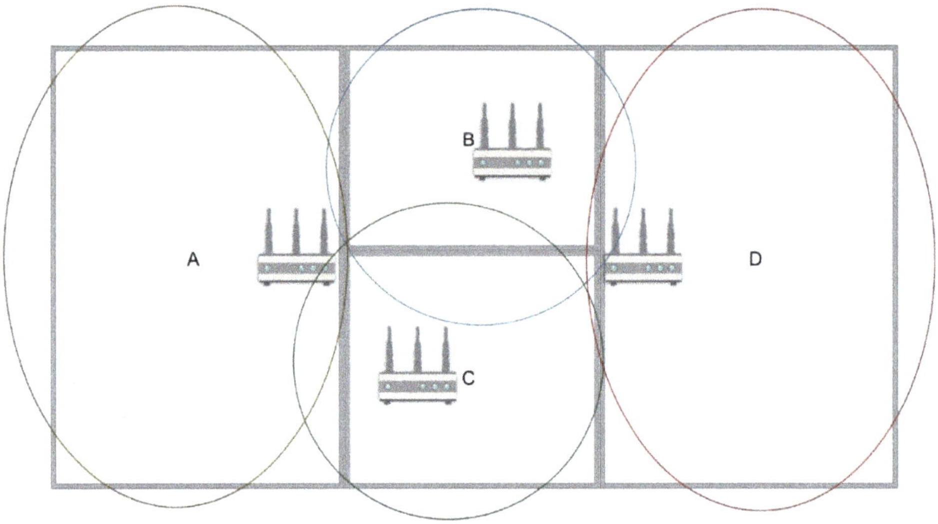

Figure 6.10: Simplified Floor Plan Needing Four Distinct Cells

Some APs provide variable output power management based on percentages, some with cryptic number systems, and others based on actual output power levels. For example, an AP may allow you to specify that the output power be 25 mW, 50 mW or 100 mW. Other APs may only allow you to state that the output power should be at 25 percent, 50 percent or 100 percent. Still others may list the power in dBm or even use a method of numeric matching where you have to look up what the numbers (such as 1, 2, 3 and 4) mean in vendor documentation. These are just examples, but it is important to know what you're looking for when you enter an AP's configuration interface.

Ethernet and Other Wired Connectivity

Unless an AP is providing WLAN services and access to a wireless-only LAN, the AP must have some interface through which it can connect to a wired LAN. In most APs, this will be an Ethernet connection. Depending on how old the AP is and the model of the AP, it may support only 100 Mbit Ethernet. Newer models should support Gigabit Ethernet, multiple gigabit ports, or even faster Ethernet standards. With an OFDM or ERP PHY, a 100 Mbit Ethernet connection is sufficient. With modern 802.11n and 802.11ac APs, a gigabit network connection is required (though the myth of requiring more than 1 Gbps is simply untrue in practically every implementation).

In addition to standard CAT5 or CAT6 cabling, some APs may support fiber connections, though these are less common today. Since fiber is rated for longer cable runs, it may provide a solution to a scenario where the AP needs to be located more than 100 meters (the limit of CAT5) from the switch port. Of course, this means the switch must support fiber, as well as the AP.

Power over Ethernet Support

Most enterprise class APs support Power over Ethernet (PoE) today. Support for PoE allows for the installation of APs in areas where no power outlets reside, but where you can run network cables to carry the power. While PoE is very popular for WLAN devices because it can provide extra features, such as power cycling the device as well as powering the device in the first place, it is sometimes more cost effective to run the power to the area. This is usually the case when only one location with no existing PoE solutions in place needs the power outlet, and the power run would only be a few feet.

Consider the implications of PoE carefully before deciding against it. You often hear that the primary benefit of PoE is the ability to install APs where there is no AC power outlet; however, it is certainly a major benefit to be able to power cycle (stop and start the device) an AP that is mounted high on the ceiling. Many PoE switches support the stopping and starting of power injection on the PoE ports using the graphical management tools the vendor provides. This means you can restart an AP from your desk, even if you cannot get into the management interface of the AP and even if the AP has stopped responding to other management interfaces that communicate with the device through the network layers. To me, this is an equally valuable benefit to that of being able to place an AP where there is no power outlet.

PoE support is usually not found in SOHO APs like those from Linksys or Netgear. Most enterprise APs do support PoE but check with your vendor to ensure you purchase a model that supports it if you need it. While more and more enterprise class APs do support PoE, a few still do not.

Mesh Networking Functions

Modern APs often provide a mesh networking function. The function allows the AP (AP1) to act as a client to multiple other APs (AP2 and AP3 for example) and treat the individual associations with these other APs as ports across which it can bridge traffic for the STAs associated with it (AP1). When a client needs to reach a destination that is reachable through AP2, but that client is associated with AP1, AP1 will bridge the packets across the association with AP2, on behalf of the client.

There is a limit to the number of associations these APs can make. For example, some APs can create up to three mesh associations with other APs. Others are not so constrained

Hotspot Support

More and more of the newer APs are coming equipped with hotspot support. This usually includes walled garden support and may also include connectivity to online payment-processing services, if you are providing a for-pay hotspot. Having this support built-in is also useful when you simply want to provide a "guest" network for visitors to your organization's facilities. In addition, several vendors are now supporting Hotspot 2.0, which is based on the Wi-Fi Alliance Passpoint certification. This allows clients to connect automatically to a WLAN, regardless of SSID, by learning about the network from the AP and determining if a proper configuration profile is on the client, that will allow connection to the hotspot. For more information on Hotspot 2.0, see the Wi-Fi Alliance website at Wi-Fi.org.

Security Capabilities

APs support a large pool of common security capabilities. These include:

- MAC address filtering (a common item in vendor's lists of security features, even though it is not one)
- 802.1X port-based authentication

- WPA/WPA2-Personal
- WPA/WPA2-Enterprise
- SSH and, preferably, SSH2 for management access
- HTTPS access to web-based management
- SNMP v3 for secure SNMP management
- Various EAP types (some are secure, some are not)
- Built-in firewalls
- Support for VPN tunnel endpoints and pass-through
- Content filtering

Your role as a WLAN administrator or engineer may include the selection of APs that support the security technologies required by your security policies. Today, these policies will likely specify that you cannot implement an AP that uses WEP or WPA for data encryption and you must, therefore, select an AP that supports WPA2-PSK at a minimum. More likely, in an enterprise implementation, you will be implementing full WPA2-Enterprise for standard clients, though some clients, such as VoIP handsets and barcode scanners, may warrant SSIDs supporting WPA-Personal or WPA2-Personal.

Management Capabilities

APs will provide different methods for configuration and management of the devices. These methods will vary from vendor to vendor, and from model to model within vendor's product lines. However, there are common methods utilized. These common methods include:

- Console (serial)
- Telnet (weak security)
- SSH2 (much better security)
- SNMPv3

- Custom software applications
- Web-based interfaces (should use HTTPS)

Console or serial interfaces are only provided on enterprise class hardware and even with them, many enterprise-grade APs no longer provide serial interfaces today. Many vendors that were once known as only SOHO vendors are beginning to attempt to cross over into the enterprise market, and they are not always bringing enterprise features with which we are familiar.

When using a console interface to configure an AP, you will usually connect a serial cable from your computer to the AP. You may also use a USB to serial converter, such as the one seen in Figure 6.11. Once connected, you will use a terminal program such as PuTTY, in Windows, to connect to the device. Once connected, you will use the CLI (command line interface) provided by the vendor. Each vendor's CLI will be somewhat different and sometimes they will be wildly different. This is one of the major arguments for using consistent hardware throughout your organization: You only have to learn one set of CLI commands rather than a varied set. The good news is that the CLI is usually used at initial configuration or for device reload, and the other graphical interfaces are usually used for ongoing maintenance and configuration support.

The Telnet and SSH or SSH2 interfaces will be similar to the console management method in that the CLI will be utilized. The difference is that the CLI is being utilized across the network, rather than through the console port and a serial cable. When using these management methods across the network, you should be careful to ensure that some form of encryption is in use. Otherwise, with Telnet for example, the commands being transmitted from your machine to the AP are being sent in clear text that is easily readable in any common Ethernet packet analyzer. Remote CLI is achieved by connecting to the AP's IP address using a tool like PuTTY and SSH2.

Figure 6.11: USB to Serial Converter

SNMP is widely supported among WLAN devices. Due to security vulnerabilities in earlier versions, you should choose only devices that support SNMP v3 and — eventually — higher. SNMP provides for centralized mass configuration management. SNMP is not a proprietary technology, so one centralized application can often manage multiple vendors' APs.

Custom software applications may come with the AP and are usually provided on a CD-ROM or at a download website when they do. These applications are usually designed to run on Windows clients since these clients are so popular in enterprises. The applications may provide first-time configuration only or they may provide for ongoing configuration management. Due to the proprietary nature of these applications, they provide limited value to very large-scale installations.

Finally, web-based configuration interfaces take advantage of built-in web server software in APs to allow for remote configuration through the Ethernet interface. While you may be able to enable web-based management through the WLAN interfaces, I do not recommend it. This means that an attacker can try to guess the password and then manage the WLAN device across the WLAN. He or she will not even need to gain access to your physical network. For this reason, if you enable the web-based administration interface at all, it should only be enabled for the Ethernet port. Web-based management interfaces are provided on nearly all APs whether they are built for enterprise or SOHO use.

In addition to the configuration features mentioned here, most WLAN APs also allow you to save the configuration to a file that can be downloaded from the device to a disk. This allows you to quickly and easily reload the configuration at a later point.

Mounting Options

APs may be placed on flat surfaces in SOHO deployments, or they may be mounted in many different ways in enterprise installations. Mounting locations and methods include:

- Wall mount
- Ceiling mount
- Pole mount

When mounted on the wall, screws are usually fastened into the wall, and then the AP's mounting hardware is slipped onto the screws. The screws may be tightened further and then the AP snapped into the mounting hardware, or the AP may have the mounting hardware already attached, and the mounting is complete as soon as the AP is slipped onto the screws. With a ceiling mount, the AP is usually attached to similar mounting hardware, but the fasteners must be passed through the tile or other ceiling material. Finally, the pole mount method usually includes a wrapping brace that passes around the pole and then fastens to the AP's mounting hardware.

Mounting an AP is more involved than just deciding among the wall, ceiling, pole or flat surface mount options. You should actually determine where the AP needs to be placed and then determine the mounting option available to you, based on the location. In other words, the mounting method will usually be dictated by the location. The ultimate goal is to provide the proper coverage and capacity in the proper location, and this means that mounting methods are secondary.

Another factor to consider when choosing a mounting method is access. Will you be able to access the reset button on the device if needed? Will you be able to view the power and connectivity LEDs to determine operational status? These factors should be considered carefully. If you do not have access to the reset button or the power cord for power cycling, can you implement an AP that supports PoE for power cycling? While this will not provide easy access to configuration resets (like the configuration reset button would), it will allow you to power cycle the device more easily.

When mounting APs, and other WLAN devices, outdoors, you will need to consider weather issues. For example, will the AP be protected from rain and wind damage? The National Electrical Manufacturers Association (NEMA) has established a set of standards for electrical equipment enclosures. These NEMA enclosures are available for mounting APs and other WLAN devices outdoors. The NEMA Standards Publication 205 defines the various enclosure standards and is available at www.nema.org.

Figure 6.12 shows an example of a NEMA enclosure. You can see that standard power connectors are supplied, as well as an RF amplifier for connectivity to an external antenna. Consider that the enclosure itself may cause blockage of the RF signal from the antenna, so you must mount the antenna in a way that the enclosure is not placed in the path of the RF signal between the antenna and the users. Note in Figure 6.12, the connectors at the bottom of the enclosure. These are used to connect antennas to the AP so that they are not enclosed. An AP with only internal antennas should not be used in an enclosure like this.

Figure 6.12: NEMA Enclosure

Configuration Process

Many new APs will come out of the box with the antennas detached. If this is the case, you will need to first attach the antennas before the AP will be able to radiate the RF signal. You may wish to wait until after configuration to attach the antennas, but this is really optional since you will not be connecting the AP to the LAN until you have configured it properly. However, be careful not to power on the radios until the antennas are connected, to avoid damage.

As the last sentence suggested, you should configure the AP before connecting it to the actual wired LAN to which it will provide access. This helps to remove the potential for wired-side access before the AP is properly configured and reduces the likelihood that you will provide an entryway into your LAN — though only for a short time — during the configuration window. Most APs come from the factory with little or no security set, so they can certainly provide a point of vulnerability by default.

After the AP is properly configured according to your security policies and configuration standards, you will need to connect the AP to the wired LAN via the Ethernet port. You may also need to connect the antennas if you did not connect them before configuration, or if you disconnected them during configuration for security reasons.

Finally, you should test the AP to ensure that you can connect to it with a client configured with the appropriate security and configuration standards that match with the AP. If you are using an AP model for the first time, you may also want to perform some load testing to verify whether the AP works as advertised (in relation to throughput and concurrent connection), or not. You may need to adjust the number of installed APs according to real-world performance with some devices.

VLANs (virtual local area networks) are commonly used in conjunction with different SSIDs to separate and identify different WLANs in a single AP. This allows the AP to service more than one WLAN.

In the end, APs come in many different shapes and sizes. APs usually support a common set of IEEE standards, security capabilities and mounting options. Common management interfaces include console, Telnet, and web-based interfaces directly on the APs or in controllers, the cloud, or management software, among others. Most APs that are used in enterprise installations today support SNMP for centralized management and may support custom software provided by the AP vendor. As a WLAN administrator, it is important that you understand these options and be able to choose among them effectively.

Enterprise WLAN Controllers

The edge architecture, where WLAN APs were placed at the edge and configured individually, was fine for smaller networks; however, as larger and larger WLANs were implemented it became apparent that configuring each AP at the edge was no longer feasible. Vendors rushed to create their own solutions to this enterprise network dilemma. The result was the creation of WLAN controllers (initially called WLAN switches). The only major difference between a WLAN switch and a WLAN controller is that the WLAN switch has the controller functionality built into it, and the WLAN controller may be in a switch, a router or some other device. Today, most WLAN controllers are dedicated devices. For this reason, I'll refer only to WLAN controllers here.

A WLAN controller contains all or part of the functionality of one or more virtual APs. At first glance, a WLAN controller may look like any other router or network appliance. Some have many switch ports, and some have only one or two. Figure 6.13 shows the Ruckus Wireless ZoneDirector 1200 and Figure 6.14 shows the Cisco WLAN controller. Notice that the ZoneDirector 1200 has only two Ethernet ports. This is because the intention is to connect the ZoneDirector to a switch and connect the managed APs to the Ethernet network separately, rather than connecting them directly to the controller.

Figure 6.13: Ruckus Wireless ZoneDirector 1200

Figure 6.14: Cisco 5508 Controller

In Figure 6.14, you can see that the Cisco 5508 controller includes 8 gigabit uplink ports. Still, these are not intended for direct AP connections, as this controller can support up to 500 APs. Instead, each port can link to a different switch, so that various groups of APs have access to the controller without having to share a single, or just two, Ethernet ports in the controller.

Of course, every WLAN vendor says their WLAN controller solution is best. To be certain, each solution has its benefits and drawbacks. As a WLAN administrator, you must analyze the features offered and then choose the best solution for your implementation. This analysis usually means looking through the vendor literature thoroughly, and sometimes requesting test equipment to work with during the analysis phase of your WLAN implementation project. Some vendors will provide the test equipment free of charge, and others will come in and perform a demonstration of the equipment for you. The reality is that smaller organizations are less likely to get free sample devices, and larger organizations are more likely to get them. If you are in a smaller organization, the product manuals, which are usually available for free download from the vendor websites, may suffice for your analysis.

When looking through the vendor literature, pay close attention to the IEEE standards that are supported, as well as the proprietary ways in which the WLAN will be implemented. Larger vendors usually remain in business for long periods of time or are consumed by other vendors who continue to support their hardware.

Many WLAN controllers include built-in site survey capabilities that are either assisted or automated in nature. The assisted site surveys will require that you walk around within the facility, after a pool of access ports have been installed, with a compatible client that can send signal information back to the controller. The automated site surveys will simply configure the WLAN according to guidelines you can generally manage centrally at the WLAN controller. The latter method usually requires more overengineering (placing more APs than are absolutely needed), and the former usually requires less; however, many switches support both.

Common Features

Because many of the features of WLAN controllers were already covered in the AP section, I will only list the common features here. Remember, a WLAN controller can centralize the "AP" processing into the switch and away from the AP with the exception, of course, of actual frame transmission and reception. For this reason, WLAN controllers often implement the features that are traditionally found in thick or autonomous APs. The following features are common:

- PoE injection into the Ethernet ports (may only be supported on a subset of the ports)
- Built-in firewall capabilities
- Port filtering and MAC address filtering
- Standards-based and proprietary WLAN security technologies such as WPA, WPA2, EAP and IEEE 802.11i

- VPN tunneling
- Common management interfaces (web, Telnet, CLI, SSH, console, etc.)
- Configuration file management
- Activity monitoring and logging
- Built-in RADIUS servers for EAP authentication types
- Redundant Ethernet ports for greater uptimes and easier maintenance
- Controller teaming or hierarchical implementation for configuration management and redundancy
- Rate limiting for the various managed WLANs; this feature is very convenient for setting up two WLANs in the same area — one for VoWLAN (no rate limits) and the other for data (rate limited)
- Hotspot support including IP redirect to map connections to a specific "starting" page (aka, captive portals)
- RBAC (role-based access control) or identity-driven management (IDM) to provide different levels of access to different users depending on RADIUS settings
- Voice prioritization for VoWLAN
- CAPWAP or LWAPP compatibility
- Wireless client roaming management and assistance
- QoS including IEEE 802.1p and IEEE 802.11e
- Internal DHCP server
- Built-in Wireless Intrusion Prevention System (WIPS)
- Built-in wireless monitoring solutions

For more information on any of the features listed here, or features not listed, be sure to visit the various vendor websites listed below and download the product manuals for their WLAN controllers and APs. These manuals will go into the details of how each vendor implements the WLAN differently and help you understand the general use of WLAN controllers in modern wireless networks. Consider visiting the following websites, at a minimum, to be exposed to various controller-based vendors, and you will quickly see that what one vendor calls a controller does not automatically match what another vendor calls a controller:

- Cisco: www.cisco.com
- Ruckus Wireless: www.ruckuswireless.com
- Aruba Networks: www.arubanetworks.com
- EnGenius: www.engeniustech.com

Configuration Process

The configuration process will vary depending on the controller vendor you choose; however, the process is generally similar when considered from a less-detailed level. The process usually looks something like this:

(1) Perform the initial controller configuration

(2) Configure WLANs in the controller

(3) Connect access points to the controller

(4) Ensure access ports are properly enabled and configured

The first step is to perform the initial controller configuration. This usually entails specifying which port will be used for WLAN access point connectivity, and which port will be used for WAN uplinks (may be a LAN link if it is only used locally and not connecting to the Internet). If the WLAN controller contains multiple ports for connections to access points or wired devices, you may configure the proper use of each port.

Next, you will need to determine if you are going to support one virtual WLAN or multiple virtual WLANs. Some controllers will support multiple WLANs with one access point, and others will require multiple access points to support multiple WLANs. However, today, the vast majority of AP vendors support multiple SSIDs

(WLAN profiles) on their APs. You will need to determine the security settings and other configuration options for each WLAN, or allow the controller to automatically select some, or all, of these features. You may also need to specify VLANs for the separation of the different logical WLANs that run on the same physical APs and controllers.

Now, you are ready to connect the access points and have them detected by the WLAN controller. Some systems will support autonomous APs as well, but they must be converted to behave as thin APs. This may be an automatic process of the WLAN controller, or you may have to perform some configuration changes manually. The APs will need to locate the WLAN controller, and this is usually accomplished in one of four methods:

- DHCP using various DHCP options
- DNS host names
- Broadcast messages
- Pre-stages controller IP addresses in the APs

Finally, ensure that the access points are working properly, and that you have the needed WLAN access in the needed locations. This will involve inspection through the WLAN controller's management interface first. You will need to be sure everything "looks" right in the controller. Second, you will need to use a laptop or some other WLAN client device to connect to the WLAN or WLANs in the various locations, to ensure that the network is functioning as you need it to function.

> Remember, each vendor's installation procedure will be different. Check with the vendors to see how their installation procedures fit into the above generic installation process. You will usually find that they simply require specific and different steps within each of these four phases.

Mesh APs and Routers

WLAN *mesh APs or routers* are devices that use proprietary Layer 2 protocols to form a mesh network or 802.11s protocols in some available options. These networks are usually said to be self-forming, self-healing and self-configuring. They are self-forming because they scan for other nearby mesh routers and form associations with them automatically. They are self-healing because they will automatically discover new routes to a destination if previously used routes become unavailable. They are self-configuring because they configure their routing tables, and many other settings dynamically, based on the discovered network topology.

The protocols that mange this self-forming, self-healing, and self-configuration process are the proprietary Layer 2 routing protocols, or they are based on the mesh BSS (MBSS) concepts defined in the 802.11s amendment and part of the 802.11-2016 standard. The proprietary protocols (meaning they are unique to each vendor) determine routing paths and other variables by inspecting signal strength and quality from neighboring mesh routers. The 802.11s protocol can perform similar functions.

Don't let the term router confuse you here. WLAN mesh routers route through the radio interfaces. They may use one radio interface to both receive and retransmit frames or they may have dual radio interfaces (or more) and have the ability to receive on all interfaces and transmit on all interfaces. Some vendors call them mesh routers while others call them mesh APs. Figure 6.15 shows the architectural components of an 802.11s-based mesh network.

The components include Mesh Gates, Portals and a mesh BSS. The Mesh Gates allow for connections between the mesh BSS and other mess BSSs. The Portals allow for connections between the mesh and another network, like an 802.3 Ethernet network. A Mesh Gate can be a dedicated AP in this role, or it can be an AP that also acts as a Mesh Gate (called a co-located Mesh Gate).

Mesh networks can be used to cover large areas and they are often used to create multi-hop backhaul links in place of short distance (a few hundred meters) bridge links.

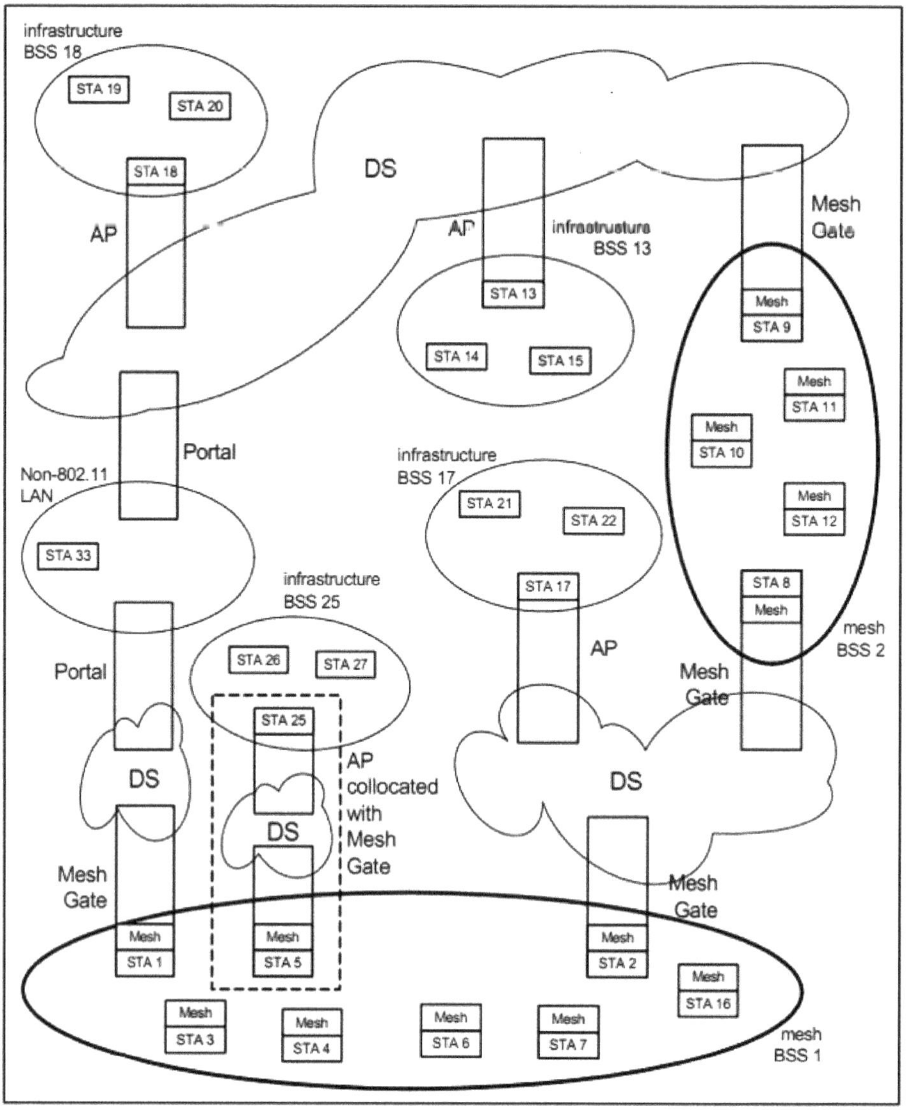

Figure 6.15: 802.11-2016 (802.11s) Mesh Networking

Remote Office Controllers and APs

Remote Office WLANs can be implemented using small WLAN controllers, which are less expensive than full-featured WLAN controllers. They allow remote and branch offices to be managed from a single location, and they simplify AP deployment and management in branch/small offices. Additionally, remote office

WLANs may be implemented using business-class wireless routers. These are devices similar to residential wireless gateways (home wireless routers) but add more advanced business features. Figure 6.16 shows an example solution for remote office deployments by Aruba Networks.

Remote Office WLAN controllers may feature:

- Integrated Router
    - NAT
    - DHCP
- Integrated Firewall
    - Stateful Packet Inspection
- Wireless Controller
    - Limited number of thin APs supported
    - Provides AP management and security
- Power over Ethernet

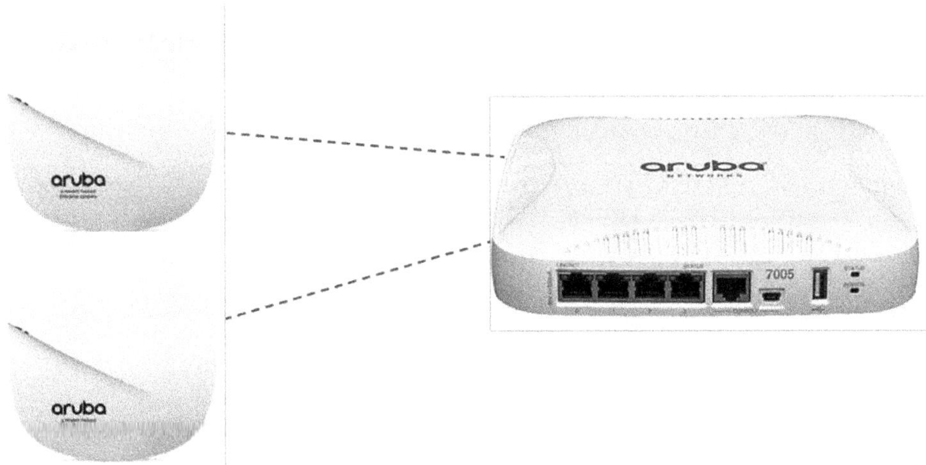

Figure 6.16: Aruba Networks 7005 Remote Office Controller Solution

Power over Ethernet (PoE)

One of the most important tasks you'll have to complete when implementing any system that required electrical power is power provisioning. You really have two options today: direct attached power (through traditional power outlets) or Power over Ethernet. Maybe in the future, we'll be able to power our APs with wireless power, but that will be a topic for the future. For now, let's explore Power over Ethernet.

PoE Injectors and Switches

I have mentioned PoE multiple times already in this chapter and in the preceding chapters. This is because the technology has proliferated the WLAN market and many APs, switches, bridges and other WLAN devices now support it. PoE is a method used to deliver DC voltage to a device over CAT5 cable. This DC voltage is used to power the device instead of a standard AC power outlet. (Most devices come with converters that convert AC power to DC power. PoE sends the power directly as DC power.)

CAT5, CAT5e and CAT6 cables have four pairs of wires in them. Only two pairs are used to carry the data. This leaves the other two pairs for other purposes. In the case of PoE, the purpose is to carry power to the device being powered. In WLANs, these devices include APs, bridges, repeaters and possibly other devices. Some implementations use the same pairs of wire that carry the data to carry the DC voltage and some implementations use the extra pairs in the cabling to carry the DC voltage, separate from the data.

As I stated earlier in this chapter, one of the most common reasons for using PoE is to power a WLAN device where no AC power outlets are available. The other benefit of implementing PoE is the ability to cycle the device being powered from remote. This latter feature is usually only available when the PoE is being provided by a managed switch. The management interface of the switch will allow you to turn of the power on a given PoE-enabled port, and then turn it back on. Power cycling is not supported by all PoE-enabled switches.

Yet another advantage of using PoE is that a licensed electrician is not usually required to install it. This is because the voltage is so low that is running across the cabling. Most any tech can run the cables and use PoE. There will likely be no building codes that will dictate specific guidelines for running the cabling and powering the end location. Figure 6.17 shows the way PoE would be utilized with an inline PoE power injector and with a PoE-enabled switch.

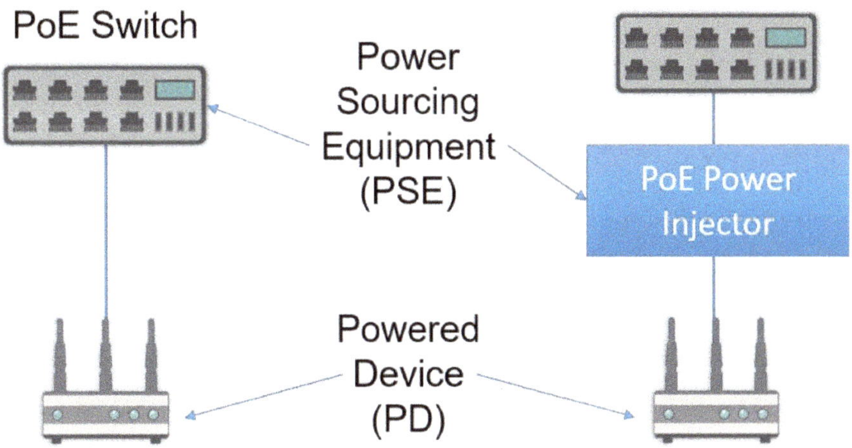

Figure 6.17: PoE with an Injector or PoE Switch

It is important that you know that PoE-enabled switches do not always, in fact they seldom do, provide power through all ports. Sometimes half of the available Ethernet ports are PoE-enabled and sometimes fewer than half can provide DC power to devices. Sometimes all ports can be enabled, but the power budget (the total amount of PoE power available) in the switch would not allow it. Be sure to check the vendor's documentation to verify the number of PoE ports being provided by the WLAN switch or standard Ethernet switch you are implementing.

Common Features

There are different types of devices that can provide voltage through Ethernet cables, which power PoE-enabled devices. These types include single-port DC voltage injectors, multi-port DC voltage injectors and PoE-enabled switches.

The single-port PoE injectors will have a single input port and a single output port. The input port is where you connect the Ethernet cable that connects to the network's switch or hub, and the output port is where you connect the Ethernet cable that connects to the device to be powered. When using a single-port PoE power injector like this, the power injector itself must be plugged into a standard power outlet. This means you will likely place the power injector in the closet (or location) with the switch or hub, and not closer to the device being powered. Due to the number of power outlets required, single-port power injectors are only recommended when one or two devices need to be powered.

A multi-port PoE power injector is really just a group of Ethernet input ports that pass through a power injection module, and then pass on to a matching group of Ethernet output ports. These devices are usually also installed closer to the switch or hub and farther from the powered device. This is due to the likelihood of having a power outlet where the switch or hub is located, since the switch or hub will need power as well. Due to the fact that multi-port power injectors can power multiple devices while they only require one power outlet connection, they are recommended in medium-to-large installations that require from three to 20 (though this is not an absolute cutoff point) powered devices. When more devices require power, you will likely opt for a PoE-enabled switch.

Large enterprises and networks with more than many PoE-powered devices will likely choose to move up to PoE-enabled switches. These switches include power injection in the same unit that is the Ethernet switch. It means powering fewer devices through standard power outlets and reducing the number of components that can fail at any given moment. When you use a single-port power injector with an access point, you introduce multiple points of failure. Image there are 20 APs that you need to power in this way. You would need 40 Ethernet cables (20 from the switch to the power injectors, and 20 from the power injectors to the APs), 20 power injectors, 20 power cords, 20 APs, and at least one switch. This means a total of 101 individual components that could fail and statistically increase the likelihood that you will have a failure at any given time. If you use a switch that can provide PoE power injection on up to 48 ports, you reduce the components involved to only 41 components. You've eradicated the need for 20 Ethernet cables, 20 power cords and 20 power injectors. The likelihood of a failure at any given moment has now been greatly reduced.

In addition to the failure probability reduction, you are gaining the benefit I discussed previously of being able to power cycle the APs from a central location. With the single-port power injectors, you would still have to go to the physical location where the power injector is located and unplug it and then plug it back in (or flip an on/off switch if it is available).

An additional benefit of PoE-enabled switches is that you do not usually have to enable PoE on all ports. For example, you can use some of the ports for wired devices, or non-PoE APs and bridges, while you use the other ports for PoE-enabled devices. This provides you with flexibility and is a valid argument for purchasing a switch that supports PoE from the factory, or at least purchasing one that can have PoE support added at a later time.

Power-over-Ethernet (PoE) (802.3-2015, Clause 33)

IEEE 802.3-2015 merged the older 802.3af PoE amendment into the core standard document, as well as the newer 802.3at amendment. The old amendments are now known as clause 33 in the 802.3-2015 document. Many, even most, vendors — at this time — are still referencing the standard as 802.3af or 802.3at, but you should know that it has been rolled into the primary standard now. If you download or access the 802.3-2015 standard in sections, clause 33 is in the section two PDF file.

The standard defines a Powered Device (PD) and Power Sourcing Equipment (PSE). The APs we've discussed that support PoE would be examples of PDs. The power injectors and PoE-enabled switches would be examples of PSEs. The clause specifies five elements:

- A power source that adds power to the cabling system
- The characteristics of a powered device's load on the power source and cabling
- A protocol allowing the detection of a device that requires power
- An optional classification method for devices depending on power level requirements

- A method for scaling supplied power back to the detect level when power is no longer requested or needed

The standard then spends the next several pages providing the details of this system. You will not be required to understand the in-depth details of PoE for the CWNA exam, but this IEEE document can act as your source for more information. You should, however, be familiar with the following two terms: midspan and endpoint power injectors, and you should be aware of the power provided by 802.3af versus 802.3at.

The standard specifies that a PSE (power injector) located coincident (in it) with the switch (technically, the data terminal equipment or DTE in the standards) should be called an *endpoint PSE*. It also specifies that a PSE located between the switch and the powered device should be called a *midspan PSE*. WLAN controllers and LAN switches with integrated PoE support would qualify as endpoint PSEs, assuming they are 802.3-2015-compliant. Multi-port and single-port injectors would qualify as midspan PSEs, assuming they are 802.3-2015-compliant.

802.3bt-2018 is an amendment to 802.3 that adds Type 3 power usage of 60 W at the PSE and Type 4 power usage at 90-100 W at the PSE. Devices requiring more than 25.5 watts should use 802.3bt PSE equipment.

Fault Protection

One note about PoE: fault protection is very important. Fault protection does the work of protecting the devices that are being powered by power injection, or that are providing the power injection. A fault occurs when a short-circuit or some other surge in power occurs in the PoE chain. Faults can occur for the following reasons:

- A device does not support PoE and uses the extra two pins used by PoE, or for some reason short circuits the pins.
- An engineer connects an incorrectly wired Ethernet cable.

Due to the nature of things, the last cause seems to be the most common cause. I know I have inadvertently "miswired" an Ethernet cable a time or two in my time. It's fairly easy to do because you're dealing with small wires using big fingers, and crimpers that haven't been upgraded or improved significantly for a few decades. When a fault occurs, the power injector should shut off DC injection onto the Ethernet cable in the path of the fault. Depending on the power injection device, you may need to manually reset the power injector, or it may monitor the line and automatically reset when the fault is cleared.

PoE Power Levels

PoE provides differing power levels depending on the class of the PD and the capabilities of the PSE. A PSE compliant with 802.3af only, can provide 15.4 watts of power onto the wire, but due to attenuation, only 12.95 watts is guaranteed at the PD. A PSE compliant with 802.3at can provide 30 watts of power onto the wire, and guarantees up to 25.5 watts of power at the PD. This information is summarized in Table 6.1.

Class	Usage	Max Power Consumed	Class description
0	Default	0.44 W to 12.95 W	Class unimplemented
1	Optional	0.44 W to 3.84 W	Very low power
2	Optional	3.84 W to 6.49 W	Low power
3	Optional	6.49 W to 12.95 W	Mid power
4	Type 2 devices	12.95 W to 25.5 W	High power
5	Type 3	Up to 51 W	Very high power
5	Type 4	Up to 71 W	Very high power

Table 6.1: PoE Classes and Max Power

Wireless Clients

Wireless clients have grown in both number and type significantly in the past five years. Before that time, the clients were mostly laptops, tablets, mobile phones, and a specialty devices. Today, there hundreds of devices that connect to 802.11 networks, and it would take several hundred pages to cover them all. This book and the CWNA 108 exam covers only the most common device types seen in enterprise WLAN deployments.

USB Adapters

USB adapters have become very common for both laptop and desktop computers. They come in two primary implementation models. The first is a dongle-type adapter that plugs directly into the USB port and the second is a device that connects to the USB port through a connector cable.

The greatest advantage of USB devices is that they are fairly universal (after all, the term USB stands for Universal Serial Bus). Saying that the USB device is universal is a reference to the fact that USB devices can be used with desktops, laptops, tablet PCs and any other device that supports the USB interface, and provides proper drivers for the WLAN NIC.

USB adapters may be 2.0 or 3.0 devices. Two important considerations should be made when selecting between 2.0 and 3.0 devices. First, USB 2.0 devices have a maximum data throughput of 480 Mbps. If the USB device is an 802.11ac device, it may not be able to achieve the highest data rates possible on USB 2.0. USB 3.0 overcomes this with data rates up to 5 Gbps. However, the second issues is with USB 3.0 devices. They can create significant RF interference in the 2.4 GHz band. In fact, the USB 3.0 circuitry in the 802.11ac USB 3.0 adapter can raise the noise floor by 5-20 dB, reducing the higher data rates to all but impossible.

Some engineers have had luck acquiring USB 3.0 hubs that connect via USB-C connectors without significant interference. I have personally seen interference coming from USB adapters themselves. So, you will have to test different devices and USB hub combinations to find those that cause the least interference.

Ultimately, a client device is a device that includes or is connected to a wireless network interface card (NIC). This card is comprised of a wireless chipset, which determines the physical layer (PHY) capabilities of the card, and is connected to one or more antennas. The antennas may be on board with the device, or external to the device. For example, USB adapters often include antennas on board and they may support external antenna connections. Figure 6.18 shows the internal board of a TRENDnet USB adapter. Note the two onboard antennas and the two connectors for use with external antennas, though the USB adapter provides no external interface to use them. The external shot of the same adapter is shown in Figure 6.19.

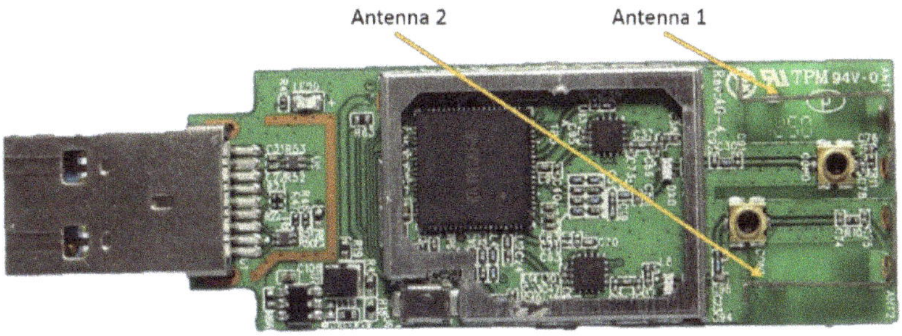

Figure 6.18: TRENDnet USB adapter (TEW-805UB)

Figure 6.19: TRENDnet USB Adapter External (TEW-805UB)

Card-Based Adapters

Card-based adapters are used in laptops and desktops, and in some tablet use cases. They include the following:

- PCI — these adapters are used mostly in desktop computers and can function all the way up to four stream 802.11ac adapters today.
- Mini-PCI, Mini-PCIe, and Half Mini-PCIe — these adapters are used mostly in laptops but may see some cases of use in tablets as well. The Half Mini-PCIe is the smallest form factor. Figure 6.20 shows a Half Mini-PCIe 802.11ac adapter.

Figure 6.20: 802.11ac Half-Mini-PCIe Adapter

Laptops, Tablets and Mobile Phones

Laptop computers are some of the earliest 802.11 wireless clients to exist. First with add-on PC Card adapters, and eventually with internal adapters, laptops are now assumed to have wireless access capabilities, though not all laptops are created equal.

In most cases, today's laptops will be equipped with either 802.11n or 802.11ac adapters. Those with 802.11n adapters may be 2.4 GHz-only, but many are dual-band. Those with 802.11ac adapters are always dual-band. When a laptop is single band, it is almost universally a 2.4 GHz-only device. It is extremely rare to see a single band, 5 GHz-only laptop.

> Remember that devices supporting 802.11n also support 802.11b/g, if they operate in 2.4 GHz. Devices supporting 802.11ac also support 802.11a/n in 5 GHz. Newer PHYs support older PHYs in the same band.

Laptops use wireless adapters that have external antenna connections. The connections are usually MC or MCX connectors. The antennas are often placed in the laptop screen enclosure to increase receptivity and they are also often located toward the back of the laptop base. Antenna location, in laptops, is very important, but remember that users move them around and use them in odd positions — antenna location can only accommodate for so many scenarios. The actual adapter may be on the laptop motherboard, but more often, today, the adapter is a mini PCI Express (PCIe) adapter or half mini PCIe (half-height mini PCIe) adapter that can be replaced. Older laptops may use the mini PCI form factor (as opposed to PCIe), but these are less common today.

Figure 6.21 shows the location of antennas in a MacBook Pro 2015 model laptop. In this case, the antennas are located in the rear of the laptop where the screen and base meet.

Newer laptop computers are shipping with 802.11ac adapters, but many 802.11n-based laptops are still in use in organizations, and some new units still use the older 802.11n chipsets as well. It is important, during laptop acquisition, to ensure that the wireless interface is appropriate for your needs. If the wireless networks managed by the organization are 802.11ac networks, it is best to purchase laptops with 802.11ac support. However, 802.11n-based laptops will work as well, though they will consume more airtime due to lower data rates used in transmission and reception.

When it comes to channels supported, older laptops that work in the 5 GHz band may support only channel 36 to 64 and channels 149 to 165. Most newer laptops will support all 5 GHz channels, with the exception of 802.11n-based laptops, which will not support channel 144 that was introduced in 802.11ac. All 2.4 GHz-based laptops will support channels 1 to 11, and laptops designed or configured to operate in the appropriate regulatory domain may support channels 12-14 as well.

Channel width is another factor in channel support. The simplest ways to increase throughput are to use more spatial streams (covered next), or wider channels. However, the use of wider channels also comes with pain points. When using wider channels, frequency reuse becomes more of a challenge. Frequency reuse is the strategy used to build capacity networks. Wider channels reduce the total number of available channels and result in the need to reuse channels within shorter distances. The result is more co-channel interference (CCI). However, in typical office spaces, 40 MHz channels are acceptable for standard deployments in 5 GHz. In 2.4 GHz, only 20 MHz channels should be used.

Figure 6.21: MacBook Pro 2015 Antenna Locations

Remember that 802.11n devices will not typically support channel 144 because it was introduced with 802.11ac. Aside from this, most newer laptops support all other 5 GHz channels.

Most laptops support at least 40 MHz channels with newer 802.11ac laptops supporting both 80 MHz and 160 MHz channels. At this time, 160 MHz channels should not be used in enterprise deployments. The clients will use whatever maximum channel width the APs support, up to the maximum width supported by the client. Again, the client will determine the link reality if the AP is using the latest chipset, and the client is using either an older chipset, or a less capable chipset.

Laptops support the entire range of possible spatial streams from 1 to 4. A device can have more transmit and receive chains than the number of supported spatial streams. In such cases, the extra transmit or receive chains are often used for diversity, to improve receptivity and transmission quality.

For example, if a laptop is said to be 2x2:2, it indicates that it supports 2 transmit and receive chains, and it supports 2 spatial streams. Typically, when a device supports more spatial streams, it can achieve higher data rates than another device with fewer spatial streams. The highest number of spatial streams in clients as of late 2017 is 4x4:4 or 4 spatial streams.

> 802.11n supported four spatial streams in the standard, but no chipsets were released supporting this number of spatial streams before 802.11ac was ratified. The highest number of supported spatial streams was 3x3:3.

The final issue to consider with laptops is security features. Because laptops can run varying operating systems (OSes), the OS will actually determine the security features available. Security is not built into the chipsets, but rather is part of the client supplicant. To determine the security options available, you must know the OS used. The most popular OSes for laptops are Windows, macOS, Chrome OS, and Linux.

> A ***supplicant***, in wireless networks, is a software component that allows for Layer 2 (and higher) connectivity to the WLAN. It provides features like roaming and security and, in most cases, is built into the OS used.

All modern OSes will support WPA and WPA2 security with a pre-shared key (PSK). With this security, a passphrase is used to configure the clients and the passphrase is passed through an algorithm (defined in the 802.11 standard) that generates the PSK. All four referenced laptop OSes support WPA- and WPA2-PSK.

Tablet devices use wireless chipsets ranging from 802.11n single stream to 802.11ac dual stream capable. Some older tablets may still have 802.11g chipsets in them or dual-band 802.11a/g chipsets, but they are becoming less common today.

Most tablets run iOS or Android OSes, with the exception of the few that run a version of Windows.

Because tablets are mobile devices by nature, users are more likely to roam around the facility while using them. For this reason, and for the mobile phones discussed next, WLANs must be designed to facilitate roaming. This design criteria is one of the most challenging to accomplish and requires expertise beyond the level of the CWNA, but the CWNA should be aware of the issues. In general, a sufficient number of APs should be deployed with the proper output power settings to facilitate the coverage and capacity to allow for roaming. For this reason, it is very important to follow the design specifications, if they were created by an expert in WLAN design, when installing and configuring APs. Mount them where the design indicates they should be mounted and configure them with output power settings according to the design. Mounting or configuring the APs differently can break the design and cause coverage, capacity and roaming problems.

As an example tablet device, consider the iPad Pro 9.7-inch model A1674 or A1675. It includes wireless antennas at the top and bottom of the tablet (when held in portrait position) and uses a two-stream, 802.11ac, dual-band chipset. Figure 6.22 shows this tablet with the antenna locations indicated.

Mobile phones, for the most part today, are the same as tablets except they have smaller screens (usually) and cellular chipsets in them, as well as Wi-Fi chipsets (though some tablets have cellular capabilities as well). For this reason, the same issues exist with mobile phones. However, mobile phones usually roam even more than tablets, and are often used for voice calls on a more frequent basis. Therefore, roaming is just that much more important for these devices.

Figure 6.22: iPad Pro 9.7 A1674 Tablet with WLAN Antenna Locations Identified

Mobile phones are commonly single- and dual-stream devices, and this is not likely to change soon. A balance must be accomplished between battery life and wireless speeds. By limiting wireless speeds, through the use of fewer spatial streams, the manufacturer can more easily create a device with extended battery life.

> When only one spatial stream is used, it has no impact on channel widths. While often confusing to the beginning wireless technician, the channel width and spatial streams are two separate things. A single spatial stream 802.11ac client can still support a 160 MHz channel, for example; however, it will be limited in data rate by the use of a single spatial stream because MIMO is not available in its multi-stream form.

As an example device, consider the LG K20 model LG-VS501 Android mobile phone. This mobile phone, shown in Figure 6.23, uses the Qalcomm SnapDragon 425 processor, which includes 802.11ac functionality. The wireless chipset supports:

- 802.11ac in 5 GHz

- 802.11n in 2.4 GHz
- MU-MIMO
- 1 spatial stream
- 80 MHz channel width max

Given these specifications, the maximum possible data rate for this phone in 5 GHz is 433 Mbps, based on an 80 MHz channel and an 802.11ac connection. Because real-world enterprise deployments rarely use 80 MHz channels, the real maximum data rate will be 200 Mbps. This kind of information is key in preparing for capacity in a WLAN. A limited amount of airtime exists, and slower devices use more of that airtime to transmit the same amount of data. Therefore, slower devices bring the entire channel down to lower throughput levels. Additionally, with 200 Mbps as the maximum data rate, in the real world, most of the phones will have data rates closer to 90 or 120 Mbps because the signal quality will not be good enough for the highest data rates. Figure 6.24 shows the chipset on the LG K20 circuit board.

Figure 6.23: LG K20 Smartphone

An additional factor with mobile phones is EAP method support. iOS phones will support the same EAP methods as iOS tablets and Android phones commonly support EAP-TLS, EAP-TTLS, PEAP and EAP-SIM.

Figure 6.24: LG K20 Qualcomm Chipset

802.11 VoIP Handsets

VoIP handsets are a unique use case as they place significant demands on the wireless network. Among these demands are:

- Effective QoS to ensure voice packets get through the network quickly. This requires 802.11e or WMM on the VoIP handsets and APs.
- Proper security. Some handsets only support WPA2-Personal and, at times, this is beneficial for faster roaming; however, it requires the implementation of a dedicated SSID for the VoIP handsets.
- Proper signal strength. VoIP handsets usually come with vendor recommendations of a minimum signal strength of -65 dBm, -67 dBm or -70 dBm. The network must be designed to accommodate this requirement.
- Single-stream devices. In most cases, to conserve battery life, and because very small packets are sent, single-stream chipsets are used in 802.11 VoIP handsets.

Considering these factors, the following guidelines should be considered:

- 802.11a/b/g-only devices are not uncommon
- Several 802.11n handsets are now available

- Knowing receive sensitivity and design specifications is essential
- Roaming support is also important
- WPA2-Personal vs. WPA2-Enterprise
- Mostly single-stream devices
- Do not require high data rates
- Do require fast access to the medium

Specialty Devices

Specialty devices are those sometimes unexpected devices that use Wi-Fi. These include the items pictured in Figure 6.25 and more.

Figure 6.25: Common Specialty Devices

It is important to know that these specialty devices often come with limiting constraints:

- 2.4 GHz-only support
- 802.11b/g-only
- 802.11n-only
- WPA2-Personal-only

Knowing these constraints can help you in the decision of acquisition, and also help you to design the WLAN to support them if you must.

Additional devices to consider, that also often come with such constraints, are:

- Handheld scanners: low throughput requirements, but often use very old PHYs (some are still in use with 802.11b).
- Push-to-talk devices: similar requirements to VoIP.
- Various IoT devices: often use cheap chipsets to reduce costs, which means older technologies (PHYs) or limited features.

Configuring Windows, Linux, Chrome OS and macOS Clients

Various operating systems provide different and yet similar configuration options. Today, nearly all operating systems used with laptops and desktops provide GUI interfaces for WLAN configuration. As a CWNA candidate, you will not be required to perform actual configuration tasks during the exam. But you should be familiar with the configuration tools available to you in the various operating systems in common use today.

Windows

In the Windows environment, WLAN settings can be configured in the GUI, or from the command line with NETSH. The GUI may be the built-in Windows configuration interface, or third-party utilities that are sometimes installed with adapter drivers. Figure 6.26 shows the GUI and NETSH interfaces in Windows 10.

Linux

Linux is well-known for its shell (command line) interface. Tools like iw, iwconfig and ifconfig are commonly used there when configuring WLANs. However, most Linux distributions also include a graphical Network Manager that can be used to connect to wireless networks. Both such tools are shown in Figure 6.27.

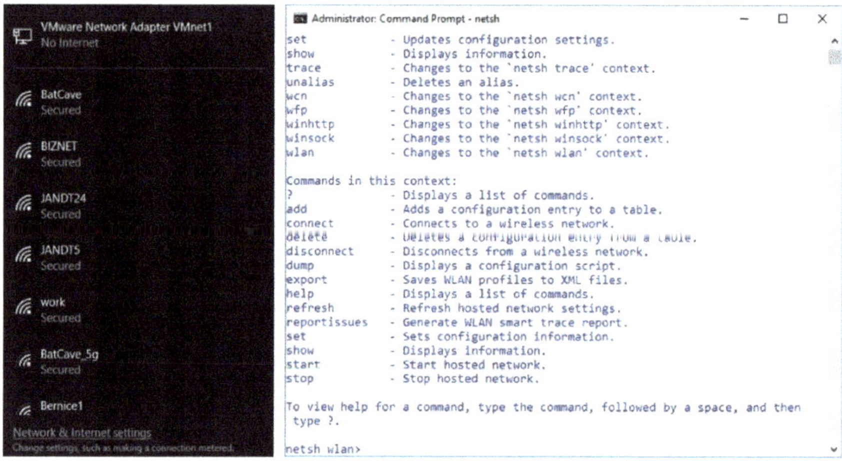

Figure 6.26: Windows Configuration Interfaces

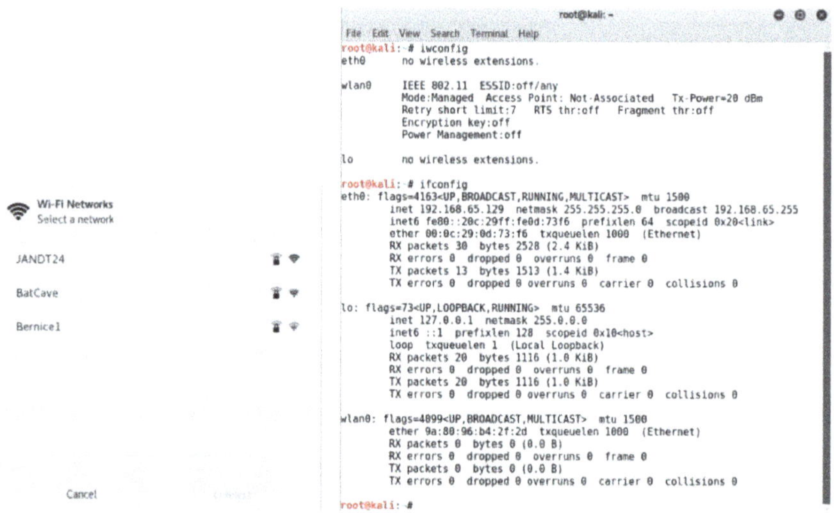

Figure 6.27: Linux Configuration Interfaces

Chrome OS

In Chrome OS, the GUI is the primary interface used to configure WLAN connections. It is similar to the Windows and Linux interfaces. Figure 6.28 shows the Chrome OS WLAN configuration options.

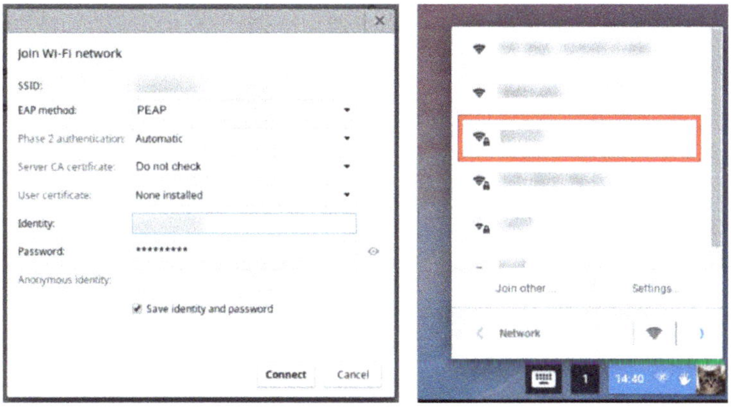

Figure 6.28: Chrome OS WLAN Configuration

macOS

Finally, the macOS, formerly Mac OS X, operating system provides a command line tool that may be used for some WLAN configuration options called airport. To use it in the newest versions of macOS, you must now create a symbolic link to it first with the following command:

```
sudo ln -s
/System/Library/PrivateFrameworks/Apple80211.framework/Versio
ns/Current/Resources/airport /usr/local/bin/airport
```

After execution, you can use the airport command directly. Figure 6.29 shows the macOS configuration interface in the GUI.

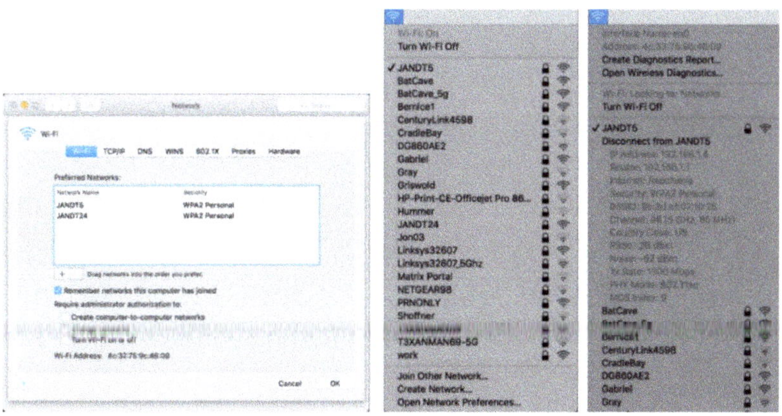

Figure 6.29: macOS GUI Configuration

Chapter Summary

In this chapter, you learned about infrastructure and client devices used on WLANs. First you learned about access points, controllers, cloud-based systems and distributed systems. Next you learned about PoE options, and finally, you explored the various client devices in use on today's enterprise networks. In the next chapter, you will begin to explore the details of 802.11 MAC operations.

Points to Remember

Remember the following important points:

- APs can operate in root/access, bridge and repeater modes.
- Using an AP in repeater mode will typically reduce the throughput by 50% for clients connected to the repeater and also reduce the available air time for the root/access AP to which the repeater is connected.
- External antennas are not used as frequently with indoor APs today, as the typical installation requires more APs for capacity anyway.
- Dual-band APs are typically concurrent, meaning they can run a 2.4 GHz and 5 GHz SSID at the same time.
- Some APs come with a configurable radio that can be configured for either 2.4 GHz or 5 GHz, while the other radio is typically dedicated to 5 GHz.
- WLAN controllers can support hundreds of lightweight APs and the APs receive their configuration and firmware from the controller.
- APs can use DNS, DHCP, broadcasts, or pre-configured controller IP addresses to locate the controller.
- When remotely managing APs or controllers, it is best to do it across the wired interface.
- HTTPS and SSH2 should be used for controller and AP management, and not HTTP and Telnet.
- With PoE, the PSE delivers the power and the PD receives the power.

- 802.3af provides 15.4 watts of power onto the wire, but only 12.95 watts are guaranteed at the PD.
- 802.3at provides 30 watts of power onto the wire, but only 25.5 watts are guaranteed at the PD.
- An endpoint PoE injector is a switch with PoE capabilities.
- A midspan PoE injector is placed in line between the switch and the PD.
- USB 3.0 adapters may cause interference in the 2.4 GHz band, raising the noise floor between 5 and 20 dB.
- Laptops typically used Mini-PCIe or Half Mini-PCIe adapters today.

Review Questions

1. What is the primary difference between an autonomous AP and a lightweight AP?
    a. The autonomous AP is heavier than the lightweight AP
    b. Autonomous APs are self-sufficient and lightweight APs require a controller
    c. Autonomous APs cannot support 802.11ac, but lightweight APs can
    d. Unlike lightweight APs, autonomous APs do not support 802.1X/EAP

2. In what way is a software-configurable radio usually able to be configured?
    a. As either a 2.4 GHz radio or a 5 GHz radio
    b. As either a three-stream or single-stream device
    c. As either an internal antenna or external antenna device
    d. Through the use of Telnet only

3. You are configuring an AP, and all of the following options are available. Which one should you use?
    a. Telnet
    b. HTTP
    c. SSH1
    d. HTTPS

4. In what band is a USB 3.0 device likely to cause interference?
    a. 60 GHz
    b. 2.4 GHz
    c. 5 GHz
    d. Sub-1 GHz

5. What command can be used to configure the wireless settings on a Linux client?
    a. iwconfig
    b. netsh
    c. airport
    d. ipconfig

6. How much power is provided to the PD on a 100 meter cable with 802.3at?
    a. 15.4
    b. 30
    c. 12.95
    d. 25.5

7. You have installed an AP and used two Ethernet cables to connect it to the wired network. One cable runs from the switch to an injector. The other runs from the injector to the AP. What kind of inject is used?
    a. PoE splitter
    b. Mini-switch
    c. Midspan
    d. Endpoint

8. In addition to DNS or DHCP, what can a lightweight AP use to locate the controller?
    a. Broadcasts
    b. WINS
    c. RADIUS
    d. LDAP

9. What command is in the macOS shell that can be used to configure some WLAN settings?
    a. iw phy
    b. airport
    c. netsh
    d. None of these

10. What kind of WLAN adapter is most likely to be in a laptop computer today?
    a. Half mini-PCIe
    b. PCI
    c. CF
    d. SD

Review Answers

1. **B is correct.** Autonomous APs can be called self-sufficient as they handle all 802.11 processing internally. Lightweight APs rely on the controller for many functions.

2. **A is correct.** When an AP has a software-configurable radio, it means that the radio can be configured for 2.4 GHz operation or 5 GHz operation.

3. **D is correct.** You should configure the AP with HTTPS.

4. **B is correct.** USB 3.0 devices are known to cause significant interference in the 2.4 GHz band.

5. **A is correct.** The iwconfig command is used in Linux, when available, to configure wireless settings.

6. **D is correct.** 802.3at can provide 30 watts of power, but given the long cable run specified, a maximum of 25.5 watts can be expected at the PD.

7. **C is correct.** A PoE injector placed in line between the switch and AP is called a midspan injector.

8. **A is correct.** Lightweight APs often support DNS, DHCP, broadcasts and pre-staged controller IP addresses to locate the controller.

9. **B is correct.** The airport command may be used in macOS to configure some WLAN settings.

10. **A is correct.** Laptops today use mostly mini-PCIe or half mini-PCIe.

Chapter 7 — 802.11 MAC Operations

In this chapter, you will be introduced to additional 802.11 terminology and specifics of 802.11 network frames. Additionally, you will learn about the processes involved in locating and connecting to WLANs. This information is important for the CWNA-108 exam, but it is also important as real-world knowledge that will prove valuable for network troubleshooting.

802.11 MAC and PHY Terminology

Before we dive deep into 802.11 frames, it might be useful to understand some key terms used in 802.11 communications. These include:

- MSDU, MPDU, PSDU and PPDU
- A-MSDU and A-MPDU
- Interframe Spaces
- Guard Interval
- Fragmentation
- PHY Preamble

MSDU, MPDU, PSDU and PPDU

The data units that are passed down through the layers have specific names. These names are used to distinguish the frame at one layer from the frame at another layer, and to distinguish the pre-serviced frame from the serviced frame at each layer. The service data unit (SDU) requires service as it comes into a lower layer from a higher layer. The protocol data unit (PDU) has been processed by the current layer and is ready to be passed down to the next layer. The important terms for 802.11 communications are: MSDU, MPDU, PSDU and PPDU. Figure 7.1 illustrates these terms in relation to OSI Model layers.

The first, MSDU, stands for *MAC Service Data Unit*. The MSDU is that which is received from the upper layers (OSI layers 7-3 via the LLC sublayer) to be managed and transmitted by the lower layers (OSI layers 2-1). It is the data accepted by the MAC layer to be transmitted to the MAC layer of another station on the network. MSDUs are included in all wireless frames that carry upper layer data; however, IEEE 802.11 management frames do not contain MSDUs, since there is no upper layer data to transfer.

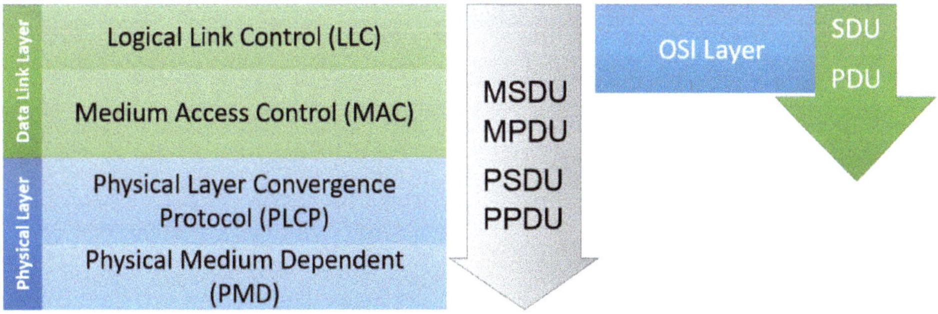

Figure 7.1: MAC and PHY Terminology

The MPDU, *MAC protocol data unit*, is that which is delivered to the PLCP so that it can ultimately be converted into a PPDU and transmitted. The MSDU is that which is received by the Data Link Layer, and the MPDU is that which comes out of the Data Link Layer and is delivered to the Physical Layer, and specifically is delivered to the PLCP. Another way of saying this is to say that the MSDU is received by the MAC from upper layers, and the MPDU is provided by the MAC to the lower layer.

The PSDU is the *PLCP service data unit*. The PSDU is that which the PLCP receives from the MAC sublayer. While the MAC sublayer calls it the MPDU, the Physical layer references the exact same objects as the PSDU. The PLCP adds information to the PSDU and provides the result to the PMD as a PPDU.

The PPDU, *PLCP protocol data unit*, is what is actually transmitted on the RF medium. The PPDU is that which the PMD receives from the PLCP. Ultimately, the PPDU is the culmination of all that has happened to the data, from the time it left the application starting at Layer 7 of the OSI model, to the time it is actually transmitted on the RF medium by the PMD at Layer 1.

A-MSDU and A-MPDU

Frame aggregation was first introduced in 802.11n and continues to be supported in 802.11ac.

The first type of aggregation is aggregated MSDU (A-MSDU). This is aggregation of MSDUs within a single MPDU. With A-MSDU frame aggregation, MSDUs are jammed together as a single payload, and a single MAC and PHY layer header are added to the set of MSDUs, making a jumbo-sized MPDU. Figure 7.2 illustrates both A-MSDU and A-MPDU framing structures.

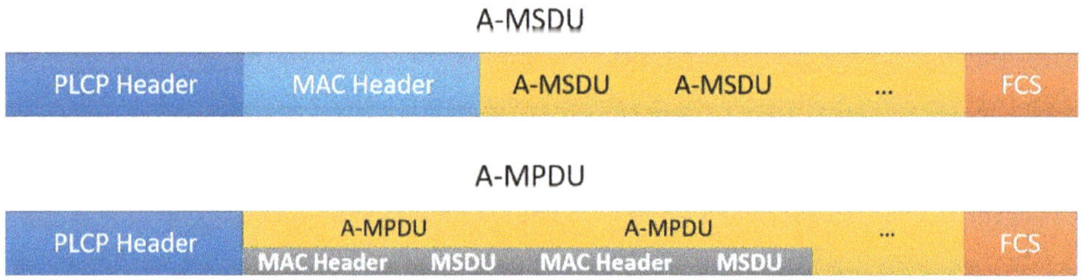

Figure 7.2: A-MSDU and A-MPDU Framing

This method of data transmission is especially efficient when many small frames are being transmitted across the wireless medium. Consider that in many cases, the MAC and PHY headers could be larger than the MSDU, and so, packing many MSDUs together (up to the maximum MSDU size) can cut medium contention and 802.11 protocol overhead by as much as 80%, in some cases. This allows for substantial throughput increases, in HT networks.

Only MSDUs that are using the same priority value may be aggregated into a single A-MSDU. All MSDUs in an A-MSDU are encrypted together, as a single encrypted payload.

Another form of frame aggregation that was specified by the 802.11n MAC layer changes was aggregated MPDU (A-MPDU). This form of aggregation is different from A-MSDU in that MSDUs are not crammed into a single MPDU, but rather multiple MPDUs (each with their own MAC header) are crammed together in a single PPDU. Think of this like putting cars onto an auto-transport 18-wheeler and sending it out onto the highway. A-MPDU can be implemented in concert with A-MSDU for additional efficiency in 802.11n devices.

With A-MPDU, each payload is encrypted/decrypted individually, since each MSDU is kept separate from others. A-MPDU requires the use of block acknowledgements, since multiple MSDUs are arriving at the same time at the destination station.

802.11ac PHY devices use only A-MPDU format frames when communicating with other 802.11ac PHY devices. By making all frames A-MPDU, instead of just multiple MPDU transmissions, it increases efficiency in the network.

Interframe Spaces (IFS)

Interframe spaces are brief idle periods on the wireless medium between frame transmissions. The purpose of an IFS is to regulate the flow of conversations, and to provide contention priority for certain stations and applications. For example, a SIFS (short IFS) is shorter than a DIFS (DCF IFS), so if one station must wait a SIFS and another station must wait a DIFS, the station waiting for only a SIFS will transmit sooner. Acknowledgements are an example of a frame that uses a SIFS. After the data frame is received, the receiving STA waits a SIFS, then transmits an ACK. This prevents other stations from cutting off the conversation by transmitting another data frame before the ACK has been sent.

There are six different types of interframe space of which you should be aware. This slide illustrates the four most common IFS types. The other two are explained at the end of this notes section. Four key interframe spaces are illustrated in Figure 7.3.

DIFS (DCF interframe spaces) are used for the first attempt at data and management frame transmission in a DCF network. The length of a DIFS is PHY-specific.

AIFS (Arbitration interframe spaces) were introduced with 802.11e to provide QoS priority to applications. AIFS are used in the same way as DIFS but are used only by QoS stations under the EDCA contention mechanisms. The length of an AIFS is both PHY-specific and Access Category-specific. Depending upon the QoS queue of the frame, the length of AIFS will vary.

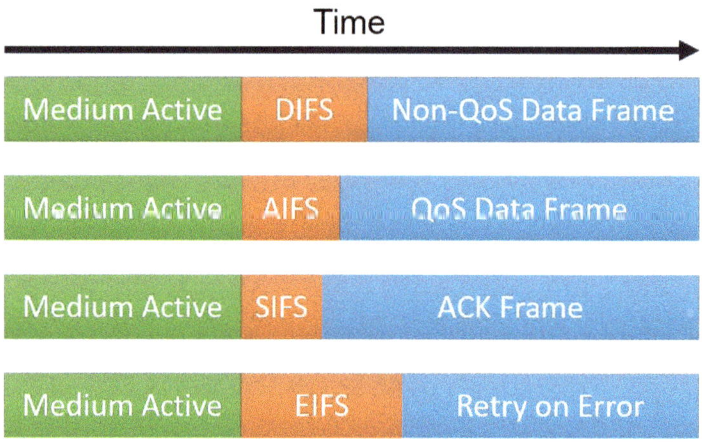

Figure 7.3: DIFS, AIFS, SIFS, and EIFS

Prior to 802.11n, SIFS (Short interframe spaces) were the shortest interframe space. SIFS are used when STAs have seized the medium and need to keep it for the duration of the frame exchange sequence. Using the smallest gap between transmissions within the frame exchange sequence prevents other STAs from attempting to use the medium, since they are required to wait for the medium to be idle for a longer gap. This gives priority to completion of the frame exchange sequence in progress.

EIFS (Extended interframe spaces) are a sort of protection mechanism. In previous sections of this course, we've talked about the Duration field and the carrier sense mechanisms. These functions allow each STA to determine when the wireless medium is busy, and when it is idle. However, virtual carrier sense mechanisms won't work properly, unless a frame is received properly, so that the stations can extract the Length and Duration information. When a station receives a frame, but it contains errors, the station cannot trust the Length or Duration values. Thus, it must have a sort of failsafe mechanism so that it doesn't interfere with other stations. The EIFS is that failsafe. When a received frame is corrupt, and the station doesn't know how long the medium will be busy, it waits an EIFS. Because of the purpose of an EIFS, it is the longest interframe space, and will decrease performance on your WLAN. Anytime a STA receives a corrupt frame, it must wait an EIFS unless a properly received frame resets the NAV before the duration of the EIFS expires.

A PIFS (PCF interframe space) is an interframe space that is used with the PCF mode of operation. Since it isn't used today, PIFS are largely unused. There is one exception: the channel switch announcement (Action frame). Since a channel switch announcement will typically follow a DFS event, it is important for the AP to be able to communicate with its BSS, so that it can switch channels. For that reason, a channel switch frame takes priority over other data frames. PIFS are not tested on the CWNA-108 exam.

Finally, 802.11n introduced the RIFS (Reduced interframe space), which is the shortest interframe space. In an 802.11n transmission burst, a very short interframe space is acceptable. RIFS is not used in 802.11ac and is not tested on the CWNA-108 exam.

Guard Interval

A *Guard Interval,* often called just a *GI,* is a period of time between symbol transmissions that allows reflections, which are caused by multipath, from the previous data transmission to settle, before transmitting a new symbol. The signal content inside the GI, called Inter-Symbol Interference (ISI), is rejected by receivers. 802.11a/g OFDM transceivers use 800ns GIs. This value was chosen by standards designers because the maximum multipath echo time is typically considered to be 800ns in an indoor environment. Outdoor GIs are typically higher, because objects are generally spaced further apart.

The HT PHY introduced two guard intervals: 400ns (called short) and 800ns (called long). Support of the 400ns SGI is optional for transmit and receive. The VHT PHY also specifies a short and long guard interval of the same timing. Figure 7.4 illustrates the short (400 ns) versus long (800 ns) GI.

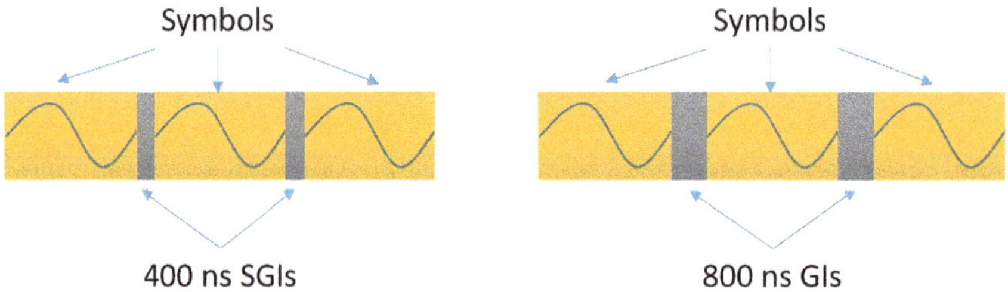

Figure 7.4: Short and Long GI

The time used to transmit a symbol is comprised of two parts: Fast Fourier Transform time, often called just FFT time, and Guard Interval time. FFT is a signal processing technique, so FFT time is the time during transmission or reception when signal processing is accomplished. The 802.11a/g OFDM symbol rate is 250 kHz, corresponding to a symbol period of 4 μs. This means the Guard Interval is 0.8 μs and the FFT is 3.2 μs in an 802.11a/g symbol.

Delay spread is the difference in time between the arrival of the first copy and the last copy of the same transmission when multipath occurs. The Guard Interval should be 2-4 times higher than the delay spread, so in most environments, delay spread should not exceed 200ns. Most offices and homes have a delay spread of approximately 50–100 ns. Choosing a Guard Interval that is too short means that Inter-symbol Interference will increase, and throughput will decrease. Choosing a Guard Interval that is too long means increased overhead due to unnecessary idle time on the wireless medium. Choosing a proper Guard Interval time period is crucial to optimizing throughput.

Choosing a 400 ns Guard Interval adds approximately 9-10% to the achievable data rate over the 800 ns Guard Interval. This fact can be seen in the 802.11n (HT) and 802.11ac (VHT) MCS tables. While 400 ns may seem like an insignificant amount of time to humans, it is a very important amount of time in wireless communications, when it occurs as frequently as the GI occurs.

Fragmentation

Fragmentation is a MAC level function that allows transmission of large frames as smaller frames, according to a user-definable length. Fragmentation improves performance in congested RF environments because:

- Larger frame sizes have a greater chance of collision, but require less processing overhead
- Smaller frame sizes have a smaller chance of collision, but require greater processing overhead

Fragmentation can be disabled to avoid the processing overhead in normal RF environments. When frames are fragmented on the RF medium, they are reassembled on the AP or controller, before being forwarded on to the rest of the network.

PHY Preamble

Frames created at the MAC layer are prepended with a Physical Layer Convergence Protocol (PLCP) header and a preamble or training fields before radio transmission. The PLCP header provides information about the data rate used to transmit the MPDU and other details about the channel.

The preamble or training fields are used to synchronize the receiver for proper reception of the MPDU frame. 802.11b and earlier PHYs use the term preamble, and 802.11a/g and later PHYs use training fields. However, for backward compatibility in 2.4 GHz, when using ERP-DSSS mode, a traditional 802.11b preamble is used, though a long and short preamble are available. ERP-OFDM mode uses the term preamble, and indicates it is precisely the same as the OFDM (802.11a preamble). However, OFDM was a sort of transitional PHY, and while it uses the term SYNC preamble, the definition of it uses the phrase OFDM training structure. So, the transition to the concept of a training field (as the terminology) instead of a preamble was in place.

Whether called a preamble or training field, the purpose is the same: to synchronize the receiver for the reception of the incoming frame. Figure 7.5 shows the PLCP header and PHY training for HT devices. Three formats are supported.

An interpretive legend for understanding the PLCP headers in Figure 7.5 is provided in Table 7.1. The important factor to note is that when HT devices use the Non-HT Legacy format, they are effectively operating as ERP or OFDM devices. The mixed format is most common, and it allows for faster HT communications of the actual frame data. HT Greenfield is rarely used, due to the existence of down-level client devices.

Acronym	Definition
L	Legacy (non-HT)
STF	Short Training Field
LTF	Long Training Field
SIG	Signal
HT	High Throughput
GF	Greenfield

Table 7.1: Interpretive Legend for Figure 7.5

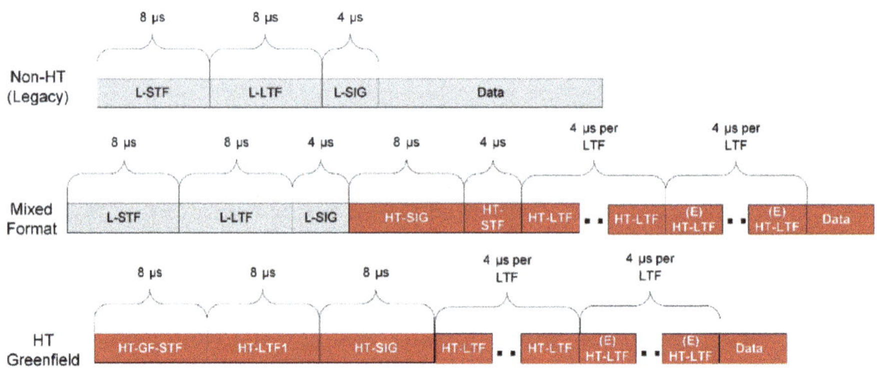

Figure 7.5: HT PLCP Header and PHY Training

802.11ac improves on this significantly by just providing a single PPDU format (though it can communicate using the three HT formats), as depicted in Figure 7.6. Notice that the first part of the 802.11ac PLCP header/training is exactly the same as the Non-HT Legacy PPDU format from 802.11n, and of course, 802.11a. This portion provides ample communications to legacy (OFDM and HT) STAs of the impending frame transfer. This information is followed by the VHT-specific PLCP header/training information and is used for MU-MIMO or SU-MIMO transmissions, based on VHT modulation and coding tables.

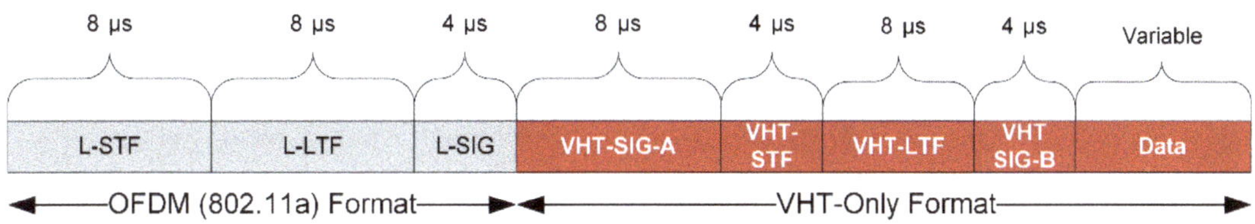

Figure 7.6: 802.11ac (VHT) PPDU Format

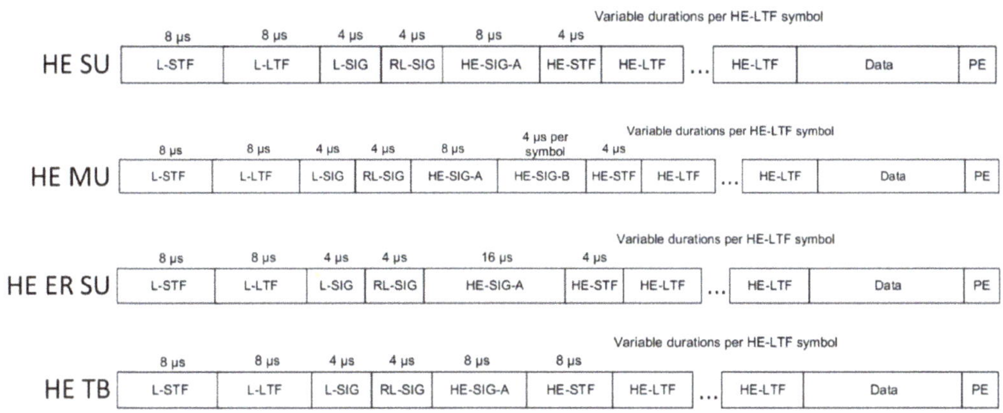

802.11ax (HE) PPDU Formats

802.11 Frame Types

The 802.11 frames provide the communication mechanism between MAC entities (Layer 2 connections) on the WLAN. This section provides an introduction to network framing and specific information about 802.11 frames of importance to the CWNA candidate.

Network Framing Overview

Network frames are a collection of organized or meaningful bits. Both devices (the sender and receiver of the frame) must understand the meaning of the bits. This mutual understanding is what we mean by the term *protocol*. In computer networking, a protocol is a standardized set of bits and communication procedures used to transfer information between two devices. The bits may be standardized by an industry organization like the IEEE or IETF, or they may be standardized in a proprietary manner by a vendor. Either way, they are meaningfully standardized and can be used for communications.

The phrase ***meaningful bits*** simply means that each bit or grouping of bits represents something. It may be an address, a mode of operation, an indicator of errors, or something else, but the bits have meaning.

Many frames are simply carriers of desired information on the network. The frame is sent in order to transfer the body of the frame (when considering data frames). The point of sending a data frame is not to send the frame itself, but the data contained in the frame. However, some organized method of sending that data must exist, hence we have frames. Other frames are meant to communicate information about the network itself, and do not carry user data. These are called management or control frames, in 802.11 terminology.

I find it helpful to begin with a simple example of a fictitious frame. Imagine that you want to have a way to send words between two devices. Words like "horse," "cat," and others. However, you have to define the target device and the source device to do so. In this simple example, we'll assume that's primarily what you have to do. Furthermore, assume that in this simple example, no more than four devices can exist on the network. Therefore, we need only two bits for the source and two bits for the destination based on the fact that two bits (for example, 01 or 10) can represent up to four values (0, 1, 2 and 3) and therefore four devices. Additionally, we need to specify how many characters are in the word. We will allow for up to 255 characters, so 8 bits will be enough for that part of the header (and we will use the extra bit creatively later). Our frame header and data would look like this (showing the actual word as text instead of binary bits for simplicity at this point):

```
SRC     DST     NUM_CHARS   DATA
##      ##      ########    word
```

Where SRC is the source address consisting of two bits and DST is the destination address consisting of two bits and NUM_CHARS is the number of characters in the data payload. Now, assume the following devices are on this simple network:

- Computer1 – 00
- Computer2 – 11
- Computer3 – 01
- Computer4 – 10

If Computer1 desired to send the word "horse" to Computer4, the frame would look like this (showing the actual word as text instead of bits for simplicity and including spaces for readability):

```
00 10 00000101 horse
```

At the Physical Layer, the network adapter would need to generate the signal for 0 twice, then the signal for 1 once, and then the signal for 0 again, followed by the signals for the bits in the NUM_CHARS field and those representing the word horse. Note that the binary value 00000101 is equal to decimal 5, which is the number of characters in the word horse. The receiving devices would all be listening for bits three and four in the frame to see if it is for them. Computer4 would see that bits three and four are equal to its own address (10), and then receive the rest of the data, in this case, the number of characters (to know how long to listen) and the word "horse." Computer2 and Computer3 would see that bits three and four are neither 11 nor 01 and know that they can ignore the rest of the data.

The benefit of knowing the source device is that the receiving device could respond with an acknowledgement frame to indicate that the transmitted frame was received as expected. That is, Computer4 could send back a standard acknowledgement message to Computer1. In our simple example, let's say that an acknowledgement is simply a set of four ones after the SRC and DST bits, and the NUM_CHARS bits set to all zeros. Computer4 would send the following frame:

```
10 00 00000000 1111
```

Notice how we used that extra value in the NUM_CHARS field creatively. We are saying that 00000000 is indicative of an acknowledgement frame. To take it one step further, if the word received was not recognized, the receiver may assume corruption has occurred, and respond with a frame indicating such. Let's say that a corrupt data notification is simply a set of four zeros after the eight zeros in the NUM_CHARS field. Computer4, in this case, would send the following frame:

```
10 00 00000000 0000
```

And there you have it. A very basic communication system with meaningful bits defined. Needless to say, 802.11 frames are more complicated than this, but this example should begin to help you understand the concept of network framing. This simple example illustrates the concept of a protocol — a standard way to communicate on the network. While this scenario is not as complicated or capable as protocols used in either Ethernet (802.3) or Wi-Fi (802.11), it illustrates the true simplicity behind frames and their use on the network. With this basic understanding, you can go further and easily understand the more detailed frame formats in Wi-Fi.

> This binary concept reminds me of my favorite T-shirt which reads, "Binary is as easy as 01, 10, 11!" and another favorite that reads, "There are 10 kinds of people: those who understand binary and those who don't!" The point, here, is that, if you don't know basic binary, you can't understand computer math and communications. If you need a refresher, read the book called "How Computers Do Math" by Clive Maxfield

Before we move on, a few terms should be understood, as they are often used when discussing frames and packets and the meaning of the bits used.

- **Most significant bit (MSB):** The bit having the highest value in binary notation. Also called the leftmost bit as it is usually the bit in the left position in binary notation (though this is not always true in the standards that define communication bits). The MSB is also called the high order bit. For example, in the 802.11 standard, the subtype field for frame type identification is specified with "the most significant bit (MSB) of the Subtype field, b7, is defined as the QoS subfield." This simply means that bit b7 (the identifier of the bit based on position), is equal to 1 for all QoS subtypes, and it is equal to 0 for all non-QoS subtypes in data frames or, stated differently, this bit determines if it is a QoS data frame or not. For example, all data frames are defined with a Type field value of 10, but the subtype field value of 0000 is standard data, and the subtype field value of 1000 is QoS data.

- **Least significant bit (LSB):** The bit having the lowest value, and the one that determines even or odd value when converted to decimal. Also called the rightmost bit as it is usually the bit in the right position in binary notation.

- **Most significant bit first (MSBF):** Indicates that, when receiving bits, the MSB is received first and the LSB is received last. Both 802.3 and 802.11 transmit the least significant bit first, instead. The opposite is LSB first (LSBF)

Here is an important example of these terms from the IEEE 802.11-2016 standard:

In control frames of subtype PS-Poll, the Duration/ID field carries the association identifier (AID) of the STA that transmitted the frame in the 14 least significant bits (LSB), and the 2 most significant bits (MSB) both set to 1. The value of the AID is in the range 1–2007.

This statement means that the two MSBs of the DurationID field determine if the field represents a duration or an AID. If it represents an AID, the two bits (remember, the MSBs) are set to 11. If it carries the duration of the frame, the bit (in this case, the single MSB) is set to 0. Further study of the standard reveals that the two MSBs can be set to 01 to represent PCF, but this will never be seen in production networks, as PCF is not used. Interestingly, the MSBs are bits 14 and 15 with bits 0-13 being the LSBs in this case, therefore, in this case, the MSBs are the rightmost bits and not the leftmost bits. However, 802 standards typically define bits from LSB to MSB and state that the LSB is transmitted first and the MSB is transmitted last, such as in 802.3-2012 Ethernet, clause 3.3.

Figure 7.7 shows the very high-level view of an 802.11 data frame. Consider the three primary components: header, payload, and footer.

The header includes 802.11-specific information. This information includes destination addresses, source addresses, delivery information (Is this a retry? Is the frame fragmented? etc.), QoS parameters and more. The payload is the upper layer information, including IP headers, TCP/UDP headers, and actual application data. The footer is used for error correction against the rest of the frame. In 802.11 networks, this is the frame check sequence (FCS).

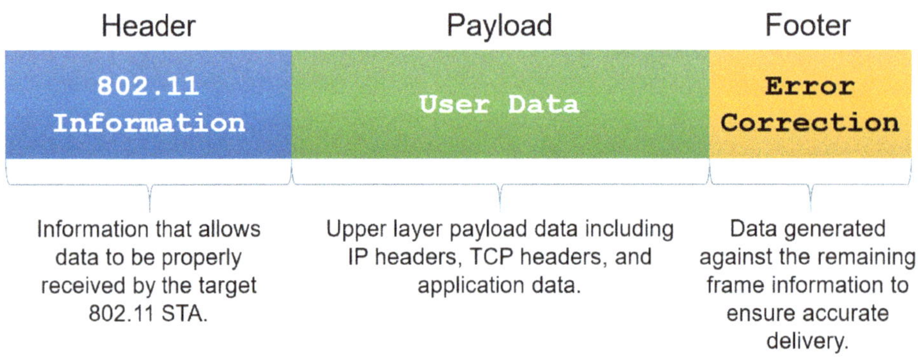

Figure 7.7: High Level View of an 802.11 Data Frame

802.11 General Frame Format

Now that you understand framing concepts in general, you can better understand 802.11 frames. This section explores the general frame format used in 802.11 framing. Figure 7.8 shows the 802.11 General Frame format, as it exists in 802.11-2016.

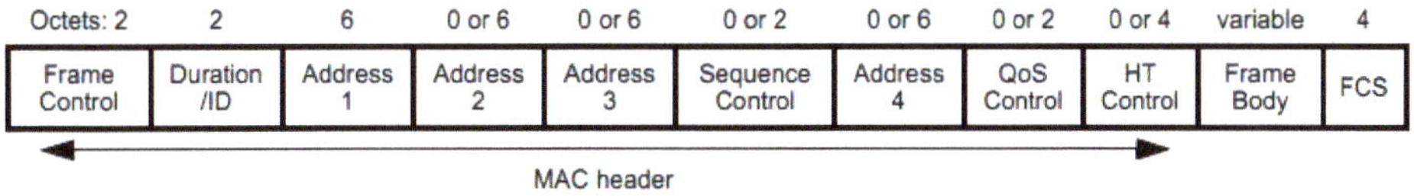

Figure 7.8: 802.11-2016 General Frame Format

The first field in the frame is the Frame Control field, which itself has many subfields. Figure 7.9 shows the Frame Control field as defined in 802.11-2016. The Frame Control fields sets important parameters for the frame. These parameters include the frame type and subtype, as well as the direction of the frame in a BSS.

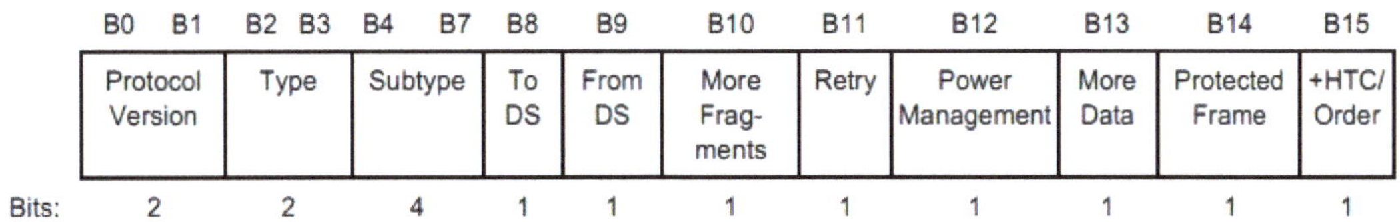

Figure 7.9: Frame Control Field

The Protocol Version bits are always set to 00, at this point, indicating that no incompatible version has been developed. If, in the future, an incompatible version is released, these bits can be used for that notification.

The Type and Subtype fields define the frame type (management, control or data), and the subtype. Table 7.2 lists the important valid values for these bits. 802.11-compatible protocol analyzers decode the frame type and subtype bits (subfields) and display the most appropriate of the three types and many subtypes in the decode view.

Type	Bits	Subtype	Bits
Management	00	Beacon	1000
Management	00	Association Request	0000
Management	00	Association Response	0001
Management	00	Authentication	1011
Management	00	De-authentication	1100
Management	00	Action	1101

Management	00	Action No Ack	1110
Control	01	Control Wrapper	0111
Control	01	Block ACK Request (BlockAckReq)	1000
Control	01	Block ACK (BlockAck)	1001
Control	01	PS Poll	1010
Control	01	RTS	1011
Control	01	CTS	1100
Control	01	Acknowledgement (ACK)	1101
Data	10	Standard Data Frame	0000
Data	10	Null Data Frame	0100
Data	10	QoS Data	1000
Data	10	QoS Null Data Frame	1110

Table 7.2: Type and Subtype Frame Control Subfield Values

The next subfields are the To DS and From DS bits. One bit each, they determine whether a frame is transmitted from a STA to the AP, from the AP to a STA, from one STA to another in an IBSS, or using the four-address MAC header format. The four-address format is used, per the standard, in a mesh BSS.

Figure 7.10 shows the To DS and From DS values appropriate, as defined in the 802.11 standard. While the direction of a frame can be defined by the source and destination address (MAC addresses), if you know the AP MAC address, the From DS subfield can be useful as a quick reference. If it is set to 1 and the four-address format is not in use, you know that the frame is traveling from the AP to a client STA. Additionally, any time you see a frame with both the To DS and From DS bits set to 0, you know it is a frame operating in an ad-hoc, or IBSS, network. This is useful in troubleshooting network problems.

To DS and From DS values	Meaning
To DS = 0 From DS = 0	A data frame direct from one STA to another STA within the same IBSS, a data frame direct from one non-AP STA to another non-AP STA within the same BSS, or a data frame outside the context of a BSS, as well as all management and control frames.
To DS = 1 From DS = 0	A data frame destined for the DS or being sent by a STA associated with an AP to the Port Access Entity in that AP.
To DS = 0 From DS = 1	A data frame exiting the DS or being sent by the Port Access Entity in an AP, or a group addressed Mesh Data frame with Mesh Control field present using the three-address MAC header format.
To DS = 1 From DS = 1	A data frame using the four-address MAC header format. This standard defines procedures for using this combination of field values only in a mesh BSS.

Figure 7.10: To DS and From DS Subfields (802.11-2016)

The More Fragments subfield is used to indicate whether the current frame is part of a fragmented frame or not. Fragmentation occurs based on the fragmentation threshold setting in the AP or client device. Fragmentation is used to increase the probability that a transmitted frame will get through in a high contention environment with hidden node issues or interference-laden environment. Sending a smaller frame results in a greater likelihood of the frame getting through before interference occurs. The fragmentation threshold defaults to 2346 to accommodate the maximum frame size without fragmentation. Interfaces allowing adjustment of this value provide the option to set it between 256 and 2346, per the standard. It should only be enabled in high-retry environments. You know fragmentation is being used when you see the More Fragments bit set to 1 in some frames.

The Retry field is useful in tracking frame transmission errors. If a frame is transmitted and the transmitter does not receive an ACK frame in response, the transmitting station will resend the frame using contention processes. When retransmitting, the frame will include the Retry field set to 1. This bit is used by the receiving STA to eliminate duplicate frames, but it can also be useful for tracking retries on the network, to see if they are causing performance issues.

The Power Management field is a 1-bit field indicating whether power management is used by the STA. The value of this field determines the mode in which the STA will operate after the completion of frame transmission. The Power

Management field is always set to 0 by an AP with its transmissions, as it does not enter power save mode.

The More Data field is used by the AP (or another STA in an IBSS) to indicate that more frames are buffered for that STA, so that it will not enter sleep mode. When set to 1 it indicates that the AP or STA is holding more frames for the STA to which the current frame is targeted. Additionally, when a STA sends a frame to the AP, and that frame includes the More Data ACK subfield of the QoS capability element (discussed more later) set to 1 and the AP has frames buffered for the STA with Automatic Power Save Delivery (APSD) enabled, the AP will set the More Data field to 1 if the ACK frame it sends back to that STA, so that the STA knows the AP has frames buffered for it.

Additional fields are defined beyond the Frame Control field, but these explanations are sufficient, as an introduction to WLAN framing. For the CWNA, it is most important to know the different categories of frames and frame type, and when and how they are used.

Basic Frame Types

802.11 frames of importance to the CWNA candidate include control, management, and data frames (of which there are two types: standard data frames and QoS data frames). This section provides an overview of the important frame types.

Control frames include:

- **Acknowledgement (ACK) frame** – used to acknowledge the receipt of another frame to the transmitter from the receiver.
- **BlockACK frame** – used to acknowledge a group of frames sent in a Block Acknowledgment negotiation with the AP.
- **RTS frame** – used to request access to the medium for a needed amount of time. Used as a protection mechanism.
- **CTS frame** – used to grant access to the medium for a needed amount of time. Also used for CTS-to-Self. Used as a protection mechanism.

Management frames include:

- **Authentication frame** — used to request and grant/deny authentication using Open System authentication.
- **Association request frame** — used to request association with an AP.
- **Association response frame** — used to grant/deny association with an AP.
- **Beacon frame** — used to announce a BSS as a broadcast and the parameters of the BSS.
- **Probe request frame** — used to seek for active BSSs within a channel.
- **Probe response frame** — used by APs to respond to a probe request with similar information to that in a Beacon frame.
- **De-authentication frame** — used to remove the authentication status (and association status, if it is active) from an AP.
- **Disassociation frame** — used to remove the association status from an AP.
- **ATIM frame** — used for power management communications in an IBSS; not required in a BSS as the information is in the AP Beacon frame.

Data frames include:

- **Data frames** — standard data frames with no 802.11e/WMM QoS information.
- **QoS data frames** — data frames including 802.11e/WMM QoS information.

As you can see, many different frame types are used in 802.11 WLANs and this is just a partial listing. The full details of all the frame types can be found in the 802.11-2016 standard.

Beacon Frames

The *Beacon frame* is transmitted periodically to allow STAs to locate and identify a BSS and to display BSS parameters. The Beacon also conveys information about frames that may be buffered during times of low power operation, so that STAs can identify buffered frames awaiting them in the AP and request their delivery.

Beacon frames include sections containing the following important information:

- Signal information (actually from the radio and not in the frame)
- Beacon interval
- SSID
- Supported rates (for 802.11b/g/a PHYs)
    - Each rate is defined as an 8-bit value (one octet)
    - Bit 7 (the first bit when reading from left to right) determines if it is a basic rate (mandatory) or supported rate (not required for connection)
        - When set to 1, it is a basic rate; when set to 0, it is a supported rate
        - Therefore, basic rates always have a value that is 128 greater than the rate
        - Additionally, these rates are encoded in the frames as double the rate. For example, a 9 Mbps rate would be encoded as 18, for reasons beyond the scope of this course
    - The remaining 7 bits (0-6) provide the actual value or the data rate (simply the data rate in binary form)
- Channel in use
- Country information (for regulatory compliance)
- ERP or OFDM information (for 802.11g and 802.11a capabilities)
- RSN information element (RSNIE) for security parameters
- QBSS element to indicate whether 802.11e QoS is supported or not, and the parameters for it
- HT sections for 802.11n capabilities
- VHT sections for 802.11ac capabilities

- Vendor-specific sections, depending on the vendor implementations

Figure 7.11 shows a partial decode of an 802.11 Beacon frame from an 802.11ac AP.

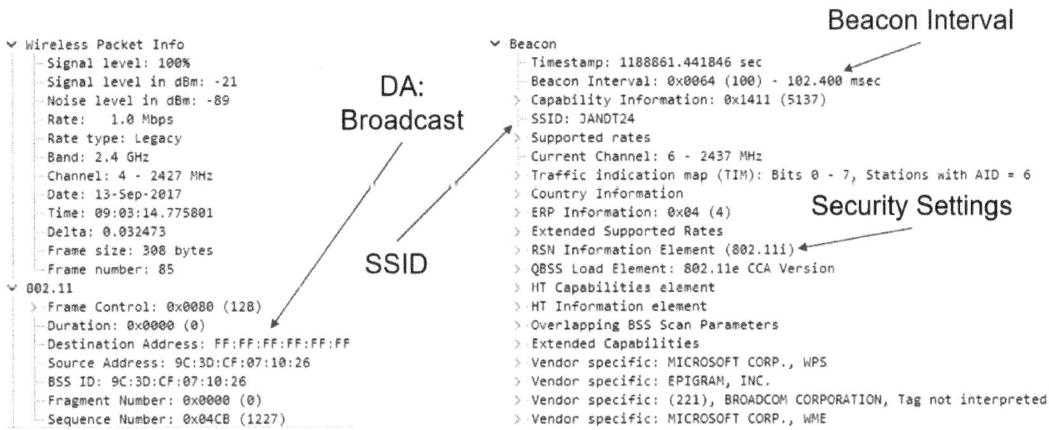

Figure 7.11: Beacon Frame Decode

Authentication Frames

802.11 authentication is either Open System Authentication or Shared Key Authentication. Shared Key Authentication is deprecated in the standard and is not secure. It should never be used today. Open System Authentication is performed before either Pre-Shared Key or 802.1X/EAP authentication.

Open System Authentication precedes 802.11 association and is a simple four-frame exchange.

- Authentication frame sent from client STA to AP
- ACK frame sent from AP to client STA
- Authentication frame sent from AP to client STA
- ACK frame sent from client STA to AP

Assuming the Open System authentication is successful, the status code in the authentication response will be 0, indicating success. In the authentication frames, the frame with a transaction sequence number of 1 is the request frame and the frame with a transaction sequence number of 2 is the response frame. Wireshark decodes the transaction sequence number field as "Authentication SEQ." Figure 7.12 shows authentication frame decodes in Wireshark.

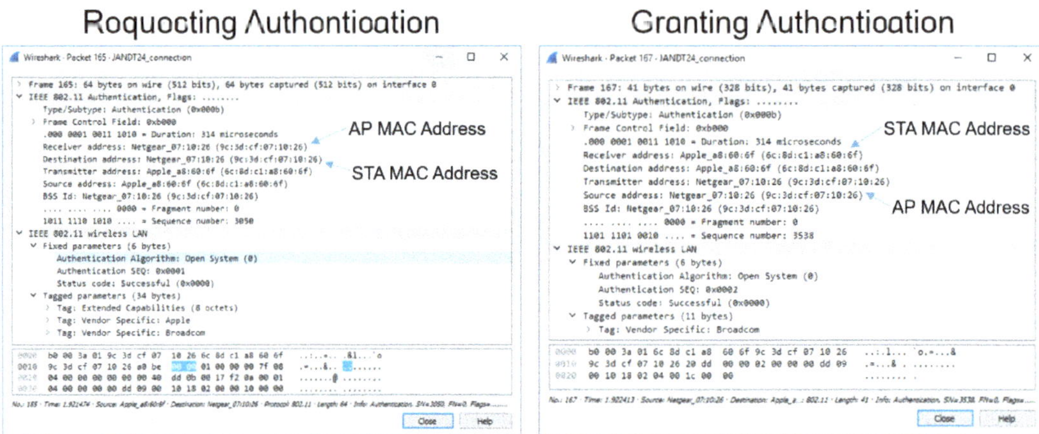

Figure 7.12: Authentication Frame Decodes

Association Request and Response Frames

The Association Request frame is used by a client station to request an association with a BSS.

The Association Response frame is used to signal the success or failure status of the Association Request to the requesting mobile station. The mobile station must have been authenticated previously for the Association Request to be "successful."

Figure 7.13 shows association request and response frame decodes in Wireshark.

ACK and Block ACK Frames

Most 802.11 frames require an ACK frame as a response to indicate successful delivery. Some frames, like Beacon frames, Probe Request frames, Probe Response frames, and other notification-only frames do not require an ACK.

Two types of acknowledgements exist in 802.11 networks: ACK frames and Block ACK frames. ACK frames indicate reception of a single frame. Block ACK frames indicate reception of a block or series of frames and require only one acknowledgement for multiple frames. Block ACK frames cannot simply be used, but must be setup between the wireless STAs using a block ACK request and response procedure.

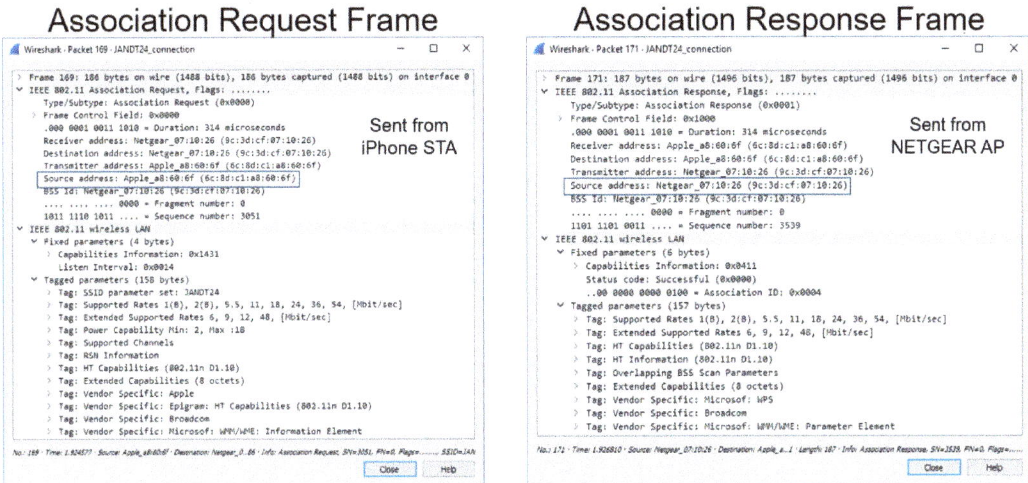

Figure 7.13: Association Request and Association Response Frame

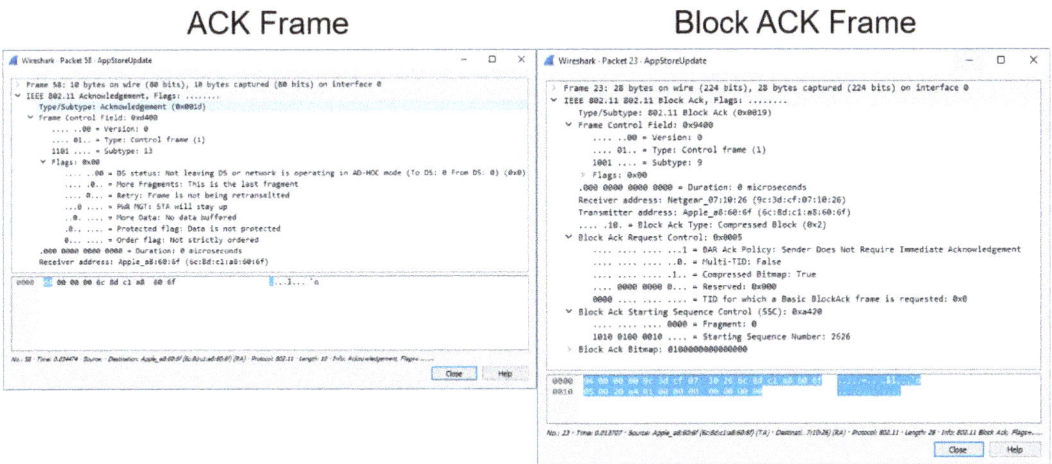

Figure 7.14: ACK and Block ACK Frames

Locating and Connecting to WLANs

Locating and connecting to WLANs is a three-step process for secure implementations. First, you must locate a target WLAN. Second, you must authenticate and associate to the WLAN with Open System Authentication. Third, you must authenticate to the WLAN using either pre-shared key or 802.1X/EAP authentication, and then install encryption keys for communications (don't worry, you don't install them, the system does it for you). In public open networks, only the first two steps are required.

Locating WLANs

Client devices use either, or both, passive scanning or active scanning to locate WLANs. Passive scanning relies on Beacon frames, and active scanning uses Probe Request and Probe Response frames, as illustrated in Figure 7.15.

When passive scanning is used, the client views the detailed information in the Beacon frame to determine if the source AP is a good target for connection. When active scanning is used, the client views the information provided in the Probe Response frame, which is similar to the Beacon frame information. Assuming the client determines a good target AP and BSS, it will move onto the Open System Authentication phase.

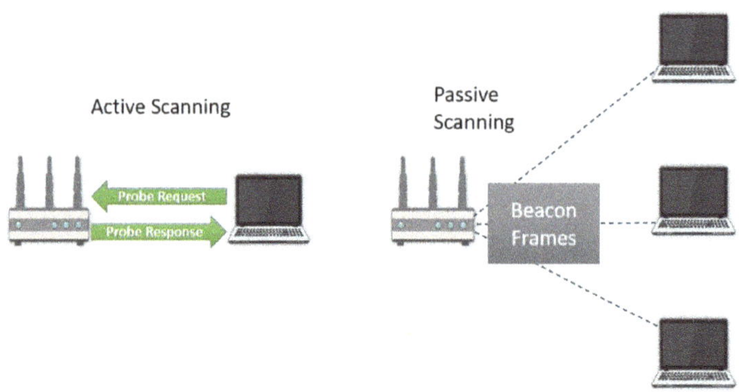

Figure 7.15: Passive and Active Scanning

Open System Authentication

Today, Open System Authentication is the only initial 802.11 game in town. Shared Key Authentication was used in the past, but it is terribly week and should not be used today. It transmits a challenge as clear text, which is then encrypted by the client with the WEP key, making cryptanalysis simple. The good news is that Shared Key Authentication is not even on option for configuration on most modern APs.

Open System Authentication, followed by 802.11 association gets the client through to State 3, in the 802.11 State Machine documented in Figure 7.16.

Open System Authentication uses a four-frame exchange to go from unauthenticated and unassociated (State 1) to authenticated and associated (State 2). Next, association request and association response frames are exchanged to go from State 2 to authenticated and associated (State 3). When you connect to an open network without WPA2-Personal or WPA2-Enterprise, the connection is complete at State 3. If Pre-Shared Key (PSK — WPA2-Personal) or 802.1X/EAP (WPA2-Enterprise) is used, you must be authenticated and complete the 4-way handshake to get to State 4 and begin using the network with security.

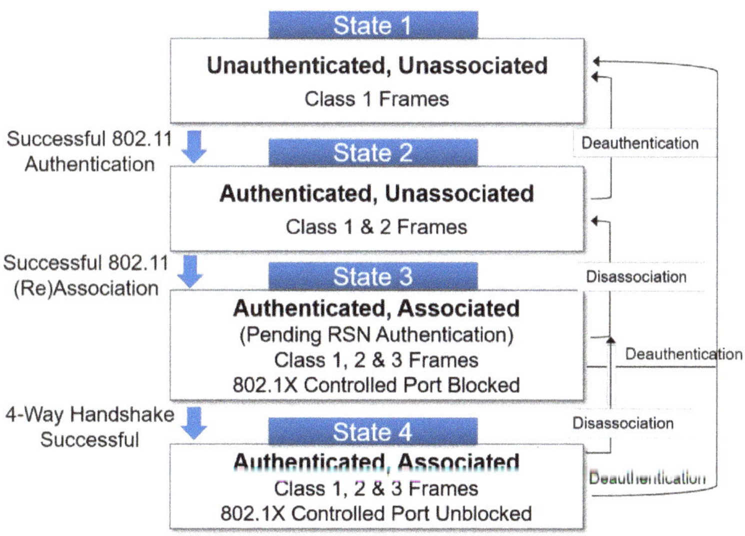

Figure 7.16: 802.11 State Machine + 802.1X/EAP or Pre-Shared Key

Pre-Shared Key or 802.1X/EAP Authentication

In addition to improved encryption and integrity algorithms, the 802.11 standard specifies the use of 802.1X port-based access control and Extensible Authentication Protocol (EAP) to provide user authentication and dynamic key distribution.

- EAP is a Layer 2 authentication protocol used by 802.3 and 802.11 as a flexible replacement for PAP and CHAP under PPP

802.1X restricts access to the network until a station has been authenticated by an authentication authority, usually residing within the wired network segment.

- Access to the network is controlled through the use of Controlled and Uncontrolled ports, which are logical entities on the same physical connection

- Prior to successful authentication, the client station may only communicate over the Uncontrolled Port

The Authentication Server (AS) is usually implemented by means of a RADIUS server. The AS may host the user database, or may communicate with an external user database, to authenticate user credentials and profiles. Two additional entities are defined:

- Supplicant — Client station wishing to join the BSS
- Authenticator — Access Point or WLAN controller, which forwards messages between the Supplicant and the AS

Figure 7.17 shows the basic 802.1X/EAP architecture with an additional fourth component — a user directory. Such directories are often used in combination with an AS.

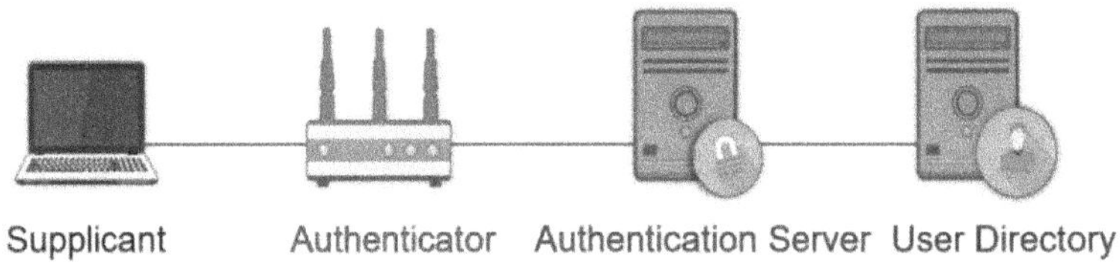

Figure 7.17: Basic 802.1X/EAP Architecture

Home and Small Businesses are not expected to have a RADIUS infrastructure, so in cases like this, a manually inserted Passphrase is allowed. The Passphrase is converted by the station to a 256-bit Preshared Key (PSK), using a standard mapping algorithm found in 802.11i, annex H.4. The Preshared Key (PSK) acts as the Pairwise Master Key (PMK), which is always 256-bits long. With this method, if the Passphrase is compromised, then the security of the wireless segment is negated. Both TKIP/RC4 and CCMP/AES can be used with passphrases. Key derivation and distribution is identical with both cipher suites.

PSK is also often used in enterprise deployments for specialty devices like VoIP handsets, barcode scanners, video cameras and more. Figure 7.18 illustrates the basic components of a PSK implementation (WPA2-Personal).

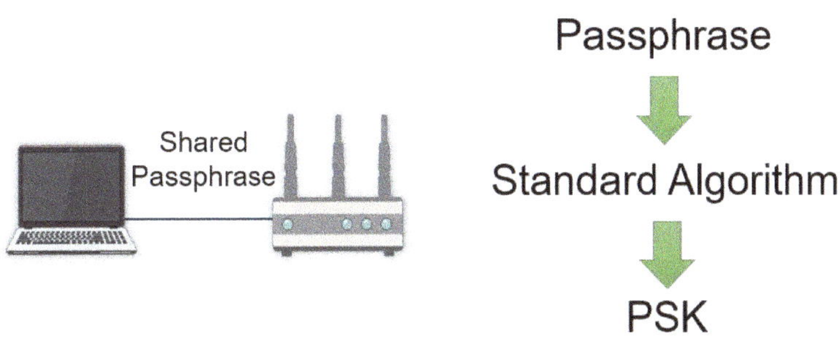

Figure 7.18: PSK Components

After either PSK or 802.1X/EAP authentication, the 4-way handshake must occur. The purpose of the 4-way handshake is twofold. First, it is used to generate and install unicast encryption keys on the AP and client. The unicast keys are used to encrypt traffic that passes only between the client and the AP. These keys are never transmitted across the wireless medium, but data (information) is transmitted to be used on each device (the client and AP), in the generation of the encryption keys. This data is simply numeric information called a NONCE (number used once). The first two messages of the 4-way handshake are used to communicate these NONCEs between the AP and client, with the AP sending its NONCE first (called the ANONCE), and the client sending its NONCE second (called the supplicant

NONCE or SNONCE). After these two messages are exchanged, the client and the AP have all the same information, and they can each generate the same unicast encryption keys.

The unicast encryption key is included in what is called the Pairwise Transient Key (PTK). The PTK includes three primary components: the Key Confirmation Key (KCK), the Key Encryption Key (KEK), and the Temporal Key (TK). The TK is used to encrypt actual data on the WLAN. The KEK is used to encrypt the group key during transmission. The KCK is an integrity check against keys in the cipher suite.

The second purpose of the 4-way handshake is to transmit the group encryption key (called the group temporal key (GTK)) from the AP to the client. In message three, the GTK is encrypted using the already-generated unicast key and transmitted to the client. In message four, the client simply confirms to the AP that the GTK is installed and ready to use. Figure 7.19 illustrates the 4-way handshake at a high level.

Figure 7.19: 4-Way Handshake

BSS Selection

Clients use various criteria when deciding to which BSS they will connect. These include:

- **Signal strength from the AP** — stronger signals are usually preferred.
- **AP settings** — finding an AP with the right SSID and supporting the data rates and other parameters matching the client is key.
- **2.4 GHz versus 5 GHz** — the client may be configured to prefer 5 GHz networks over 2.4 GHz.

Additional factors may be considered but given that these client algorithms are so tightly guarded by vendors, it is impossible to know every factor used in the decision process.

Chapter Summary

In this chapter, you learned the basics of 802.11 frame and initial connection to the WLAN. To understand this, you had to learn some important 802.11 terms, like MSDU, MPDU, PSDU and PPDU, and also learn the 802.11 framing structures. In addition, you explored the process used to locate and connect to a WLAN, including scanning, authentication, association, 802.1X/EAP or PSK authentication, and BSS selection. Now that you're connected to a network, in the next chapter, you will learn how devices communicate on that network.

Points to Remember

Remember the following important points:

- The MSDU and MPDU are associated with the MAC sublayer of the Data-Link layer (Layer 2) in the OSI Model.
- The PSDU and PPDU are associated with the Physical layer (Layer 1) of the OSI Model.
- The Guard Interval (GI) is a space of time between symbols within in 802.11 frames.
- The short GI is 400 ns and the long GI is 800 ns.
- Interframe spaces (IFS) are used between frames.
- SIFS is always used before ACK frames.
- DIFS is used before Non-QoS data frames
- AIFS is used before QoS data frames.
- Frames may be fragmented before transmission to reduce the likelihood of interference with the transmission.
- The 802.11ac amendment provides a single PHY header that is compatible with all other 5 GHz PHYs (both HT and OFDM).
- The 802.11 general frame format includes a header, data payload (when required) and frame check sequence (FCS).

- Beacon frames are transmitted by APs to announce the presence of a BSS and the parameters of that BSS.
- Probe request and probe response frames are used in the active scanning process.
- Association request and association response frames are used to associate with an AP and move to State 3 of the 802.11 state machine.
- Authentication frames are used for Open System authentication with the AP.
- After reaching State 3 of the 802.11 state machine, if 802.1X/EAP or PSK is in use, they must successfully authenticate, and then perform the 4-way handshake for encryption key generation and installation.

Review Questions

1. What is the term used for the information received by the PHY and ready to be processed by the PHY (PLCP) for transmission?
   a. PSDU
   b. PPDU
   c. MSDU
   d. LSDU

2. What does a PPDU have that a PSDU does not?
   a. Destination MAC address
   b. A PLCP header and PHY preamble or training information
   c. Frame Control field
   d. Data payload

3. The guard interval is a small space between what other 802.11 components?
   a. Frames
   b. IFS
   c. Symbols
   d. DIFS

4. What was the length of the original guard interval before it was shortened with the ratification of 802.11n?
   a. 800 ms
   b. 800 ns
   c. 400 ms
   d. 400 ns

5. What is the name of the first field in the 802.11 general frame format?
   a. Duration/ID
   b. Frame Control
   c. FCS
   d. None of these

6. What kind of frame must be transmitted by a STA when that STA successfully receives a data frame?
   a. Beacon
   b. Authentication
   c. Probe response
   d. ACK

7. When active scanning is used, what frame types are involved in the exchange?
    a. Beacon and Acknowledgement
    b. Probe Request and Probe Response
    c. Beacon and Authentication
    d. Beacon and Association

8. How is State 2 of the 802.11 state machine described?
    a. Unauthenticated and associated
    b. Authenticated and associated
    c. Authenticated and unassociated
    d. None of these

9. After 802.1X/EAP or PSK authentication is successful, what needs to transpire?
    a. Open System Authentication
    b. 802.11 association
    c. 4-way handshake
    d. BSS selection

10. Which one of the following is not commonly used by clients in BSS selection, within known algorithms or metrics?
    a. AP vendor ID
    b. AP Beacon signal strength
    c. Desired SSID
    d. Band preference

Review Answers

1. **A is correct.** The PHY, more specifically, the PLCP receives the PSDU, though the MAC passed down the MPDU. That is, the PSDU is the MPDU received in the PLCP for processing.
2. **B is correct.** A PLCP header and PHY preamble or training information.
3. **C is correct.** The GI is a space between symbols and not between frames.
4. **B is correct.** 800 ns. It was shortened to 400 ns for devices that support the SGI.
5. **B is correct.** The Frame Control field is the first field in the general frame format.
6. **D is correct.** Acknowledgement (ACK) frame.
7. **B is correct.** Probe Request and Probe Response frames are used in the active scanning process.
8. **C is correct.** Authenticated and unassociated.
9. **C is correct.** The 4-way handshake should occur immediately after 802.1X/EAP or PSK authentication succeeds.
10. **A is correct.** AP vendor ID is not used. The others commonly are.

Chapter 8 — 802.11 Channel Access Methods

802.11 networks use RF, which is an open shared medium. For this reason, some access method must be used that allows devices to share the medium. The access method used in 802.11 networks are the Distributed Coordination Function (DCF) and Enhanced Distributed Channel Access (EDCA). This chapter provides details on these access methods. Additionally, channel width operations, HT and VHT operation modes, HT and VHT protection mechanisms and power management options are explained.

Distributed Coordination Function (DCF)

Wired networks, such as Ethernet, can detect collisions. Therefore, they use Carrier Sense Multiple Access with Collision Detection (CSMA/CD), if required. In reality, with the modern use of switches, full duplex implementations do not require CSMA/CD as a dedicated link is used for traffic up and down, to and from switch ports, and the switch controls, when and how traffic moves out of the switch. Figure 8.1 illustrates CSMA/CD.

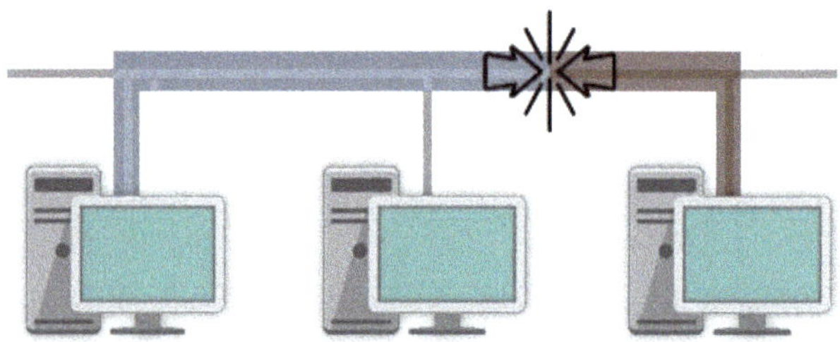

Figure 8.1: CSMA/CD

Wireless networks cannot detect collisions. Therefore, while they still perform carrier sense and have multiple STAs accessing the medium, they must attempt to avoid collisions. This is accomplished through a distributed algorithm, based on information seen on the medium by the STAs. The general concept is CSMA with Collision Avoidance (CSMA/CA) and is illustrated in Figure 8.2.

DCF is the basic implementation of CSMA/CA in 802.11 networks. EDCA is the enhanced version that also provides for QoS capabilities.

Using the foundational DCF access method, the same coordination function logic is active in every station (STA) in a basic service set (BSS), whenever the network is in operation. Stated differently, each station within a DCF follows the same channel access rules. This method is contention-based, which means that each device competes with other devices in the same channel to gain access to the wireless medium. After a transmission opportunity is obtained and observed, the contention process begins again. As the original 802.11 network access method, DCF is the simplest channel-access method; however, being the first access method, it lacks support for quality of service (QoS). In order to maintain support for non-QoS devices in QoS-enabled networks, support for DCF is required for all 802.11 networks.

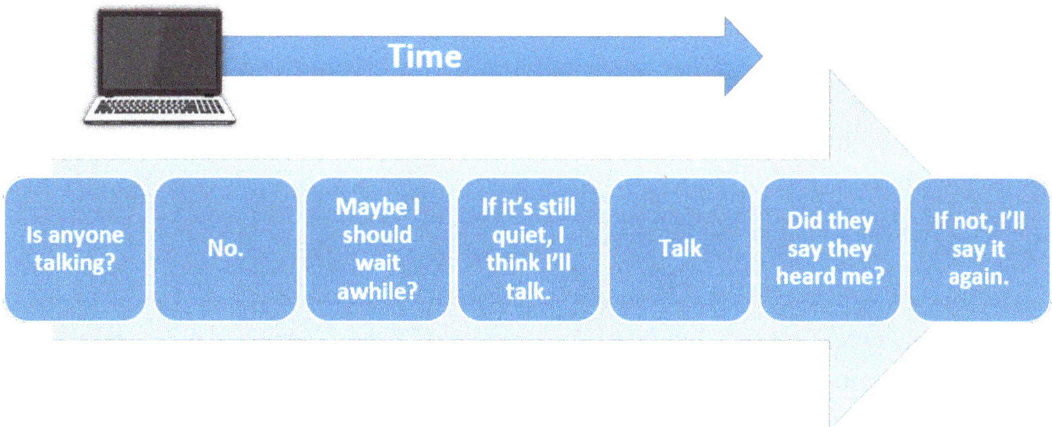

Figure 8.2: CSMA/CA

Figure 8.3 shows the basic components of DCF from a single STAs perspective. It includes interframe spaces, backoff timers and carrier sense throughout.

Figure 8.3: DCF Single STA Perspective

You've learned about interframe spaces in earlier chapters. They are brief delays that must be implemented before contending for access to the medium. You could say they are not part of contention, but they precede it. The different IFS durations (among DIFS, SIFS, EIFS, etc.) allow for higher priority frames, like ACK frames, to be transmitted before lower priority frames, like data frames.

You may wonder why an ACK frame should have a higher priority than a data frame. The answer is that if the station that sent the frame requiring an ACK (for delivery verification) does not receive the ACK, it will retransmit the data frame, wasting precious air time. Therefore, with DCF, ACK frames are preceded by a SIFS, and data frames are preceded by a DIFS. SIFS is shorter than DIFS, and so the ACK frame gets priority.

Even while waiting for the IFS to expire, STAs perform carrier sense. If an 802.11 frame is seen while waiting for the IFS to expire, the STA can set the network allocation vector (NAV) timer based on the Duration value in the frame being transmitted, wait for the current frame to complete and then count down the NAV timer before it continues. Figure 8.3 does not list the NAV timer, but it can kick in anything carrier sense is processing.

Carrier sense uses both physical carrier sense and virtual carrier sense. Physical carrier sense is called Clear Channel Assessment (CCA). It is used to see if the wireless medium is busy with either an 802.11 frame, or significant energy from a non-802.11 transmitter.

Carrier Sense (CS) is the part of CCA that detects and measures the signal strength of 802.11 frames. If the frame is below a specific threshold, which the CWNA need not memorize, the STA can ignore the frame. If the frame is above that signal threshold, the STA must read the Duration (if possible) and set its NAV. If it cannot read the NAV (because the frame is too weak to decode the data rate), it must still wait for the medium to become silent.

Energy Detect (ED) is the part of CCA that detects and measures the raw RF energy seen on the channel. If this energy is above a specific threshold (which is higher than the CS threshold), the STA must wait until the energy is gone to transmit.

As stated, the ED threshold is higher than the CS threshold. That is, an 802.11 STA will attempt to transmit a frame, even if RF energy is detected at a significant level. Consider that the STA is sending the frame to a different location, and it is likely that the RF energy that is local to the STA will not be local to the receiver location at the same levels. Therefore, it is logical to transmit in the presence of non-802.11 RF energy, while it is not acceptable to transmit in the presence of 802.11 RF energy at the same levels.

Virtual carrier sense is achieved using the Duration value in 802.11 frames and a local STA timer called the NAV. The duration value in the MAC header indicates the time required to complete the transmission opportunity after the current (the frame in which the duration value resides) frame is completed.

The Network Allocation Vector (NAV) is set in 802.11 STAs, based on the duration values seen in 802.11 frames. For example, if a frame is seen with a duration value of 143 microseconds, the STA knows that it should be silent for 143 microseconds after the current frame is completed. This allows for interframe spaces and acknowledgement required to complete the entire frame transmission process. Figure 8.4 shows the Duration value (of 48 microseconds) in an example 802.11 frame.

If physical and virtual carrier sense indicate that the channel is clear, a STA may proceed through the IFS, but it will then land on a random backoff timer, as indicated in Figure 8.3.

During contention, and after the interframe space, stations must wait a randomized period of time before they can transmit. The "random" time that must be waited is dictated by a range of potential values called the *contention window*.

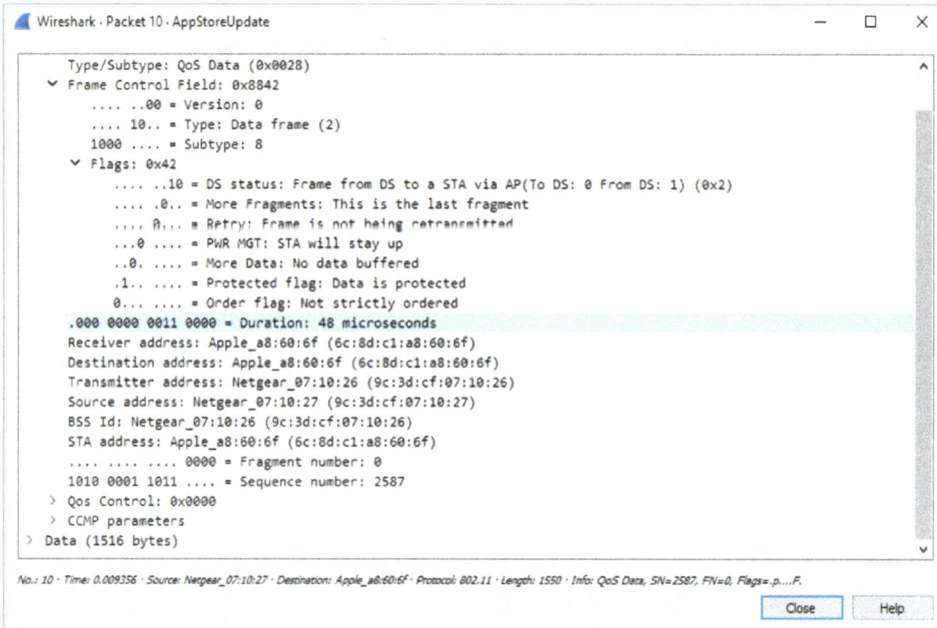

Figure 8.4: Duration Value in an 802.11 Frame

The contention window (CW) is a range of values from which STAs randomly pick a number. The random number represents the length of time that the station must wait before it can transmit. Of course, it's more detailed than that.

The CW (range of numbers) always begins at 0. The range of numbers is always 0-x. The x changes based on a number of variables, including the PHY technology, whether this is the first attempt at a transmission, or a retry (and how many previous retries there have been), as well as QoS priority when EDCA is in use.

802.11b has the largest default contention window, while 802.11a/g/n/ac have similar CW values.

It is important to understand that the upper limit of the CW is not the actual value that will be selected by a STA. For example, if the CW range is from 0-31, the STA may select 1, 6, 13, 17, 25, 31, or any other number within that range. It selects this number randomly to attempt to avoid collisions with other stations. The selected number represents the number of "slots" or slot times (as a point of reference,

802.11b has a 20μs slot time) that the wireless medium must be idle before the STA can transmit a frame (i.e. wins contention).

The series of events are as follows:

- When a STA has a frame to transmit, it randomly picks a value from the CW.
- STAs begin to countdown that value, which is called the backoff time.
- When the WM is idle (as determined by carrier sense mechanisms) for a slot time, the backoff timer decrements by one.
- When the WM is busy, the STAs must wait, and start the process over (carrier sense, interframe space, backoff countdown (but the back off counter doesn't increase)).
- When the backoff timer reaches zero and the WM is idle, STAs may transmit.

In the end, DCF offers four main components:

- Interframe space
- Duration/ID
- Carrier Sense
- Random Backoff Timers

DCF allows fair access to the medium for all stations. DCF does not prioritize access for stations based on applications, and DCF offers fair use of the medium (only one transmission at a time). Before being allowed to transmit data frames, stations must have:

- DIFS = expired | interframe space
- NAV = 0 | Duration/ID
- CCA = idle | carrier sense

- Slots = 0 | random backoff timers

Enhanced Distributed Channel Access (EDCA)

First, there is some confusion over whether this is called EDCA, EDCF or EDCAF. To clear up the confusion, first know that there is no such thing as EDCF in the standard. That removes one point of confusion. EDCA and EDCAF are referenced in the standard.

EDCA is the replacement for DCF. It is the overall process defined for channel access. EDCA Function (EDCAF) is more granular in that each access category (discussed more later) has an EDCAF to determine when a waiting frame in that access category queue may be transmitted on the wireless medium. EDCAF uses the EDCA process to ultimately gain access to the medium for frames in its queue.

The 802.11e amendment originally introduced EDCA, and also introduced HCCA, which is not implemented in WLANs. HCCA was a Hybrid Coordination Function (HCF) Controlled Access (HCCA) solution that used a central control for medium access. EDCA was contention based, like DCF, and so it become the popular method implemented in 802.11 devices and incorporated into the WMM certification by the Wi-Fi Alliance.

With DCF, waiting frames are in a kind of First-In-First-Out (FIFO) queue within the local STA, and no priority is given to one frame type over another in a typical sequence. EDCA added priorities for different frame types. Applications can tag IP packets with priorities, and these priorities can be used to assign Layer 2 MAC 802.11e priorities to the frames for transmission. For example, voice frames can be tagged with a higher priority than a frame that simply contains e-mail transmissions. Video frames can be tagged with a higher priority than a frame that simply contains FTP transmissions.

EDCA uses four independent EDCAFs to provide priority queues for transmitted traffic. These four EDCAFs are defined by access categories (ACs). The defined ACs are:

- AC_VO – voice traffic
- AC_VI – video traffic

- AC_BE — best effort traffic
- AC_BK — background traffic

BEYOND THE EXAM: Wait, Six Queues?

A lesser-known fact of the EDCA operations in the 802.11-2016 standard is that the EDCA implementation can use either four queues or six queues. When four queues are used, there is a one-to-one mapping with the four ACs. When six queues are used, two alternate queues are created. One for voice and the other for video. For example, the AC_BK queue is simply called BK. The AC_BE queue is called BE. However, if alternate queues are used, AC_VI has one queue called VI and another called A_VI (A is for alternate). Also, AC_VO has one queue called VO and another called A_VO.

Within the standard, a parameter called dot11AlternateEDCAActivated exists. If this parameter is set to true, the alternate queues are used. If the parameter is set to false, only the four main queues are used. When the alternate queues are in use, the Intra-Access Category Prioritization (IACP) is used to provide fine-grained prioritization in the communications. When six queues are used, only four EDCAFs exist. One EDCAF decides between the two AC_VO queues, and one decides between the two AC_VI queues. The goal of IACP, for example, is to indicate that not all voice frames are the same priority, and not all video frames are the same priority. Therefore, within the two higher priority EDCAFs (voice and video), you can sub-prioritize into two additional priorities each. That is, within the two ACs (intra-access category), you can perform prioritization.

So, why do so few know about this? The answer is simple. It is not implemented in many (if any) production solutions, and it was introduced in the 802.11aa-2012 amendment for MAC Enhancements for Robust Audio Video Streaming. Sometimes those IEEE folks can sneak up on you.

In production 802.11 WLANs today, the four ACs map to four queues that are managed by four EDCAFs, as seen in Figure 8.5. The MSDUs are processed and tagged with user priorities that map to the ACs. There are eight user priorities (0-7) that map to the ACs. The user priority is assigned to the MSDU in the layers above the 802.11 MAC, and the 802.11 MAC maps them to the ACs. The mapping is as indicated in Table 8.1.

User Priority (same as 802.1D UP)	AC	Designation
1,2	BK	Background
0,3	BE	Best Effort
4,5	VI	Video
6,7	VO	Voice

Table 8.1: User Priority to AC Mapping

In the end, EDCA works by placing frames in the different queues based on their priority. The queues have priority from lowest to highest as BK, BE, VI and VO. Therefore, voice frames have the highest priority and background frames have the lowest. The priority is achieved by using different IFS delays for each AC, and by using narrower contention window ranges for higher priority ACs. The end result is a probabilistic priority mechanism that makes it more likely that voice frames will be transmitted before video, video before best effort, and best effort before background.

Figure 8.5 shows the architectural overview of EDCA. Note that, in this diagram, the traditional four queues exist: VO, VI, BE, and BK. An EDCAF is assigned to each queue and implements the EDCA parameters (AIFSN, CWmin, CWmax) for each queue. In a given TxOP, any queue may win the contention based on random backoff timers, but the VO and VI queues are more likely to win more often; hence, a probabilistic QoS mechanism.

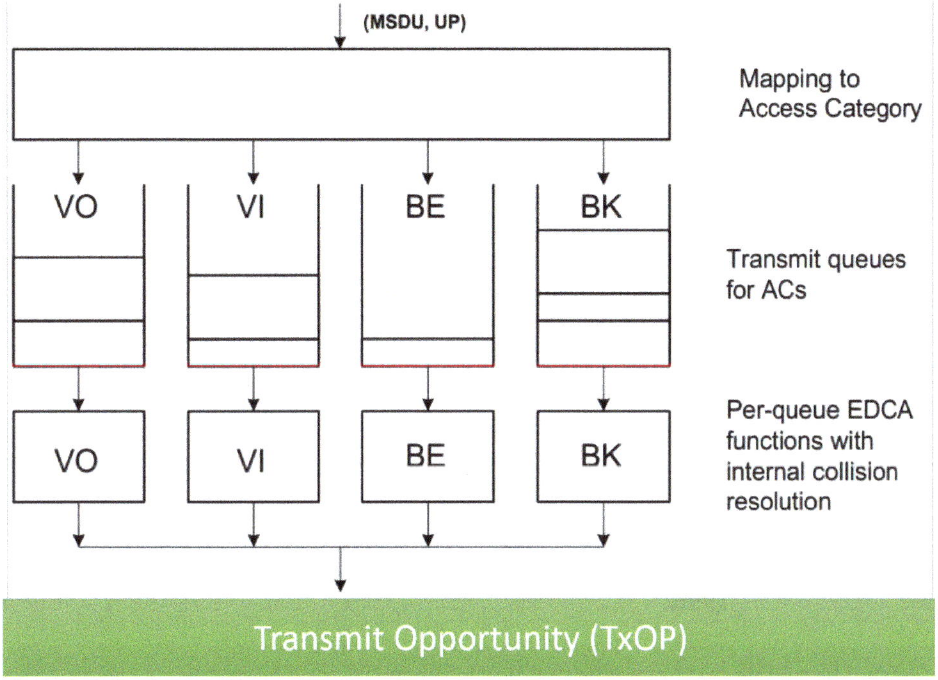

Figure 8.5: AC Queues

Channel Width Operations

HT devices support 20 MHz and 40 MHz channels. The 40 MHz channels are called bonded channels and should not be thought of as one 20 MHz channel and then the next non-overlapping 20 MHz channel (for example, in 2.4 GHz), because they actually just use the next 20 MHz of frequency space. Of course, in 5 GHz, this is the next non-overlapping channel. HT 40 MHz channels are often defined as +1 or -1 configurations. This simply means that the secondary channel is either the next 20 MHz (+1) or the preceding 20 MHz (-1) around the primary 20 MHz channel. Figure 8.6 illustrates the 802.11n channel width operation options.

Figure 8.6: HT (802.11n) Channel Width Operations

802.11ac supports wider channels of 80 and 160 MHz. I will not address 160 MHz channels in-depth here, as there is no practical use for them in the vast majority of current WLAN deployments. Figure 8.7 illustrates the 802.11ac channel width operation options.

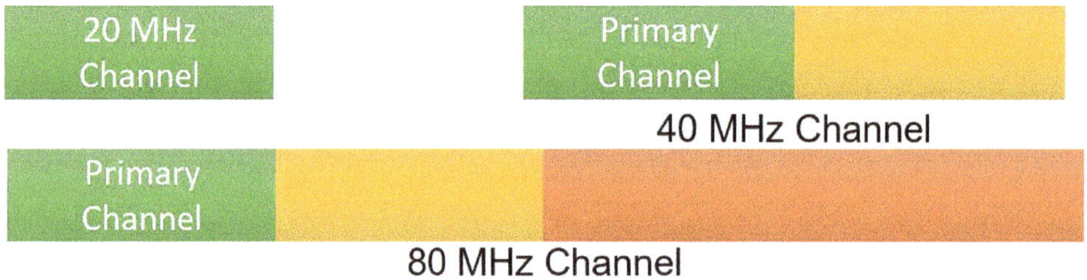

Figure 8.7: VHT (802.11ac) Channel Width Operations

It is important to know that, with both HT and VHT, even when 40 MHz and larger channels are used, a 20 MHz primary channel always exists. The primary channel is used for BSS operation management, and for backward compatibility with 20 MHz-only client devices. Therefore, when looking at a spectrum capture (with a spectrum analyzer) of a 40 or 80 MHz channel, you will typically notice that 20 MHz of the channel is more utilized than the rest. That is the primary channel.

HT and VHT Operation Modes

HT and VHT devices support different modes of operation. HT devices can operate in three modes:

- **Non-HT** – These frames are formatted as OFDM (5 GHz) or ERP (2.4 GHz) frames, and use a FORMAT parameter of NON_HT. This mode is used to operate an 802.11a or 802.11g device.
- **HT-Mixed** – These frames are formatted with a preamble compatible with OFDM (5 GHz) or ERP (2.4 GHz) and use a FORMAT parameter of HT_MF. This mode is used for backward compatibility, while still allowing transmission of MPDUs at higher 802.11n data rates.

- **HT-Greenfield** — These frames are formatted to work only with 802.11n (and now 802.11ac) devices and use a FORMAT parameter of HT_GF. In Greenfield mode, down-level STAs (802.11a/b/g) are not supported in the BSS.

The chosen operating mode is specified in the AP configuration.

VHT devices support all three HT modes (for backward compatibility), and they support the new VHT frame format and use the FORMAT parameter VHT. This mode is compatible with 802.11a/n devices (remember, 802.11ac only operates in 5 GHz), because of the non-VHT first portion of the preamble.

HT and VHT Protection Mechanisms

HT and VHT devices use protection mechanisms to coexist with older PHYs. ERP also uses such mechanisms to coexist with HR/DSSS devices. The common protection mechanism is the RTS/CTS (Request to Send/Clear to Send)) exchange. Figure 8.8 shows the RTS/CTS frame formats.

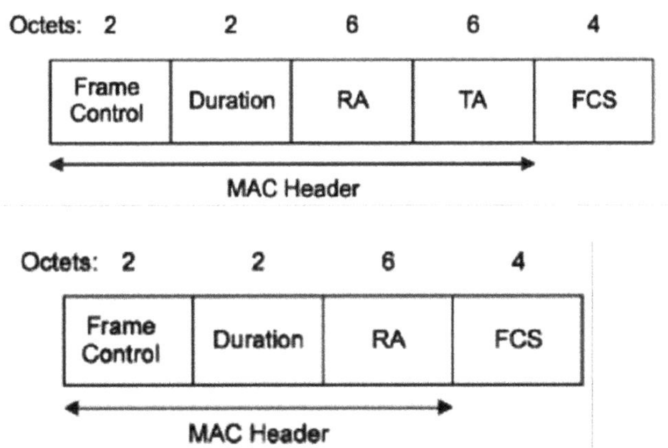

Figure 8.8: RTS/CTS Frames

The purpose of both RTS and CTS frames is to carry a duration value, which sets the NAV of receiving stations, in order to protect a subsequent frame. For that reason, both frames are as short as possible, carrying only essential information. RTS and CTS frames provide the core of the most commonly used protection mechanisms on WLANs. RTS/CTS is a possible solution for the hidden-node problems (discussed in chapter 13).

CTS-to-Self is a method used, mostly by APs, to avoid requiring the RTS frame. Simply stated, CTS-to-Self is a CTS frame used without an RTS frame.

RTS/CTS frames are sent at the lowest required data rate in the BSS — the same data rate as the Beacon frame. Therefore, every STA participating in the BSS should be able to see the CTS frame sent from the AP, even if they can't see the RTS frame sent from a client. Figure 8.9 illustrates the difference between data communications without RTS/CTS and with RTS/CTS/

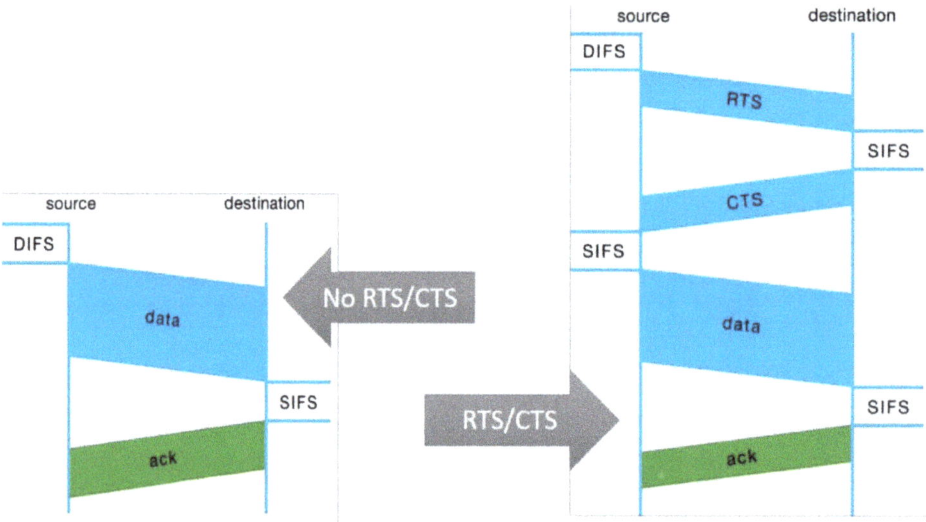

Figure 8.9: RTS/CTS Illustrated

The RTS/CTS protocol uses an exchange of control frames to force all stations in a basic service area (BSA) to defer to the upcoming RTS/CTS-protected transmission. The RTS frame is the first frame in a four-way frame exchange between the transmitter and the receiver. The purpose of the RTS is to transmit the duration information of the upcoming frame transmission to those stations in the neighborhood of the transmitter, so they may update their network allocation vector (NAV), to help avoid collisions.

The CTS frame is the second frame in the four-way frame exchange. The purpose of the CTS frame is to transmit the duration information of the upcoming frame transmission to those stations in the neighborhood of the immediate receiver (typically an AP), so they may update their NAV to help avoid collisions.

The Data frame (or data fragment) is the third frame in the four-way frame exchange. This frame carries the payload of the upper layers of the protocol stack and carries a duration value long enough to cover the ACK frame that follows.

The ACK frame is the fourth frame in the four-way frame exchange. This frame acknowledges correct receipt of the Data frame and sets the NAV of any station that can hear it back to zero (0).

BSS Color (802.11ax)

The general concept of BSS Color is that a STA can ignore an 802.11 frame from an overlapping BSS (OBSS) with a different BSS color using different thresholds than traditional DCF and EDCA. When a STA receives a frame with a matching BSS Color to its BSS, it will follow normal EDCA procedures. When a STA receives a frame with a different BSS Color to its BSS, it can ignore it (not update its NAV) and continue with contention as long as it is below the defined threshold. The difference in threshold is typically 20 dB, which is significant in reducing contention for inter-BSS frames (inter-BSS frames are frames seen by a device from a different BSS [an OBSS] on the same channel).

- Introduces in S1G and now implemented in 802.11ax (HE)
- Allows STAs to identify the BSS from which a frame originates based on number values in the frame
    - The BSS Color subfield of the HE Operation element defines the number
    - Range is from 1 to 63
- If a collision of BSS color values is detected, the BSS can define a new color
- Achieved with a BSS Color Change Announcement element in the Beacon, Probe Response, Association Response, Reassociation Response, or HE BSS Color Change Announcement frames

- Collisions may be detected directly by the AP or by the STAs and reported to the AP

Power Management Options

802.11 devices primarily use one of two power save options: legacy power save mode and unscheduled-automatic power save delivery (U-APSD).

Figure 8.10 shows the basic legacy power save process. Devices are either in active mode or power save mode. APs are always in active mode. Client STAs can enter power save mode.

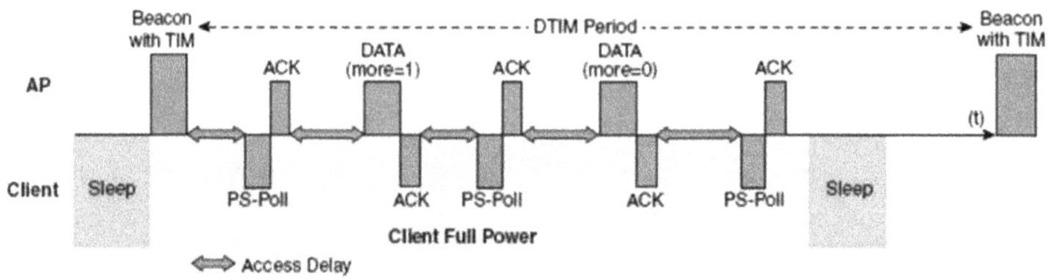

Figure 8.10: Legacy Power Save

Wireless mobile devices frequently rely on batteries as their power source. 802.11 stations have the ability to extend battery life by switching off the radio portion of the station adapter, for brief periods.

The 802.11 standard originally offered two power management modes of operation:

- Active mode — The mode where no power-saving features are enabled
- Power Save mode — The mode that allows stations to save power by being powered down (dozing) while inactive on the network. Stations awaken to receive frames destined to them

Power Save mode can be enabled or disabled on the client stations. When Power Save mode is enabled, client stations will alternate between two states:

- Awake

- Station is fully powered, and may receive frames at any time

- Doze
  - Station is not able to transmit or receive, and consumes very little power
  - Station must awaken periodically (based upon the Listen Interval parameter), to ascertain if traffic intended for it is stored at the AP

APs are required to know whether their associated client stations are using Power Save mode, but they do not track the individual state of each station moment by moment. Instead, beacons from an AP carry a Traffic Indication Map (TIM) information element, which lists the stations that have frames buffered at the AP awaiting delivery. It is the responsibility of the client to awaken and read the TIM from time to time. Since the times when beacons are to be transmitted are known by all stations, they can awaken themselves in time to read the TIM. Stations will typically awaken at least in time to receive beacons that are classified as Delivery Traffic Indication Maps (DTIMs).

- DTIMs are notifications of whether or not the AP has multicast or broadcast messages to transmit
- DTIMs are not necessarily carried in every beacon, but may be scheduled for every other, or every third beacon (for example)
- If the AP has multicast or broadcast traffic buffered, it will be transmitted immediately following the beacon containing the DTIM

With U-APSD, the trigger for retrieving traffic is the client sending traffic to the AP. The AP, when acknowledging the frame, tells the client that data is queued for it, and that it should stay on. The AP then sends data to the client typically as a TXOP burst, where only the first frame has the EDCA access delay. All subsequent frames are then sent directly after the acknowledgment frame. In a VoWLAN implementation, there is only likely to be one frame queued at the AP, and the VoWLAN client would be able to go into sleep mode after receiving that frame from the AP.

This approach overcomes disadvantages of legacy power save (e.g. latency) and is much more efficient. The timing of the polling is controlled via the client traffic, which in the case of voice, is symmetric, so if the client is sending a frame every 20 ms, it would be expecting to receive a frame every 20 ms, as well. Figure 8.11 illustrates the U-APDS power save option.

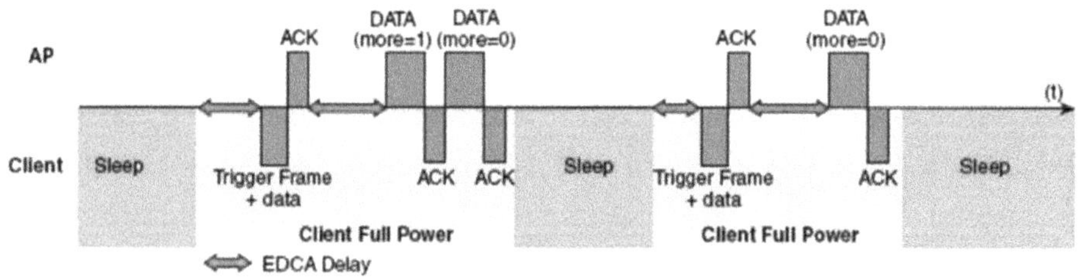

Figure 8.11: U-APSD Illustrated

Target Wake Time (TWT) is actually supported by both the S1G and HE PHYs; however, the focus of this course is on the more typical PHYs used in WLANs and, hence, 802.11ax (HE).

TWT reduces contention in the channel because the AP can define waking times for stations and, therefore, impact when they communicate. It also improves power management for the clients over traditional 802.11 power management schemes. STAs can negotiate a unique schedule with the AP when supported, which can allow clients with varying requirements to established longer periods or shorter periods between wake times. Alternatively, STAs can simply participate in a broadcast TWT schedule already existing on the AP.

- TWT scheduling AP
  - The AP that schedules broadcast TWTs and communicates them in the TWT element
- TWT scheduled STA
  - A client that follows the broadcast TWT schedule from the scheduling AP
- TWT accomplished two primary benefits
  - Reduced contention in the channel
  - Improved power management for the clients

Chapter Summary

In this chapter, you learned about the channel access mechanisms in 802.11, including DCF and EDCA. You also learned about operational modes for HT and VHT devices, and the protection mechanisms they use, particularly RTS/CTS frames. Finally, you discovered the two primary power management options in 802.11 wireless devices, including legacy power save and U-APSD.

Points to Remember

Remember the following important points:

- CSMA/CA is used in 802.11 WLANs.
- The foundational channel access function is DCF.
- Carrier sensing is both physical (802.11 frame detection and energy detection) and virtual (NAV timers set based on frame duration values).
- The Frame duration value is the time required to complete the transaction AFTER the current PPDU completes.
- EDCA enhances DCF with QoS options.
- An EDCAF is a function assigned to an AC that is used to gain access during a transmit opportunity for that AC.
- Four ACs are used in EDCA: AC_BK, AC_BE, AC_VI, and AC_VO, and they are listed here in order of lowest to highest priority.
- User priorities (eight of them) map to ACs.
- Each AC has varying IFS delays and contention window ranges. The highest priority ACs have the lowest maximum CW range value, and the shortest AIFS.
- RTS/CTS frames are used as protection mechanism so that newer PHYs can silence the channel and transmit at their higher data rates.

- APs can use CTS-to-Self, which is just a CTS frame without an RTS frame.
- The RTS/CTS process is used to set the NAV timer in STAs in the channel.
- HT devices can operate in legacy, mixed or Greenfield modes.
- VHT devices support the three HT modes, and also add a new PPDU format that is compatible with all 5 GHz PHY devices.
- Legacy power save is based on STAs awaking and looking at the Beacon frame to see if frames are waiting on the AP.
- U-APSD is based on STAs awaking and transmitting anything at all to the AP, and the AP, in turn, will let the STA know if frames are waiting in the queue.

Review Questions

1. What kind of carrier-sensing mechanism is used to detect RF power from non-Wi-Fi devices?
    a. CS
    b. NAV
    c. ED
    d. AC

2. What does the Duration value represent when it is used to set the NAV in a STA?
    a. The length of time it will take to transmit the current frame
    b. The time to complete the transaction after the current frame
    c. The time it takes to send the ACK for the current frame
    d. None of these

3. What IFS type is variable depending on the priority of a frame waiting for transmission?
    a. DIFS
    b. SIFS
    c. AIFS
    d. RIFS

4. After the IFS, what must count down before a STA can attempt to transmit?
    a. Backoff timer
    b. Carrier sense
    c. Energy detect
    d. None of these

5. When 40 MHz and larger channels are used, what portion of the channel is used for BSS operations?
    a. The second portion
    b. The first portion
    c. All of it
    d. The primary channel

6. What is used by APs to clear the channel as a protection mechanism, but not typically used by client STAs?
    a. CTS-to-Self
    b. RTS/CTS
    c. DTIM
    d. ATIM

7. When a legacy power save client awakes and sees traffic waiting based on the TIM in the Beacon frame, what can it do to inform the AP that it is awake?
    a. Send a Probe Request frame
    b. Send a PS-POLL frame
    c. Just wait, the AP will detect it
    d. None of these

8. Why can a VHT device use a single PPDU format and still be compatible with older 5 GHz devices?
    a. It actually uses the identical entire PPDU format as 802.11a
    b. It uses the Non-HT PPDU format of 802.11n
    c. The VHT preamble is backward compatible
    d. It transmits two PPDUs at the same time — one compatible with 802.11a and the other VHT-specific

9. What is the highest AC in EDCA?
    a. AC_BK
    b. AC_BE
    c. AC_VI
    d. AC_VO

10. What is the function called that processes the individual AC queues in a QoS STA?
    a. EDCA
    b. EDCF
    c. EDCAF
    d. AIFSN

Review Answers

1. **C is correct.** Energy Detect (ED) is used to sense non-Wi-Fi energy on the medium.
2. **B is correct.** The Duration value is equal to the time it takes to complete the current transaction, after the PPDU containing the Duration value is completed.
3. **C is correct.** The Arbitration IFS (AIFS) varies, depending on the AC of the frame awaiting transmission.
4. **A is correct.** The backoff timer is set based on a random selecting from the CW range.
5. **D is correct.** The primary 20 MHz channel is used for BSS operations and communications with 20 MHz-only STAs.
6. **A is correct.** The CTS-to-Self is a CTS frame without an RTS frame, mostly used by APs to clear the channel before transmitting to a high data rate STA.
7. **B is correct.** When the STA awakes, if the TIM in the Beacon frame indicates waiting frames, it can send a PS-POLL frame to the AP to trigger transmission of the frames.
8. **C is correct.** The VHT format uses a preamble that can be understood by all historic 5 GHz 802.11 PHYs.
9. **D is correct.** AC_VO (Voice) is the highest priority access category (AC).
10. **C is correct.** EDCAF (EDCA function) — one is assigned to each queue, when the traditional four queues are used.

Chapter 9 — WLAN Network Architectures

Previous chapters have covered WLAN architectures and explained some of their features. In this chapter controller-based, cloud-based, distributed, and controller-less architectures are discussed by example. A WLAN vendor's solution is briefly described to help you understand these varying architectures. Additionally, scalability and availability solutions, as well as roaming solutions are explained.

The second major section of this chapter is all about RF planning models. Two primary models exist: multiple channel architecture and single channel architecture. Both are addressed and important issues like channel selection, AP placement, CCI/ACI, and cell sizing are explained. Let's get to it.

WLAN Architectures

In this section, the four popular WLAN architectures are reviewed by exploring the models implemented by four different vendors. The vendors have been chosen as randomly as possible, and their inclusion is not intended to indicate whether they are best or worst, in that particular model. All vendor solutions have their strengths and weaknesses, and the CWNA should evaluate them all, to make the best decisions for his or her network.

The following vendors are reviewed:

- Controller-Based Model: Cisco
- Cloud-Based Model: Mojo Networks
- Distributed Model: Aerohive
- Controller-less (Autonomous) Model: EnGenius

Before we dive into the model reviews, however, it is important to review some terminology introduces in earlier chapters:

- **Centralized Data Forwarding:** a data forwarding model where all user data is passed through the AP in a tunnel to the controller or cloud (though typically not in the cloud anymore), and then forwarded on to the final destination.
    - **Core Layer Forwarding:** the data is sent to the network core, where a controller resides for destination forwarding.

- **Distribution Layer Forwarding:** the data is sent to the distribution layer controller and sent from there to the final destination.
- **Access Layer Forwarding:** the data is sent to a controller in the Access layer and sent from there to the final destination.

▪ **Distributed Data Forwarding:** a data forwarding model, where all user data is forwarded directly from the controller-based, or otherwise managed APs, to the final destination.

▪ **Control, Management and Data Planes:** conceptual planes that include different types of communications. The control plane is about network control protocols, like routing protocols and switching protocols, and WLAN solutions like radio resource management (RRM) and automated radio management (ARM). The management plane is focused on managing the devices and monitoring them, such as WLAN configuration and monitoring. The data plan is focused on user data transfer.

With the terminology review complete, let's begin our exploration of WLAN architectures. These architectures are primarily focused on how APs are managed and how user data traverses the network (for example, centralized or distributed data forwarding).

Controller Based Model

Cisco has been in the controller market longer than anyone. Back when they were called WLAN switches, Cisco acquired the company that developed them, and eventually transitioned to the term controller instead of switch (which makes sense when you think about it — no use confusing people, given that they are a leading seller of wired switches). Today's Cisco controllers implement CAPWAP, Layer 3 UDP tunnels, and a split-MAC model for WLAN operations.

LWAPP, the predecessor to CAPWAP, could operate at Layer 2 or Layer 3. CAPWAP, used in the Cisco controller-based model, uses Layer 3 tunnels and UDP packets to communicate between the controller and AP. UDP ports 5246 and 5247 are used for communications for control and data packets, respectively.

The split-MAC model is common among controller-based APs. The term split-MAC simply indicates that some of the 802.11 MAC operations are in the AP, and the others are in the controller. Table 9.1 lists the common breakdown of a split-MAC architecture, but some controller-based vendors may vary from this.

MAC Functions (Controller)	MAC Functions (AP)
802.11 authentication	Beacons
802.11 association	Probe Responses
802.11 reassociation	ACK and Block ACK
802.1X/EAP	MAC encryption/decryption

Table 9.1: Split-MAC Breakdown

WLAN profiles are created in the controllers, and the APs receive them from the controller. Controllers support AP groups and mobility groups. AP groups allow APs to be managed as a collective. Mobility groups indicate the controllers across which clients may roam to managed APs. Such solutions, by whatever name the vendor chooses, are common in WLAN controllers today.

The newest controllers that use CAPWAP can also encrypt the tunneled communications between the APs and the controllers. This provides for enhanced security on the wired side of the network.

Controllers are located using pre-provisioning, DNS, or DHCP. With pre-provisioning, APs are configured while disconnected from the network, with the IP address of the controller. This allows administrators to choose the controller to which the AP associates, with complete control.

When DNS is used, host names are created in the DNS server that map to the controller IP address. The AP will connect to the network and receive IP settings from DHCP, which will include the DNS server address. Next, the AP will use the DNS server to locate the controller. Finally, the AP will associate with the controller, and come under its management.

When DHCP is used, in addition to providing the AP with IP configuration settings, it will use DHCP options to pass the controller IP address to the AP. With Cisco APs, this is DHCP option 43.

Many controller-based APs also support broadcasts for the location of the controller. In this case, broadcast messages are sent out and the controller can be located, as long as it is on the same subnet (or broadcast domain) as the AP.

Controller-based APs must locate the controller and they accomplish this through broadcasts, pre-provisioning (pre-staging), DNS, and DHCP. This information should be remembered for the exam.

An important consideration with controller-based architectures is scalability and availability solutions. Think about it. If you have one controller and 20 APs and your controller fails, you have no WLAN, assuming your WLAN vendor does not support some kind of automatic switchover to autonomous operation by the APs, which is not exactly common. Therefore, availability must be considered.

Most controller-based vendors support an N+1 high availability model, and Cisco does, as well. N+1 allows one controller to be a backup for multiple primary controllers. In other words, it is not required that you have two controllers for every controller, in order to have redundancy and high availability. Think of it like a hot swappable drive in a RAID 5 array. You may have three active drives in the array and one hot swappable drive. You do not have to have a hot swappable drive for each active drive, as the odds of multiple active drivers failing concurrently are much lower than a single drive failing. In the same way, it is unlikely that multiple controllers will fail on the WLAN at the same time, so an N+1 high availability model is cost efficient, while still providing the needed availability.

Scalability is a separate concern. Cisco and other controller based vendors offer scalability through the addition of more AP licenses on a controller, the addition of more controllers, the addition of more APs, and the ability to manage multiple controllers from a single pane of glass. All of these provide for potential scalability.

For example, imagine you have purchased three controllers. Two primary and one as a backup for the two primary controllers. The controllers can support up to 500 APs, but you only licensed them for 100. You can therefore support 200 active APs on the WLAN. In a Cisco environment, the backup controller could handle all 200 APs, if both primary controllers happened to fail, but it would only allow for this management for 90 days. Now, suppose you need to add another 50 APs — 25 to each controller. You will simply need to acquire the additional AP licenses, and purchase the APs. No additional controllers will be required. The backup controller can still handle the now 250 APs, in the worst-case scenario, for 90 days. This is a form of scalability.

An additional factor that must be considered with controller-based architectures, is the implementation of QoS for high priority traffic, such as VoIP, and other real-time communications. Because the data traverses the network back to the controller (when centralized forwarding is used), you should understand what QoS solutions are used as it traverses the network. Two perspectives should be considered: downlink and uplink traffic. Downlink traffic comes from the network to the wireless client. Uplink traffic goes from the wireless client to the network.

For uplink traffic, the client application should properly configure QoS markings (DSCP) at Layer 3 of the OSI Model. These QoS settings will be mapped to 802.11e/WMM ACs, for priority queuing and transmission to the AP. The AP sends the 802.11 frame in a tunnel to the controller. The tunnel may not have the QoS priority you desire, but controllers often let you configure this. The controller sees the DSCP markings and applies the appropriate QoS markings before forwarding onto the destination. It is very important to realize, that any decapsulation device in the process can completely remove QoS markings, if the device is not configured properly for it. QoS must be configured in devices, end-to-end.

For downlink traffic, the originator (such as a remote VoIP phone) will mark the IP packets with appropriate DSCP markings and send the data. It arrives at the controller, and is forwarded to the AP. The AP processes and maps the DSCP markings and applies the appropriate AC for transmission to the client device. Figure 9.1 illustrates this communication process.

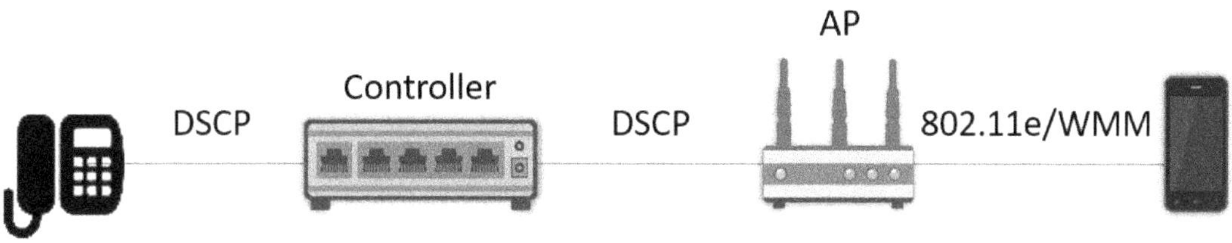

Figure 9.1: Downlink QoS in 802.11 Networks

Controller-based WLANs also support virtual local area networks (VLANs). Each SSID in an AP can be mapped to a separate VLAN, thus allowing separation of the traffic originating from the different SSIDs.

Finally, controller-based networks offer advantages in the area of roaming. Roaming can be intra-controller or inter-controller, and the latter is where mobility groups come into play. Cisco controllers support both Opportunistic Key Caching (OKC) and Fast BSS Transition (FT), which is the 802.11 standard method of roaming. Both methods perform key caching and key management methods that allow for faster roaming from one AP to another, when 802.1X/EAP is used. Of course, if PSK is used, fast roaming methods are not required, because roaming occurs very quickly with PSK, without additional infrastructure support.

Cloud-Based Model

As we cover the other networking models, the details in the controller-based section are not required. But it is important to note areas of differentiation. Mojo Networks is one of several cloud-based AP vendors. They provide AP management through the cloud service, removing the need to have local controllers. However, they do offer a local virtual machine service that can allow for GRE tunnels to be established with the virtual machine, and function in a similar way to a controller (as to user data termination), while the cloud still receives monitoring information from the APs and manages the AP configurations.

In a cloud-based model like this, the APs need an initial connection to the cloud, in order to register with the cloud service, and acquire their configuration profile. Once configured, they function from a MAC perspective like autonomous APs. However, they will communicate back monitoring information to the cloud, and

Mojo Networks suggests that they will require about 1.5 Kbps per AP deployed, for this communication. The communications occur across UDP and are encrypted with AES encryption. If the cloud should be unavailable, due to a local Internet outage or a cloud service failure, the APs will continue to provide 802.11 operations with their current configuration locally.

Because cloud-based systems like this gather so much statistical data, they can implement advanced monitoring and troubleshooting solutions, as well. For example, Mojo Aware provides data analytics to allow for such troubleshooting. For example, when clients attempt to connect, but fail, Mojo Aware detects this, automatically reports on the problem, and gives a likely solution. Additionally, it tracks performance problems and correlates them with things like low signal strength and high retry rates, which can be very helpful in the troubleshooting process.

Distributed Model

Aerohive refers to their distributed model as cooperative control. The concept of a distributed or cooperative control model is summarized in that the APs are configured by the management solution, but they also have local intelligence for configuration and operations beyond a controller-based model. In this model, the management plan involves communications with the Hive Manager (which may be in the cloud or on location), but the control and data planes stay among the APs.

What this really means is that things like fast and secure roaming, QoS, radio channel and output power management, are all coordinated and managed among the APs themselves without requiring the Hive Manager to make decisions. Everything is reported back to the Hive Manager for monitoring (remember, the management plane), but the distributed cooperative control protocols handle the control plane.

> The key takeaway is that a distributed model relies on communications among the APs for control plane management and a physical controller is not required. Cooperative control protocols are used to allow this operation.

Similar in results is the virtual controller concept implemented by some vendors (like Aruba Networks), that allows one of the APs to act as a controller. The difference is that the intelligence is not distributed, even though a physical controller is not required. Virtual controller solutions are not really distributed models as they still require a controller, even though it is a virtual device.

Controller-less Model

In this section, the EnGenius EWS360 AP, which can be configured in autonomous mode, will be used as the example device. Controller-less models fall into two basic categories: strict autonomous APs and managed autonomous APs. EnGenius offers a managed autonomous AP model, through their Neutron Series switches and the APs can also be configured as strict autonomous APs. In this section, we will step through the configuration interface of the EWS360AP and see the options typically available in autonomous controller-less models. Figure 9.2 shows the starting screen for the EWS360AP.

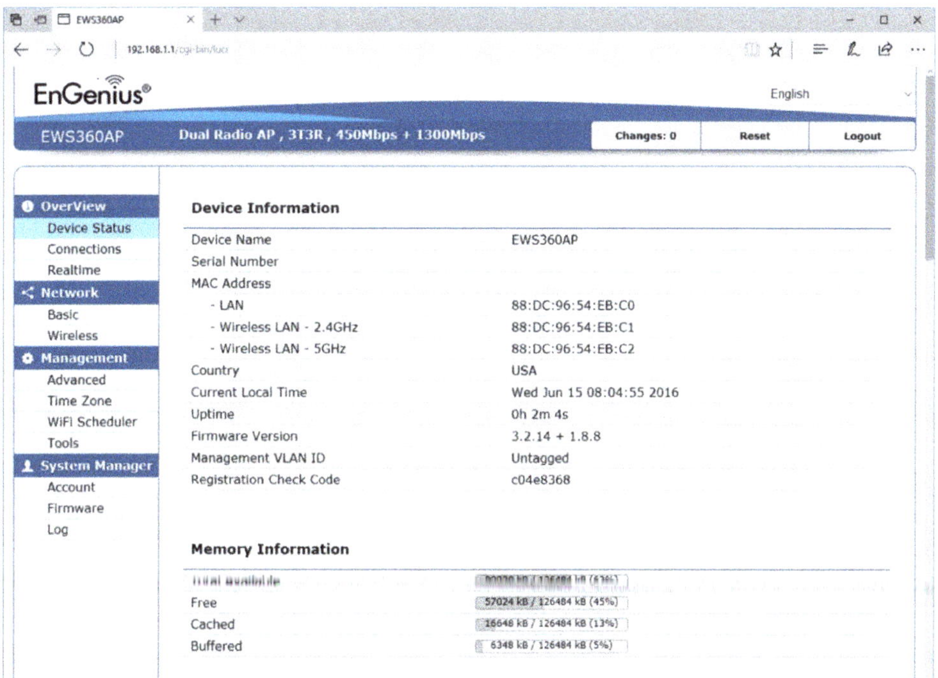

Figure 9.2: EWS360AP Start Screen (OverView > Device Status)

Given that this is an autonomous AP, in this mode, the first step would normally be to configure the logon account (admin in this case) and change the default password. I will not perform that task here, but it should be done in a production system. The first thing we need to inspect is the Network > Wireless page to see what options are available. Figure 9.3 shows this page.

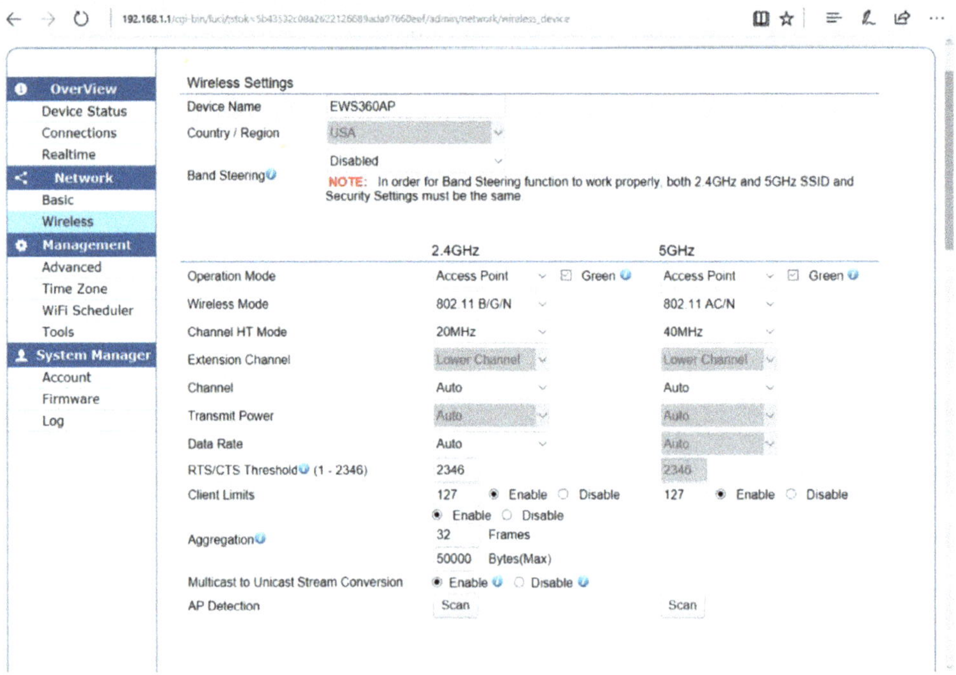

Figure 9.3: The EWS360AP Wireless Page

On this page, you can see the settings available. They are typical of an AP and include:

- **Operation Mode:** Devices may support Access Point, Bridge, Repeater and others.
- **Wireless Mode:** Used to configure compatibility, for example, 802.11b/g/n, 802.11n-only, 802.11b/g, etc.

- **Channel Mode:** Determines the channel width.
- **Extension Channel:** Determines if the upper or lower channel is used as the secondary channel.
- **Channel (primary channel used):** When set to Auto, the AP attempts to determine the best channel. Some vendors' algorithms are better than others'. You can also statically configure the channel.
- **Transmit Power:** Thankfully, the EnGenius AP simply lists transmit power levels in dBm, but some vendors may show percentages or other meaningless representations. When this is the case, consult the vendor documentation to figure out what should have already been in the configuration interface.
- **Data Rate:** Used to configure the data rates allowed in the BSS.
- **RTS/CTS Threshold:** Used to determine the size of frames that require RTS/CTS, even if protection mechanisms are not required.
- **Client Limits:** Used to limit the number of clients that can connect to the AP.
- **Frame Aggregation:** Used to determine how A-MSDU or A-MPDU will be used.

You will also notice in Figure 9.3, that Band Steering is an optional configuration item. This is explained more in the next chapter, but for now, know that it can be used to either encourage, or force, 5 GHz clients to connect on the 5 GHz radio.

The screen in Figure 9.3 is showing the settings for the first configuration profile. The EWS360AP supports up to eight profiles, or eight SSIDs, mapped to eight VLANs in this autonomous mode, as you can see in Figure 9.4. For each profile, you can determine the following:

- SSID
- Security
- Hidden SSID: Does not broadcast the SSID in Beacon frames.
- Client Isolation: Disallows communications directly between clients.

- VLAN Isolation: Traffic will be tagged, based in the specified VLAN ID.
- Layer 2 Isolation
- VLAN ID: The VLAN ID to use for traffic tagging, if enabled.

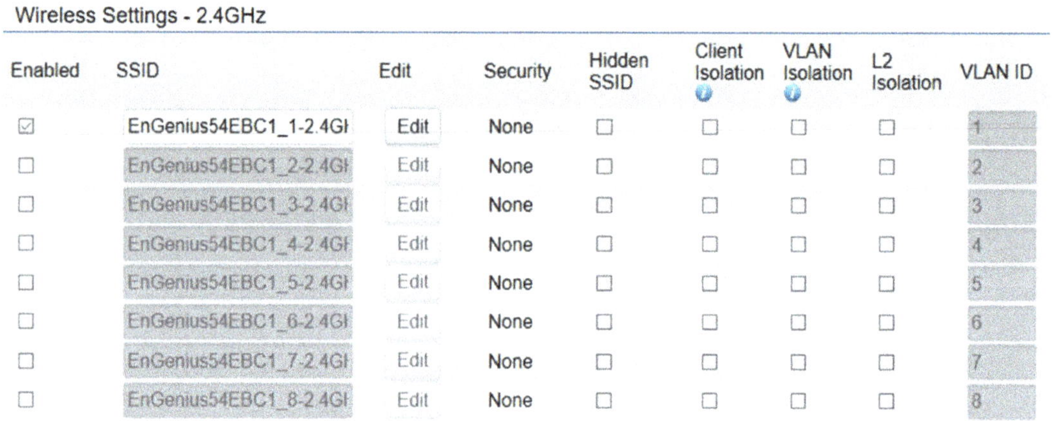

Figure 9.4: The Wireless Settings (Profiles) Section of the Wireless Page

Guest networks are also supported and can be configured to use a unique DHCP pool in the AP, as shown in Figure 9.5.

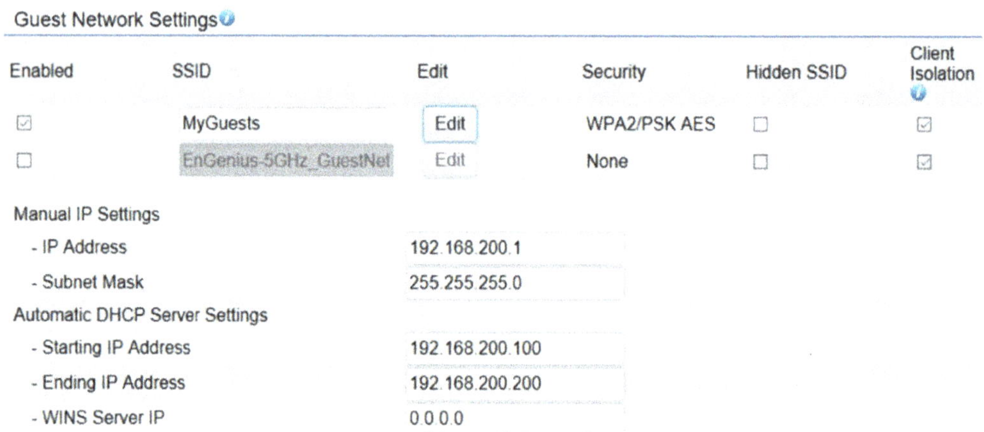

Figure 9.5: Configuring a Guest Network

As you can see, an autonomous AP allows for the configuration options most often needed. The challenge with autonomous APs is the time required to configure and reconfigure large groups of them. For this, a network management solution, which is also available from EnGenius, is more efficient.

RF Planning Models

RF planning models fall into two primary categories: MCA and SCA. This section addresses both models. It is important to understand that MCA has become more complex over time, as we now have automated MCA and static MCA. Both are addressed here.

Multiple Channel Architecture

The traditional WLAN architecture is *the multiple-channel (multichannel) architecture (MCA)*. A multichannel architecture is built with careful planning, and maintained over time, when channels are statically defined. The 802.11 PHYs that operate in the 2.4 GHz band provide three non-overlapping channels. In the United States, the non-overlapping channels are 1, 6, and 11. I'll focus on the 2.4 GHz band here to make the explanations simpler. The 5 GHz bands offer many more non-overlapping channels with 802.11a/n/ac. Strategically configuring APs to use these channels, and then staggering the channel usage throughout a coverage area, allows complete coverage of larger areas.

As an example, consider the simple floor plan in Figure 9.6. The building houses approximately 150 workers, and has very old and thick external concrete walls, and a center block wall along both sides of the long hallway. Assuming this entire single-floor building needs coverage, multiple APs will be needed. In order to provide the highest data rates to all users, APs will be installed, and power levels will be adjusted accordingly.

MCA plans are often depicted with hexagons or circles to represent the coverage of each AP. In the real world, APs and their antennas do not ever propagate the signals in a perfect hexagonal or circular shape. Today, there is no need to do such guesswork, because we have exceptional tools like iBwave Wi-Fi, Ekahau Site Survey, AirMagnet Survey Pro, and TamoGraph Site Survey. These tools use algorithmic calculations, and awareness of building materials, to generate RF coverage maps that are very close to what you will see in the actual implementation. The tools can also be used in active survey mode, in which they gather real signal metrics, and show you exactly what the implemented environment looks like. We will explore such tools from that perspective more, when we discuss post-deployment validation surveys in a later chapter.

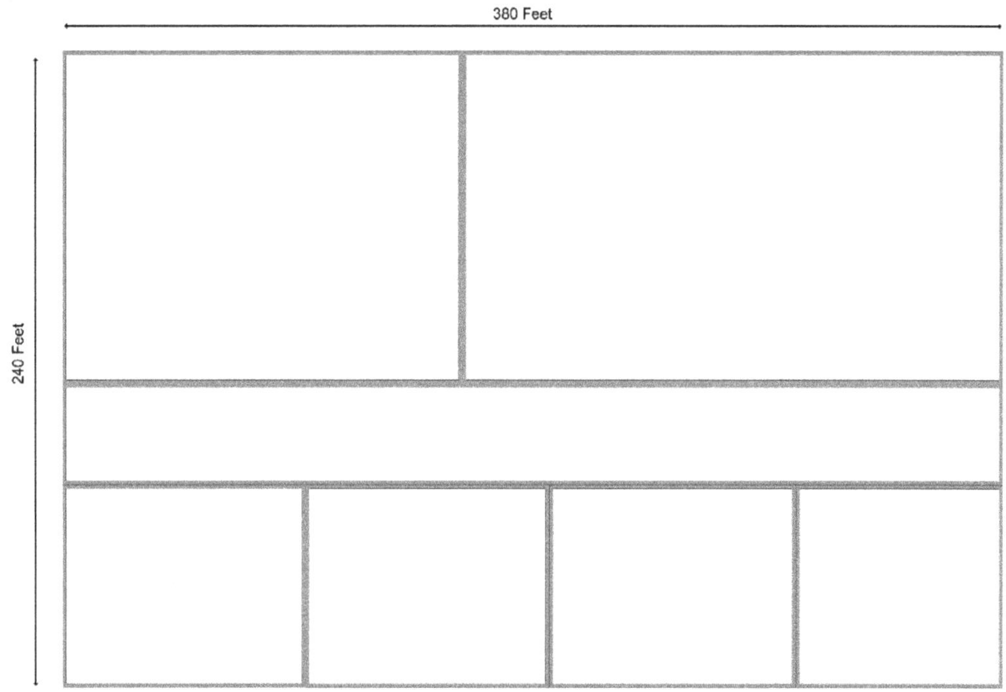

Figure 9.6: Floor plan of intended coverage area

Figure 9.7 shows a possible plan for this network. Four APs were deployed at 25 mW of output power and omni-directional antennas. Two APs were deployed (at each end of the hallway) with Yagi antennas aimed inward and 25 mW of output power. The result is that the weakest coverage areas indoors are at -71 dBm. The location of the APs is important. By placing the APs strategically, the two APs in the lower portion of the graphic can cover all four of those office spaces.

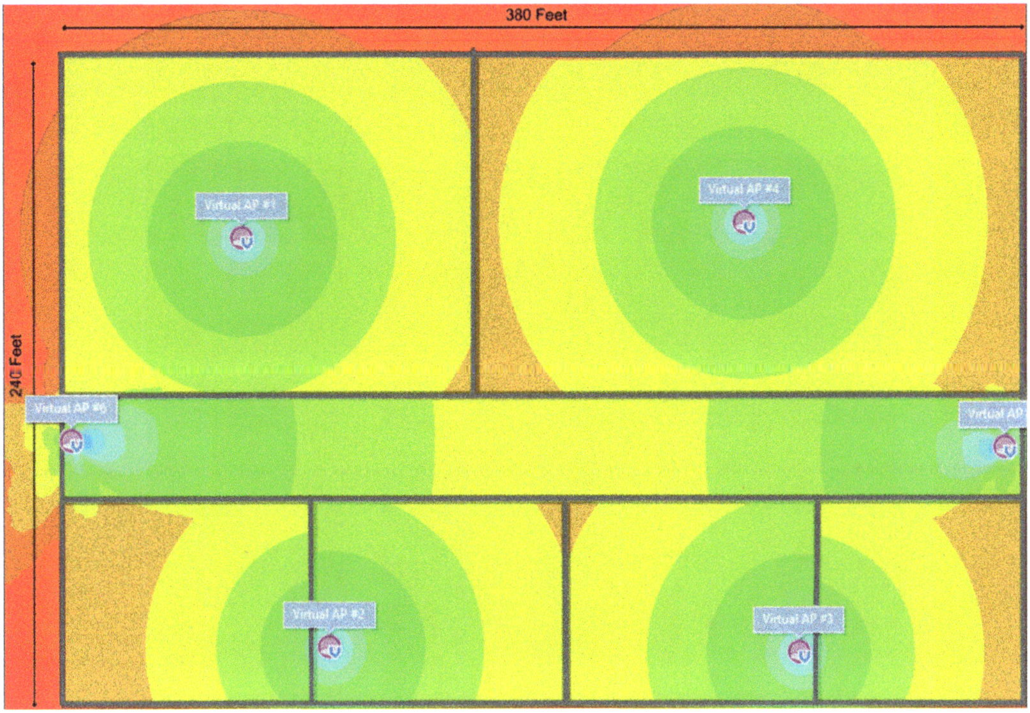

Figure 9.7: Network Plan (Created in TamoGraph Site Survey)

Figure 9.8 shows the plan with the channel map revealed. As you can see, some co-channel interference (CCI) is likely to occur on channel 11, and possibly on channel 1. But given the space, and an attempt to use 2.4 GHz, some CCI is to be expected.

The same concepts applied here would be applied in a large building, or a building with multiple floors. You must implement channel plans with the goal of providing effective coverage and capacity, while diminishing the impact of CCI. This goal is easier to accomplish in 5 GHz, due to the increased number of channels, but acceptable performance can be achieved, even in the 2.4 GHz band, within many spaces. With the advent of dual-band APs, it is not uncommon today to disable some 2.4 GHz radios in some APs, due to the increase coverage typically accomplished in 2.4 GHz.

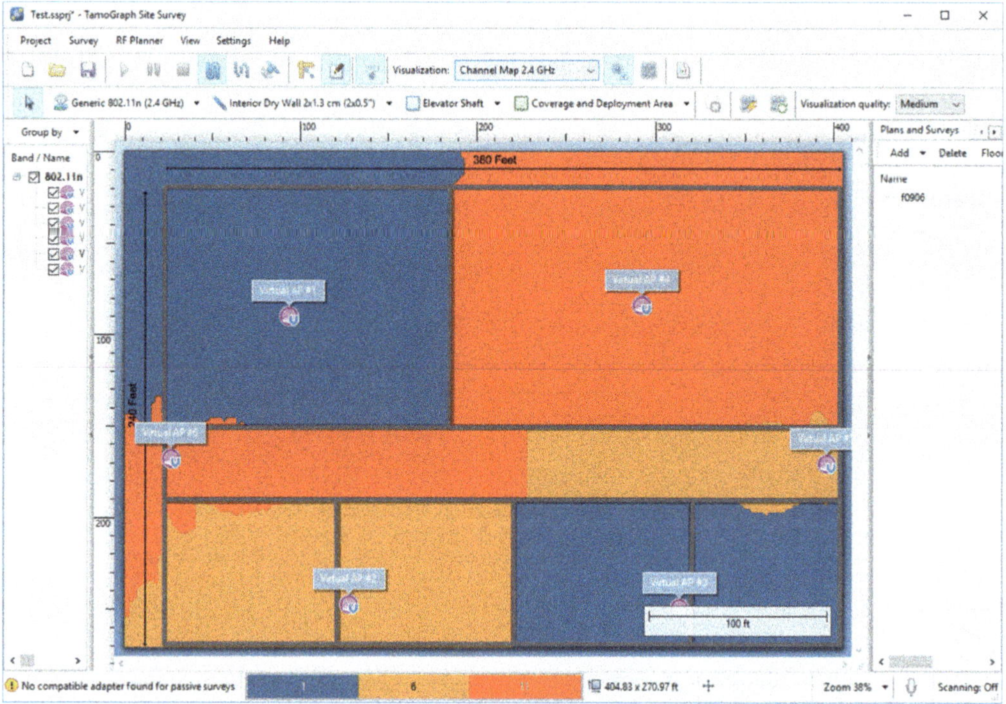

Figure 9.8: Channel Plan Shown in TamoGraph Site Survey

Several challenges are introduced with the MCA solution:

- Output power settings must vary at each AP, and this causes site surveys and WLAN design to be more difficult and time consuming.
- Adjacent-channel non-overlapping interference (interference among channels 1 and 6 or channels 6 and 11) is common, and measures must be taken to reduce it, though using realistic output power settings on the AP (under 100 mW) is usually sufficient.
- It is more difficult to implement high density areas (such as conference and meeting rooms and arenas) within the context of a larger WLAN.
- Over time, WLANs require manual or automated adjustments as the environment changes.

Single Channel Architecture

As stated previously, the two primary design methodologies in today's wireless networks are multi-channel architecture (MCA) and Single Channel Architecture (SCA). Although both appear to be self-explanatory, there is an amount of complexity in successfully deploying a single channel network or using more than one channel in your typical multi-channel network. In light of the fact that SCA can be configured to operate on a single channel, following all design principals is critical to the success of any wireless network whether it be SCA or MCA. Defining requirements, a thorough design approach, and validation are all crucial to being successful. Although most wireless networks are designed using multiple channels, the seemingly controversial Single Channel Architecture definitely has benefits and has established a foothold in multiple verticals, such as education and healthcare. Some of SCA's concepts have since been adopted by MCA vendors and, possibly, even influenced standards development such as the 802.11ax amendment.

At SCA's most basic level, a group of APs make up a virtual basic service set identifier (BSSID) that may span an entire campus or may be broken into smaller groups covering a campus. More simply stated, a client transitions through a group of APs which make up a single virtual BSSID. The client believes it is connected to a single AP, so transitions are seamless and transparent because of this "hand-off" rather than a roam. Because of this seamless transition, the client experiences little to no latency while moving from AP to AP across the virtual BSSID. The average hand-off within the virtual BSSID is roughly 2-3 ms. This is significantly better than even the most optimized 802.11 v/k/r network, where 10-15 ms can be achieved. With roaming improvements such as this, it may be easy to identify how this may be advantageous already.

Fortinet's (formerly Meru Networks) SCA offering uses the phrase "Virtual Cell" to describe a group of APs that operate using the same virtual BSSID. Without going outside of the scope of this book, Virtual Cell describes the technology which includes the coordinating function that is responsible for deciding which AP will handle each client at any time. This virtual cell or virtual BSSID is typically only broken when introducing different hardware model APs, features, band, etc. For example, you may use a combination of Fortinet AP832 and AP1020 access points in the same ESS, but their virtual BSSID will be different because they are different hardware models. Note the channel and BSSID in Figure 9.9 In this same scenario, you could also configure each one of these hardware model groups on their own channel; AP1020s using channel 11 on the 2.4 GHz band and channel 44 on the 5 GHz band, as opposed to AP832s using channel 1 on the 2.4 GHz band and channel 149 on the 5 GHz band, as shown in Figure 9.10. Although this is preferred, it is not mandatory to have different hardware/radios on separate channels.

ESS-AP Configuration (256 entries)

ESS Profile | ESS-AP Table | Security Profiles

ESS Profile	AP ID	AP Name	Interface Index	Channel	Operating Channel	Admin State	Max Calls	BSSID
RVH-LCPS	1	RVH-L400	1	1	1	Up	0	00:0c:e6:02:fa:91
RVH-LCPS	1	RVH-L400	2	149	149	Up	0	00:0c:e6:02:fb:57
RVH-LCPS	2	RVH-L402	1	1	1	Up	0	00:0c:e6:02:fa:91
RVH-LCPS	2	RVH-L402	2	149	149	Up	0	00:0c:e6:02:fb:57
RVH-LCPS	3	RVH-L404	1	1	1	Up	0	00:0c:e6:02:fa:91
RVH-LCPS	3	RVH-L404	2	149	149	Up	0	00:0c:e6:02:fb:57
RVH-LCPS	4	RVH-L406	1	1	1	Up	0	00:0c:e6:02:fa:91
RVH-LCPS	4	RVH-L406	2	149	149	Up	0	00:0c:e6:02:fb:57

Figure 9.9: A group of four Fortinet AP832 APs in the same ESS profile. Note the channel and BSSID

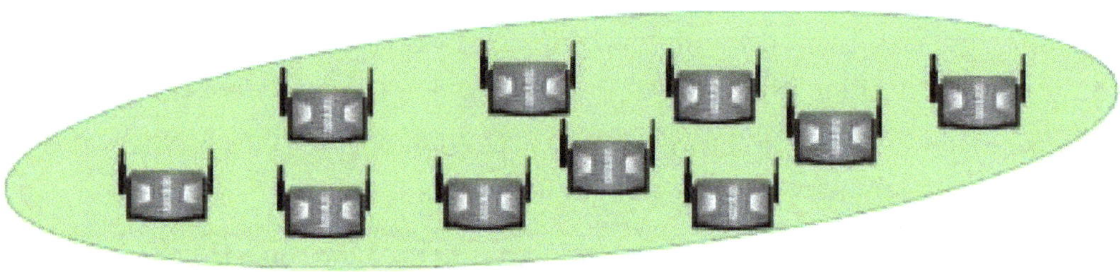

Figure 9.10: The 2.4 GHz Radio of Each AP is on Channel 1 and the 5 GHz Radio is on Channel 149

In a virtual WLAN, the controller choses which AP should respond to the client. The wireless controller is always aware of the conditions of the wireless network. The controller knows which APs are busy and which APs can see each other. Fortinet achieves this by a frame report from every access point to the controller and its coordinator. By default, these frame reports are sent every three seconds. This, like many other features of the Fortinet solution, can be adjusted via scripting, based on your deployment specific needs. These reports may need to happen more frequently in an environment with fast-moving clients; for example, forklifts or robots. This awareness of wireless network conditions allows the controller to assign the best AP to a client in the area. If a client is moving in an area, the controller will notice that the client's signal is degrading, from the perspective of the AP to which it is currently associated. The controller will evaluate the network and will assign a better AP to the client. From the client's perspective, it has not roamed because it is still connected to the same virtual BSSID. In MCA wireless networks, the client is responsible for making the majority of the decisions on when and how to roam. In contrast and as stated above, in an SCA environment, the roaming decisions are handled by the WLAN controller.

In an SCA wireless network, the controller uses proprietary algorithms to decide which access point is the best for a client to be assigned to. These proprietary algorithms along with RSSI, load, etc., are used to assign APs to clients. How each SCA vendors' algorithm works is considered intellectual property and is beyond the scope of this book.

Single Channel Misconceptions

There are many misconceptions when designing and deploying SCA wireless networks. What seems to be the biggest misconception is that you do not need to channel plan or perform a site survey. While you can technically design your virtual WLAN using a single channel across either band, you may get mixed results based on the requirements. Failure to meet the requirements that were initially set out before the design phase can make any wireless network fail, whether it be Multi or Single Channel Architecture. Most, if not all, design practices that are used for MCA wireless networks are also used for SCA wireless networks.

Another misconception of SCA is that all 2.4 GHz radios are set to one channel, and all 5 GHz radios are set to another channel. Contrary to this belief, there are different design methods to meet different requirements using SCA. A couple different examples will be discussed in the following section.

Another notable misconception is that SCA and virtual cell are one and the same. Single Channel Architecture is independent of virtual cell technology, but the misconception may stem from the fact that SCA WLANs are rarely deployed without virtual cell.

Designing for Single Channel Architecture

Wireless networks serve a dynamic range of clients, which require the network to be flexible. When designing either an MCA or SCA wireless network, there are several steps that should always be followed; define requirements, design, implement, and validate. The wireless network design steps are covered more deeply in the Certified Wireless Design Professional certification materials. With Single Channel Architecture, you have a few design options at your disposal to be successful in meeting your requirements. Simply placing each APs' radios on the same channel (in each band) can work, but results may vary based on the requirements of your wireless network. You can channel stripe and/or channel layer.

Channel striping is sometimes considered on multi-level structures, like shown in Figure 9.11 and 9.12, where you may have APs above or below one another. For example, a three-story school may use channel 36 on its lower level, channel 44 on its entry level, and channel 52 on its second level. This is often done to provide additional capacity, while also segmenting different physical and virtual layers of a wireless network.

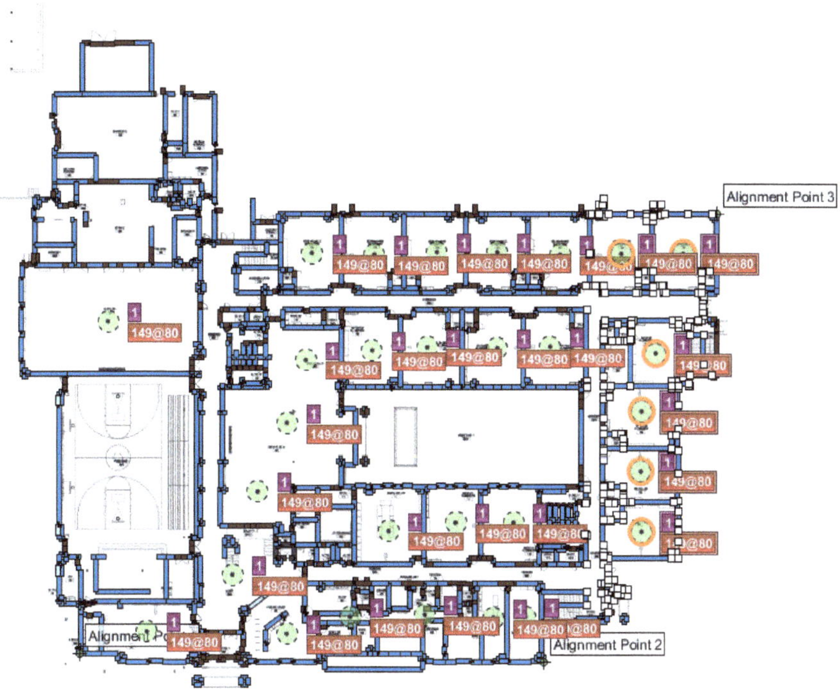

Figure 9.11: First Floor Single Virtual BSSID Channel Stripe

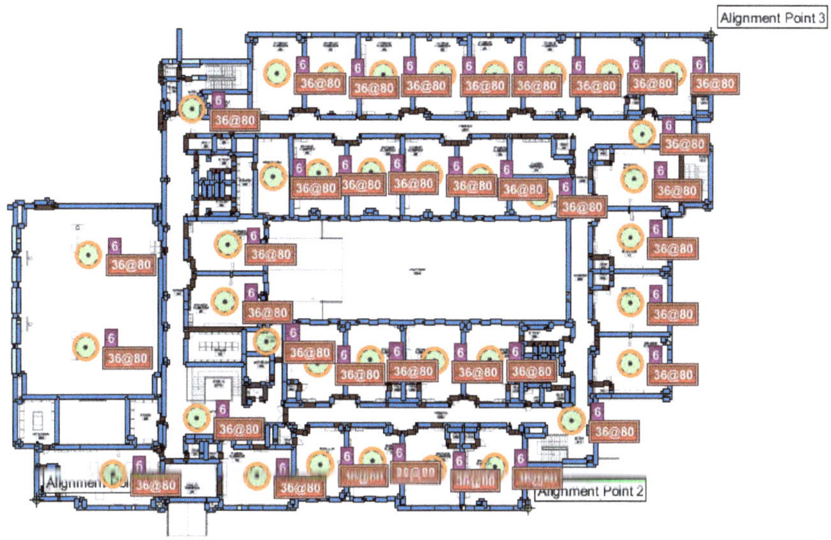

Figure 9.12: Second Floor Single Virtual BSSID Channel Stripe

Channel layering is the practice of placing multiple co-located APs in the same general area, each on different channels. Channel layering is also sometimes called channel stacking. Layering of the channels could be used in a specific high-density area, or across an entire campus. Layering channels adds capacity (per additional channel) to the network. For example, layering could be used to add capacity or redundancy, isolate different types of devices, or segregate applications. Each layer uses a unique virtual BSSID, and therefore a virtual cell, per channel, as seen in Figure 9.13. Because of the spectral efficiency of using a virtual BSSID, channel capacity can be dramatically increased. In a hospital, this could allow you to run medical records software and devices on a dedicated channel layer in the same airspace as the guest network. Designing in layers could potentially mean that even in the case of a major incident, with many people in the ER, there's no impact to service for the medical records system.

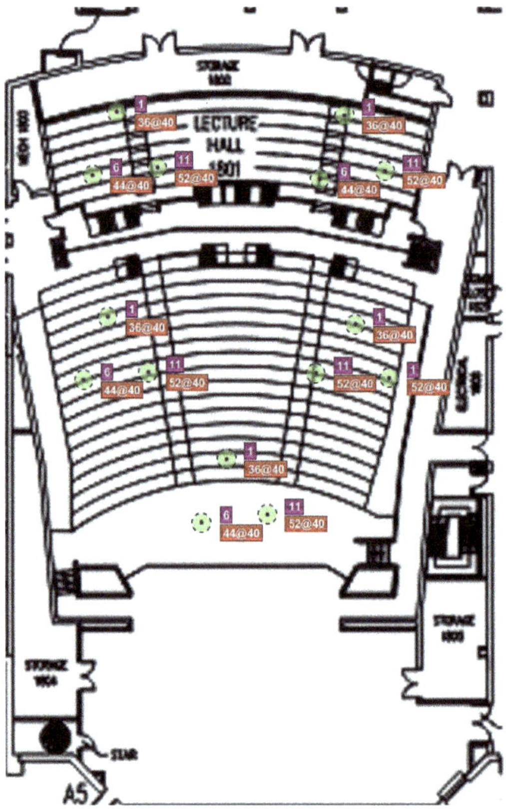

Figure 9.13: Three Layered But Separate Virtual BSSIDs

In the case of 802.11ac, 80 MHz channels can be leveraged with less concern about spectral re-use. However, narrower channels can be used for a layered design, which may require very high density. This is one of the most beneficial aspects to virtual cell\SCA technology.

It is noteworthy that some wireless deployments employ channel striping in a multi-channel design, as well. This commonly means allocating non-overlapping channels on APs dedicated to a specific function. For example, an MCA striped network could consist of channels 40, 52, 140, and 157 allocated to a voice SSID (application), and a different set of channels allocated to another, different application. Although both employ different architectures, there are striking similarities between a channel-striped SCA deployment and a channel-striped MCA deployment.

Chapter Summary

In this chapter, you learned about various WLAN architectures, in more detail than previously presented. You also learned about MCA and SCA RF channel planning. In the next chapter, you will focus on matching requirements to various WLAN features and functions.

Points to Remember

Remember the following important points:

- Centralized data-forwarding routes user traffic to the controller, and the controller forwards it to the destination.
- Distributed data-forwarding routes data directly from the APs to the destination.
- CAPWAP uses UDP for communications between the APs and controllers.
- CAPWAP is a Layer 3 protocol, and LWAPP supports Layer 2 or Layer 3.
- In a split-MAC architecture, authentication, association and reassociation are usually handled in the controller.
- Controllers are located by APs using broadcasts, pre-provisioning, DNS and DHCP.
- N+1 high availability indicates that a single controller is a backup for several other controllers.
- End-to-end QoS involves DSCP (Layer 3) markings and 802.11e/WMM ACs.
- In most cloud-based architectures, the APs continue to function, even if the cloud is unavailable.
- In the distributed model, the APs cooperatively choose channels, output power, and other settings without a controller.
- You can limit the number of clients that can connect to most enterprise-class APs.

- Client isolation is a common feature of APs, and it prevents client devices participating in the BSS from communicating directly with each other.
- The MCA RF planning model uses multiple staggered channels with multiple APs to provide coverage and capacity.
- The SCA RF planning model can use a single channel with centralized control over which APs transmit at any given moment.

Review Questions

1. In what networking plane would RRM exist?
    a. Management
    b. Control
    c. Data
    d. None of these

2. What model allows data to be sent from the AP directly to the wired destination?
    a. Distributed Data Forwarding
    b. Core Layer Forwarding
    c. Distribution Layer Forwarding
    d. Access Layer Forwarding

3. What protocol is used with CAPWAP tunnels?
    a. RTLS
    b. IPSec
    c. UDP
    d. PPTP

4. In a split-MAC model, which one of these is usually performed in the AP?
    a. 802.11 authentication
    b. Beacon frames
    c. 802.1X/EAP
    d. 802.11 association

5. What can an AP use to locate a WLAN controller when DHCP is not configured to provide the location, and the controller is on a separate network?
    a. Broadcasts
    b. WINS
    c. DNS
    d. NTP

6. When three controllers are installed as primary controllers, and one is installed as a backup for all three, what kind of high availability is described?
    a. N+1
    b. RAID5
    c. Redundant +1
    d. None of these

7. What Layer 3 QoS marking method plays a role in QoS in WLANs?
    a. TLS
    b. AC
    c. WMM
    d. DSCP

8. In a cloud-based model, what happens when the cloud is unavailable?
    a. The APs cease to function, and the WLAN is inoperable
    b. The APs function based on the last received configuration
    c. The APs switch to autonomous mode and must be manually configured
    d. The APs switch to controller-based mode with the onsite backup controller

9. What is determined by the RTS/CTS threshold?
    a. The frame size at which RTS/CTS must be used
    b. The signal strength at which RTS/CTS must be used
    c. The number of associated clients at which RTS/CTS must be used
    d. None of these

10. While no WLAN design is easy, why is MCA RF planning simpler in 5 GHz bands?
    a. Automatic channel planning actually works there
    b. Signals do not propagate as far
    c. More channels are available
    d. We know more about the 5 GHz band

Review Answers

1. **B is correct.** RRM and other automatic channel configuration solutions operate in the control plane.
2. **A is correct.** Distributed Data Forwarding.
3. **C is correct.** CAPWAP tunnels use UDP.
4. **B is correct.** Beacon frames.
5. **C is correct.** DNS is used to resolve a hard-coded host name to the IP address of the controller.
6. **A is correct.** N+1 high availability is described.
7. **D is correct.** DSCP markings are used to determine the proper AC for WLAN transmissions.
8. **B is correct.** The APs will continue to function with the last received configuration, if the cloud is unavailable.
9. **A is correct.** The RTS/CTS threshold, available in most AP configuration systems, defines the frame size at which RTS/CTS must be used, even if protection mechanisms are not required.
10. **C is correct.** MCA design is simpler in 5 GHz because there are more channels.

Chapter 10 — WLAN Requirements and Solutions

Objectives Covered

This chapter is focused on meeting requirements of WLANs. First, we address WLAN roles and applications and, in this context, by roles we mean uses in various industries and sectors. Next, we look at specific WLAN features how they can assist in meeting requirements. Finally, we explore network services that are needed by the WLAN.

WLAN Roles and Applications

In this section server WLAN roles will be considered. These roles are covered on the CWNA exam, but more importantly, they are common environments that CWNA professionals must administer and support.

Corporate data access and end-user mobility

Corporate data access and end-user mobility is the primary reason for implementing indoor WLANs in standard office environments. The WLAN acts as the point of access to the corporate or organizational network. When designing corporate data access and mobility, keep the following issues in mind:

- Ensure sufficient coverage in needed areas. You do not want users frustrated by the lack of coverage in areas where they have meetings, or where they must move around while performing their responsibilities.
- Ensure sufficient throughput. It's not enough to say that the WLAN signal can be detected at a given location. Does it provide a sufficient data rate and throughput level for the users' needs?
- Ensure proper security. You must consider authentication, authorization, confidentiality and integrity. Authentication can be provided with 802.1X. Authorization can be implemented using VLANs and role-based access control. Confidentiality can be provided with CCMP/AES encryption. Finally, integrity can be provided with integrity checks.
- Ensure availability. Availability is different than performance. Performance is about throughput. Availability simply means that the resource is there. In most cases, both performance and availability must be provided, but many cases warrant a greater focus on availability.

BEYOND THE EXAM: Availability versus Performance

I have noticed a primary focus on performance when reading the whitepapers and literature from WLAN hardware and software vendors. However, I would argue that availability is more important than performance in many cases. For example, if an organization is relying on a wireless network for mission-critical connectivity, occasional slowdowns in throughput may be tolerable, but complete loss of connectivity will not be accepted. Availability ensures that the network will always be there when the user needs it. It is certainly far more common to see a wireless network "go down," than it is to see our traditional wired networks do the same.

So, how can we build availability into our wireless networks? The answer is, as usual, redundancy. Redundancy must be built into the infrastructure first. If you are using a centralized WLAN model (a wireless controller with lightweight APs), you must implement multiple controllers in order to achieve redundancy within the infrastructure. Additionally, each coverage cell should either have APs that operate through the different controllers, or they must support automatic failover to an alternate controller, should the controller with which they communicate fail. APs may even failover to a self-contained mode when the controller fails, should the vendor implement such features.

Another redundancy consideration is the power supply to the APs. If only one AP is used to provide coverage in an area, and that AP's power source fails, the area will lose coverage. The solution may be to use APs that support PoE, but to power them with traditional power lines during normal operations. With a brief outage, the AP may be able to come back online through PoE, if a localized power failure occurs within the building. An alternate solution would be to implement two APs in the coverage area. One AP could be powered by PoE, and the other with standard power lines. With modern systems, one of the APs could be in monitor mode — for RF reporting purposes — and the other AP could be in access mode — for client network access. If the AP in access mode fails, the monitor mode AP can be switched to access mode to service the clients.

As you can see, availability options do exist, but it is up to you to select the best solution for your needs. My main point here is to remind you that availability may be as important as performance in your wireless network design. Don't forget this.

Network extension to remote areas

Network extension roles involve the use of bridges or APs acting as bridges. The network extension may also be implemented with a wireless repeater in short range extension scenarios, but it will not perform as well as a bridge link or mesh deployment. The following items should be considered when implementing network extension with 802.11 hardware:

- Ensure efficient spectrum utilization. If you have existing wireless networks in the 2.4 GHz band, consider implementing the network extension using the 5 GHz band. The bridge link between the existing network and the new coverage area does not need to support direct client connections, so any 2.4 GHz clients will not be hindered by this choice.

- Ensure sufficient throughput for the link. Consider the number of users that will connect in the new remote area. How much throughput will they require for the connection back to the existing network? Will communications mostly take place within the remote network, or does the extension exist solely for connectivity back to the existing network? If extra throughput is needed, you could implement multiple 802.11n or 802.11ac 5 GHz bridge links.

- Ensure proper security for the link. It is not uncommon for a network administrator to forget about the importance of security when implementing bridge links. Make sure that the bridges can talk to each other, and to no other wireless stations.

When implementing network extensions, remember that visual line of sight is not always required. If you only need to span 50-100 meters, you may be able to accomplish this with standard indoor APs operating in bridge mode. Use two or three such APs to created redundant links, and you have a highly available and well-performing network extension without exorbitant costs. Additionally, by using patch or panel MIMO antennas, you may be able to achieve very high data rates and avoid some levels of interference.

WLAN Bridging (Building-to-Building)

Building-to-building connectivity is very similar to network extension roles. The key difference is usually the distance of the link, and the fact that antennas and bridges may have to be installed in outdoor locations. For this reason, we have slightly different design requirements:

- Ensure proper antenna alignment. The bridge antennas must be very accurately aligned for longer distance links. In addition, the proper antennas must be used. For most outdoor bridge links that are PtP, you will use highly directional antennas, such as parabolic dish or grid antennas. If the bridge links are PtMP, you may use semi-directional antennas at the central location, or you may choose to use an omni-directional antenna.

- Ensure proper line of sight. Remember that, for long distance links, you must keep the first Fresnel zone cleared by at least 60 percent and preferably 80 percent. This will ensure a consistent connection. You may have to plan for first Fresnel zone inspections periodically over the life of the link, as foliage and new construction can interfere with the link over time.

- Ensure protected installations are used. For outdoor installations, protected enclosures should be used for two reasons. First, you want to protect the equipment from weather and exposure damage. Second, you want to prevent theft. Outdoor antennas and bridges are easy targets for thieves. Using a secured enclosure can help to protect against both problems.

Last-mile data delivery – Wireless ISP

Last mile refers to the final segment of the network that reaches from the provider's infrastructure to the subscriber's location. Unless you are the wireless ISP (WISP), you do not have to be concerned about the technical implementation of the wireless connection point on the side of the WISP infrastructure. However, you must design the internal implementation and requirements. The following should be considered:

- Ensure sufficient throughput. The wireless connection must provide enough throughput to meet your needs. Are you using the WISP for standard Internet access, or to implement a VPN-based WAN connection? How much throughput will be required of the users? Depending on the data rates available through the WISP, you may have to purchase multiple links.

- Ensure proper installation location. The location for the WISP's client device or devices should be planned carefully. Make sure it is close enough to the internal network connections so that extra cable runs are not required. If you are implementing multiple connections to the WISP, consider physically installing the different WISP clients in different areas of the building. This decision can help overcome localized (within the building section) power outage problems and also provide redundancy for connectivity to the Internet.

- Ensure the proper service-level agreement. Make sure that the vendor is committed to providing the services you need. It's not uncommon to closely read a service agreement only to find language like "data rates up to 10 Mbps," when you actually need to guarantee 10 Mbps. Know what you're buying and know that it meets your requirements.

Small Office/Home Office (SOHO) Use

In the typical SOHO environment, it is not uncommon to implement a single AP, or even a single wireless residential gateway (a fancy name for a wireless router that connects with your Internet Service Provider). The single device will usually provide Internet router, basic firewall functionality and, possibly, features like website filtering and authentication for Internet access. However, a few important issues are either unique to SOHO implementations, or more common among them. The important requirements to consider include:

- Ensure cooperative frequency usage. Be careful not to cause frustrations for neighboring wireless networks by using available channels that do not cause interference with other nearby networks. If you're implementing a brand-new wireless network, you can usually achieve this best in the 5 GHz band. However, strategic planning can also allow for non-interfering implementations in the 2.4 GHz band, but this space is becoming more congested every day.
- Ensure proper security. In SOHO implementations, it is very common for the wireless network to be wide open. This is not a good practice. Even if you adhere to the Bruce Schneier mindset that basically says, if the nodes are secure, the network is secure, you must realize that the business data traversing the wireless network has value and should be protected with encryption. For that matter, Bruce intended this only for home networks or networks that do not transfer data of value.

Mobile office networking

Mobile office networks often use similar equipment to SOHO networks. The considerations are similar, as well. However, one additional consideration must be made, and that is the Internet or WAN connectivity provisioning. With mobile offices, it is important to be able to uplink from any established location. You must select an ISP or WAN service provider that offers coverage in the target area. In most cases, this means using a wireless or satellite service for the uplink. Even with such services, you must ensure that coverage will be available where you need it, and when you need it.

Educational/Classroom use

For educational and classroom use, the glaring difference between corporate access and classroom access is in the area of regulations. The United States government, for example, requires that public schools filter their Internet access to prevent students from using the school network to access information that is deemed improper. Considerations for educational and classroom use installations include:

- Ensure a proper filtering technology is used. If you implement a wireless network in a regulatory domain that does not require filtering, this will not be a concern for you; however, in regulatory domains that require filtering, you must ensure that the hardware and software you implement can support an appropriate filtering technology.
- Ensure proper security is used. Like all other networks, classroom-use networks must implement proper security. The last thing you want is a cracker penetrating your school's network and attacking student computers, or possibly transmitting inappropriate information to their computers.

Industrial — Warehousing and Manufacturing

In a warehousing or manufacturing implementation, new concerns arise. You are now dealing with equipment that may generate incidental RF energy and cause interesting RF propagation behaviors. However, a good site survey should ensure proper operation, even in these environments. Other than the added complexity involved in the site survey, warehousing and manufacturing implementations are very similar to traditional corporate access installations. Just remember that all that shelving and metal equipment will do interesting things to your RF propagation.

Healthcare — Hospitals and Offices

When installing a WLAN in hospitals and healthcare environments, availability becomes a very important concern. Research has shown that hospitals can greatly reduce costs by implementing wireless healthcare networks. These networks allow nurses and doctors to enter patient information into central databases using mobile devices or digital notebooks in the patient rooms. However, if the network frequently fails, the medical staff tends to give up on the network and fall back to traditional records management. The result is that the potential benefits are never realized.

In addition to the need for high availability, you must consider the unique problems that may be presented in a hospital. For example, some rooms may be specially designed to block RF waves. Examples of such rooms include X-ray rooms and MRI rooms. Be sure to consider this factor and either plan for coverage within the rooms, or clearly communicate to the medical staff that coverage will not be available within these locations.

The primary concern with healthcare devices is availability and durability. The network must be available to the devices as the information they communicate is critical. This means coverage and capacity become important at all new levels. Every location must be covered by at least two APs and the throughput capacity must be sufficient, such that critical communication packets are not lost.

Most devices used in healthcare are stationary or nomadic. The nomadic devices are indeed portable but are not typically used while in motion. Rather, they are transported to a room or treatment area and are then utilized in that location. Such equipment includes X-ray cards, pulse oximeters, intravenous pumps and vital sign monitors.

The highly mobile devices that do require connectivity during motion are Wi-Fi phones, supply chain delivery robots and real-time location tracking devices. The same kinds of demands are placed on these highly mobile devices in healthcare as in other environments; however, in healthcare, like in emergency response, the stakes are higher.

> Security is a special consideration in healthcare networks. In order to comply with HIPAA regulations within the United States, healthcare providers must ensure that the patients' medical information remains confidential. You simply cannot implement an internal wireless network in a hospital without using encryption.

Municipal and Law Enforcement

Municipalities and law enforcement communications usually take place in licensed or reserved frequencies; however, emergency networks and WLANs located in headquarter buildings can still take advantage of license-free standard WLAN technology. If the network uses a licensed or reserved frequency, and you are responsible for implementing the network, you must ensure that you have the proper licenses and permissions to utilize the chosen frequency range.

Transportation Networks

Business commuters are expecting more from their travel providers. Airlines are implementing WLAN Internet access in-flight. Commuter trains already have Internet and many automobile manufacturers are looking for ways to utilize WLAN technology in their vehicles. Can you imagine pulling out of your garage only to have your car detect that you have only a half-gallon of milk available? Furthermore, can you picture how this car may detect as you are passing by a store that has milk on sale? This concept is not in use today, but the potential is very real.

The major concerns in a transportation network are similar to the concerns in a hotspot, as covered in the following section. The only additional concern is that which is similar to a mobile office — the uplink connection to the Internet or WAN. The difference between a mobile office and transportation networks is that the mobile office is nomadic — it is installed in a static location and remains there until it is no longer needed. The transportation network, however, is mobile in that the train, plane or automobile is moving and must maintain the connection. Satellite is the most common uplink solution, but WiMAX — or WiMAX-type — technologies may also be used within metropolitan areas.

Hotspots — Public Network Access

When implementing a hotspot, you will apply the typical considerations used for corporate data access. The primary goal of a hotspot is usually to provide either paid or free Internet access. However, three new considerations surface in a hotspot implementation:

- Ensure proper separation from your internal network. If your internal network uses the same Internet connection as the public hotspot uses, make sure that a firewall exists between the hotspot network and your internal network. I recommend that you install a separate Internet connection for your network, but if this is not an option, firewalls and physical network separation (a router and separate switches) can do the trick.

- Ensure payment processing is confidential. The payment processing should occur across an HTTPS connection, or some other connection that encrypts the communications, so that the users' payment information remains confidential.

- Ensure the users understand the terms of service. Now, you can't really ensure that they understand the terms of service, but you can ensure that they say they understand and agree to them. It is important that you can prove the user agreed to the terms of service, so that they cannot bring litigation against your organization for any damage done to their system while connected to your network. It's very similar to the old "I was on your property when I slipped and fell" issue. You must prove that you have not neglected your responsibility to the user, and a terms of service agreement usually takes care of this.

Hospitality

Hospitality deployments have the following characteristics:

- Require guest access, may be free or paid access

- May have soundproofing materials between rooms impeding propagation, in some instances

- Heavier capacity requirements than the past, with modern, savvy Wi-Fi users bringing laptops, tablets, and mobile phones

High Density Deployments

High density, very high density, and large public venue installations (like stadiums) require careful planning. Characteristics of high-density networks include:

- Many more APs with lower output power settings

- Implementation of solutions like band steering and load balancing to improve capacity

- Use of directional antennas in strategic locations to accommodate more APs

- Use of materials in the building or area to intentionally attenuate the signals

High density networks are those that have a high volume of users/devices connected to the network, as opposed to the networks of a decade ago, that rarely have more than 5-10 devices connected to an AP. For example, imagine a conference room with 40 people and three Wi-Fi devices each. If the room is 30x50 feet, that's 120 devices in just 1,500 square feet (140 square meters) of space. In the early part of the last decade, only 4-5 of those 40 people would even have a Wi-Fi device with them.

For even greater demands, consider the sports stadium or arena with many thousands of devices within just a few acres of property. Additionally, consider that these devices are being used to upload videos, photos and other content generated at the event. Clearly, new strategies for high density design were required.

When considering the number of client devices that can be serviced by a single AP radio, the following items must be considered:

- What data rates are supported by the clients?
- What applications will be used?
- Important characteristics include packet sizes, packet number and latency requirements.
- What usage level is expected?

Deployment Requirements and Solutions

In this section, you will briefly explore various technologies and gain an understanding for how they may assist you in meeting typical requirements of a WLAN. First, a list of common use types is provided, with issues for consideration. Then, specific technologies are defined and the solutions they provide is noted.

Use Types

WLANs are used for several purposes. Some are quite traditional, and others are created and new. As a CWNA, you should be familiar with the following:

- **Data:** Data communications are the most common communication on most networks. They vary significantly. You must analyze the applications used, their throughput requirements, the percentage of utilization, and the number of users, in order to determine the impact on WLAN design decisions.

- **Voice:** For voice communications, latency is the most important factor. This is because high volumes of throughput are not required. Voice is a lot of small packets, but they have to get there fast. Implementing end-to-end QoS is a must in networks that support VoIP. VoIP generally requires unidirectional delay of less than 150 ms.

- **Video:** For video communications, latency is important, and throughput must be accommodated as well. Depending on the codec used and the streaming technology, real-time video can require from 300 Kbps to 20 Mbps. High capacity is essential, when a significant number of video streaming devices are in use, such as in a classroom or training center.

- **Real-Time Location Services (RTLS):** RTLS is incorporated into many WLAN vendor solutions today, and it requires special design considerations. In many cases, the APs/sensors must be around the periphery of the area with additional APs/sensors throughout to accomplish proper triangulation processes. In most cases, a coverage-based WLAN design will not meet the requirements of RTLS.

- **Mobility:** Mobile devices such as tablets and smartphones are highly mobile. Therefore, roaming becomes a significant consideration. Users will likely desire to move around while they continue using their devices. Security, as it related to roaming, is covered in Chapter 11.

- **High density:** High density design is one of the most complicated WLAN implementation scenarios. It is discussed in more detail later in this chapter.

- **Guest access:** Guest access can be as simple as a separate SSID, or as complex as a separate SSID, with registration, a captive portal, separate VLANs, GRE tunnels, and more.

- **BYOD:** *Bring Your Own Device (BYOD)* requires considerations of security and data leakage. BYOD is not synonymous with guest access. In many cases, the users want to use their devices to access your applications and data. This requires, in most cases, a strong Mobile Device Management solution that can control the abilities of the devices and protect data on them.

Technologies and Benefits

The following technologies should be understood at a basic level, and the purpose or benefit of their use:

- **AirTime Fairness:** AirTime Fairness is a proprietary feature that seeks to improve the use of airtime. In normal 802.11 operation, contention processes slightly favor devices using more recent physical layer technologies. However, the advantage is somewhat slight. Further, multiple devices using the same PHY technology, such as 802.11n, may have highly disparate connectivity rates, such as 6.5 Mbps compared with 300 Mbps. As you can imagine, the device with slower data rates takes significantly longer amounts of time to transmit the same amount of data as the higher-rate devices. Airtime Fairness seeks to balance airtime usage instead of balancing the number of frames. This technique improves performance for the entire cell, while minimizing the impact on the lower-speed stations.

- **Band Steering:** Band steering can be an effective tool in all WLANs, but it is particularly beneficial in high density networks. It works by simply ignoring 2.4 GHz requests and responding to only 5 GHz requests, once the 2.4 GHz maximum connection limit is reached. Alternatively, it will always delay a response in the 2.4 GHz band to see if the client attempts to connect on the 5 GHz band. If the client does not connect on the 5 GHz band, the AP may respond to future attempts (probe requests or authentication requests) in the 2.4 GHz band.

- **Hotspot 2.0/Passpoint certification:** A Wi-Fi Alliance certification for hotspot deployments that allows members to connect automatically to hotspots that participate in any subscription service with which they have accounts. Connections are automatic and transparent to the user.

- **Radio Resource Management (RRM)/Adaptive Radio Management (ARM):** The basic features of all *RRM* solutions is dynamic channel selection or assignment, and transmit power level management. Most RRM solutions use some kind of neighbor communications that provide information to APs about their neighbors, and then this information is passed up the chain to a WLAN controller that uses the data to make decisions about channels and power levels. RRM can be useful in highly dynamic environments, or when no local administrators are skilled in channel selection. Over time it will likely improve and be more useful in even more scenarios.
- **Mobile Device Management (MDM):** The generic name for a solution that allows for the registration (enrollment), management and decommissioning or disenrollment of mobile devices such as laptops, tablets and mobile phones. *MDM* may be used to manage enterprise-owned devices, or user-owned (Bring Your Own Device (BYOD)) devices. Organizations typically treat enterprise devices differently than user-owned devices, but both may be managed by a modern MDM solution.
- **Network Access Control (NAC):** *Network Access Control (NAC)*, called Network Access Protection (NAP) by some vendors, is used to analyze a connecting device, determine if it meets minimum network access requirements, and then either grant or deny access accordingly. Many NAC solutions also provide quarantine access, which places the device on a separate network, from which they can install updated, new anti-virus definitions, etc.

Required Network Services

Several infrastructure network services are required for WLAN functionality. This section provides an overview of these services.

Dynamic Host Configuration Protocol (DHCP)

DHCP is used to provide IP configuration sets to computers on the network. It is sometimes said that DHCP provides IP addresses to computers, but this is an over simplification. In actuality, DHCP can provide the following, and more:

- IP addresses
- Subnet masks

- Default gateways (routers)
- DNS server
- Lease durations
- Configuration options

When planning for DHCP support of the WLAN, at least three things should be considered:

- Will you use an existing scope, or create new separate scopes for the WLAN operations?
- What options should be provided by the DHCP server for proper WLAN functionality?
- How long should leases endure?

The selection of a DHCP scope for the WLAN is actually very important. It can impact roaming capabilities and result in a large subnet, if you're not careful in your planning. For this reason, vendor recommendations should be considered when answering the first question. In general, it is best to place WLAN traffic on a separate IP subnet from wired traffic, so that it can be managed accordingly. IPv4 is still the most commonly used IP solution, with IPv6 very slowly gaining momentum.

Lease durations or timeouts should also be considered, relative to DHCP. This is particularly important in networks where users come and go, such as guest WLANs or mobile areas of the WLAN, like a lobby. Setting the lease duration to a few hours instead of a few days can reduce DHCP pool (scope) exhaustion and allow the DHCP operations to continue.

DHCP Option 43 can be used to provide the WLAN controller IP address, so that IPs can locate the controller when they first connect to the network. The client sends an Option 60 request to receive information in some instances.

Domain Name System (DNS)

DNS is used to resolve domain names to IP addresses. It is used in WLANs to service clients for host name resolutions, but it is also used for specific WLAN functions, including locating controllers. For example, when a Cisco AP receives its IP configuration from the DHCP server, if option 43 is not specified, the AP will try to resolve the DNS hostname *CISCO-CAPWAP-CONTROLLER.localdomain*, where *localdomain* is the domain configured by DHCP.

Network Time Protocol (NTP)

NTP is used to configure the time on a device using a network time server. Accurate time configuration is important for logging and other operations. Most WLAN controllers and APs can have their internal clock set using a network time server.

ACL Management

Access Control Lists (ACLs) are used on network routers to control the traffic that is enabled for pass-through to other networks. Like a firewall, the routers must allow the communications through required by the WLAN. In fact, ACLs on routers are more likely to cause problems than firewall rules, as the controllers and APs are more frequently inside of the firewall. Check the vendor literature for the ports (both UDP and TCP) used for various communications and ensure that the appropriate ACLs are configured.

VLAN Management

Virtual LANs (VLANs) are used to segregate network traffic into separate broadcast domains, even when they traffic originates in the same physical switch or switch aggregate. Additionally, VLANs, through tunnels, can span multiple physical subnets. In WLANs, VLANs are used to separate wireless traffic from wired traffic, and they are used to separate one wireless network from another, such as separating an enterprise WLAN from a guest WLAN using the same APs.

The VLAN can be configured in relation to the SSID within the AP configuration profile. It can also be configured through information provided by the RADIUS server at the time of authentication. Additionally, VLANs can simply be assigned, based on the switchport configuration to which the AP is connected. That is, with an autonomous or cloud-managed AP, the switchport can determine the VLAN on which all traffic coming from the AP is placed.

RADIUS and LDAP

RADIUS is used for authentication, and LDAP is a protocol used to access a directory service. A directory service contains network objects, like user accounts and groups. In many deployments, RADIUS servers are used for authentication, but they often depend on large-scale directories for account information, such as Microsoft Active Directory. LDAP is used to make the connection from the RADIUS server to the Active Directory Domain Controller.

From a design perspective, you can recommend the use of a RADIUS server that fits well with the existing network solutions. For example, in a Microsoft Active Directory environment, using the Microsoft RADIUS server makes sense. In a Linux/Unix environment, FreeRADIUS or a commercial RADIUS product may be recommended, or the organization may choose to use a RADIUS server from the network infrastructure provider like Cisco.

Public Key Infrastructure (PKI)

Because many EAP types rely on certificates, and some depend on client certificates, a PKI may be required. A PKI is comprised of certificate authorities (CAs) that either authorize other CAs to distribute certificates or generate and distribute (issue) certificates themselves. As a WLAN designer, you will have to determine whether a PKI will be available or not, so that you can select the appropriate EAP type. If a constraint is in place requiring an EAP type that must have client certificates, you should be sure to document the requirement of a PKI in the design plans, though the details of the PKI can be left up to the individuals who will implement that project.

Wired Network Capacity Requirements

As WLAN speeds go up, so do the demands on the wired link available to the AP. However, many myths are floating around about 802.11ac and 1 Gbps wired links. Due to varied data rates of clients, MAC overhead, and other factors, even 4x4:4 802.11ac APs are unlike to read the full capacity of a 1 Gbps wired link before 2020, in the vast majority of installations. Will we ever get to the point where every AP needs a 10 Gbps link or greater? Sure, we're just not there yet.

Cable Types and Lengths

For today's cable runs, CAT6 cables should be used. They provide the best quality link and support full 802.3at PoE capabilities. Remember that you should not run an Ethernet cable beyond 100 meters, or approximately 300 feet. This can be an important constraint during AP installations.

Chapter Summary

In this chapter, you explored various use core roles of WLANs and some of the requirements they impost. You also considered features of WLAN technology and the capabilities they provide. Finally, you considered the network infrastructure services needed to make your WLAN work. In the next chapter, you will explore security issues and solutions in 802.11 WLANs.

Points to Remember

Remember the following important points:

- It is important to understand the SLA, when subscribing to WISP services.
- With modern devices supporting 5 GHz radios, it is a good idea to use 5 GHz APs, even in SOHO installations.
- K-12 classroom networks often require filtering in compliance with local government regulations.
- Security is important in healthcare deployments because very specific privacy constraints are required for health regulation compliance.
- High density deployments require more APs, strategic placement of APs, directional antennas, and the use of existing materials to intentionally attenuate the signals.
- VoIP typically requires less than a 150 ms unidirectional delay.
- Video is sensitive to delays but is even more demanding on throughput.
- RTLS deployments typically require APs around the periphery of the facility, and throughout.
- AirTime Fairness attempts to give more airtime to higher data rate devices.
- Band steering attempts to direct STAs to 5 GHz, when they support it.
- RRM and ARM provide automatic channel selection and power settings on an enterprise scale.

- When DHCP is used for guest networks, the leaser duration is usually set to a shorter window than on enterprise networks.
- A PKI may be required if you use 802.1X/EAP methods that require client certificates.
- CAT6 cables should be used for most modern deployments and should not exceed 100-meter cable runs.

Review Questions

1. What two important factors must be accommodated in all WLAN deployments?
    a. Coverage and Capacity
    b. RTLS and AirTime Fairness
    c. Coverage and Low Latency
    d. DMG PHY and TVHT PHY

2. What must be carefully reviewed, when subscribing to a WISP service?
    a. 802.11ac spatial streams supported
    b. Service Level Agreement
    c. Mounting options for customer equipment
    d. Types of antennas in use

3. What type of room in a hospital is a common area of challenge for Wi-Fi coverage?
    a. Lobby
    b. Cafeteria
    c. X-ray rooms
    d. Office space

4. What kind of materials are sometimes used in hospitality construction, that can cause increased attenuation for RF signals?
    a. Carpet
    b. Windows
    c. Doors
    d. Soundproofing

5. Where MUST APs or sensors be placed in RTLS deployments?
    a. In the center of the area
    b. Around the area
    c. On the roof
    d. In the basement

6. For what scenario is an MDM solution a good choice?
    a. As an alternative to a RADIUS server
    b. BYOD implementation
    c. RTLS deployment
    d. VoIP deployment

7. What proprietary solution attempts to grant more transmission time to devices with higher data rates?
   a. AirTime Fairness
   b. Band Steering
   c. RRM
   d. ARM

8. What solutions typically use neighbor communications to gather information for a controller, to make effective decisions in channel assignment for APs?
   a. Band Steering
   b. RTLS
   c. RSSI threshold
   d. RRM

9. What is often used to ensure that connecting clients meet minimum security requirements?
   a. MDM
   b. RRM
   c. NAC
   d. ARM

10. What can be used to segregate network traffic into separate broadcast domains?
    a. VLANs
    b. RBAC
    c. RADIUS
    d. Band Steering

Review Answers

1. **A is correct.** Coverage and capacity are required design specifications for all WLAN deployments.
2. **B is correct.** The SLA should be reviewed to ensure you are subscribing to the service you require.
3. **C is correct.** X-ray rooms often have lead-lined walls that hinder RF propagation.
4. **D is correct.** At times, hotels will install special soundproofing materials between rooms during construction. The materials can cause high levels of RF attenuation.
5. **B is correct.** RTLS APs/sensors are installed throughout the area, but they typically must be installed on the periphery of the coverage area.
6. **B is correct.** BYOD deployments can benefit significantly from MDM solutions.
7. **A is correct.** AirTime Fairness gives more airtime (in downlink communications) to higher data rate clients.
8. **D is correct.** RRM and ARM provide automatic channel selection and output power settings in enterprise deployments.
9. **C is correct.** NAC is used to validate that clients meet minimum requirements of the network, including security requirements.
10. **A is correct.** VLANs are used to segregate network traffic into separate broadcast domains.

Chapter 11 — Security Solutions for WLANs

Securing a wireless network is a primary concern for the wireless administrator. Wireless networks introduce new threats against which we must defend. Wired and wireless networks also have shared security concerns. For this reason, we'll begin this chapter by addressing general security principles that would apply equally to wired or wireless networks. Next, we'll cover the weak security methods that should not be used, followed by effective security solutions and security enhancements. Finally, we'll explore secure management protocols.

Security Principles

Both wired and wireless networks are susceptible to similar attacks. While wireless networks introduce new attack methods that we must prevent or mitigate, we must still protect them against traditional attacks. This reality means that the traditional security concerns of wired networks all apply to our wireless networks, and we also have to deal with unique wireless threats.

Network security has focused mostly on wired networking for much of the history of networking. A cursory glance at any book on the topic of network security, reveals that wireless security is an add-on or afterthought, at best (unless the book is actually on the topic of wireless security). I would suggest that the mobility of the wireless nodes makes them a more appealing target for the attacker. For example, it's much easier to steal a laptop, than it is to steal a desktop. Of course, it's much easier to eavesdrop on the network communications of a wireless node, than it is a wired node; wireless nodes transmit the data through the air, where any attacker can eavesdrop on it. For these reasons alone, I feel that more energy should be expended to secure the average wireless node, than the average wired node. Now, I'm not suggesting that we should ignore the wired nodes. I'm only suggesting that greater caution must be taken when implementing a wireless node, and more is required to secure a typical mobile device.

With that said, accomplishing wired security levels is really the first step to securing a wireless node. On wired networks, we've traditionally focused on authentication, authorization and accounting. In some scenarios, communications confidentiality and integrity are also considered. And of course, availability is important. Implementing redundant routes through the network, for example, can help to mitigate the impact of a denial of service (DoS) attack.

You may have recognized the six areas I mentioned in the preceding paragraph as the AAA and CIA security models. The triple A (AAA) security model focuses on authentication, authorization and accounting. The CIA security model focuses on confidentiality, integrity and availability. I'll review each of these six areas of protection that have been the focus of security on wired networks in the following sections, to lay a foundation of security principles.

Authentication

One of the most important components of a security strategy is an identity management system (IMS). An IMS provides a storage location for identity objects, typically called user accounts, and one or more methods for connecting to that storage location and proving identity ownership — a process known as *authentication*. User accounts are objects that identify users and are owned by users. The user accounts provide properties for use by authentication systems and network operating systems. Besides user accounts, certificates, biometrics, tokens and other credentials may also be used for authentication or identity management.

Without a clear understanding of authentication and identity management, you will have difficulty installing a secure system. Both basic and advanced authentication systems exist, and many systems include the ability to support both. Windows Server systems allow for advanced authentication mechanisms through the Network Policy Server (Microsoft's RADIUS implementation) and allow for basic authentication using passwords against the Active Directory database. Each method serves a valid purpose and is best for certain scenarios. When you determine which method is right for your scenario, you have taken the first step to secure authentication.

Once you've selected the appropriate advanced or basic authentication method, you must determine whom to authenticate. Will you only authenticate known or identified users, or will you allow some level of anonymous access?

Authentication should not be confused with authorization. Authentication can be defined as proving a person or object is who or what he or it claims to be. Authorization is defined as granting access to a resource by a person or object. Authorization assumes the identity has been authenticated. If authentication can be spoofed or impersonated, authorization schemes fail. From this, you can see why authentication is such an integral and important part of network and information security. When an attacker breaks your authentication system so that he is seen as an authenticated user, the authorization become irrelevant. Authentication must be strong, if authorization is to serve its purpose.

Advanced authentication systems generally utilize stronger credentials and better protection of those credentials, than basic authentication systems. The strength and protection of the credential is determined by the effort it takes to exploit it. A password-protected credential is usually considered weak, when compared with biometric-protected credentials. This, in some cases, is a misconception, because strength of authentication really depends on how the authentication information (the credential and proof of ownership) is sent across the network. If you were to implement a biometric system, such as a thumb scanner, and the client sent the credentials and proof of ownership (a unique number built from the identity points on the user's thumb) to the server in clear text, it would be no more secure than a standard password-based system; however, I am not aware of any biometric authentication system that sends the authentication data as clear text.

The key element, which will provide a truly strong authentication pathway, is the encryption or hashing of the user credentials, or at least the proof of identity information (for example, the password). This can be accomplished with Virtual Private Networking (VPN) technology, or with well-designed authentication systems. One example of a well-designed authentication system is 802.1X with a strong EAP method. 802.1X and EAP methods are used to secure both wired and wireless connections, at the network access level.

Remember that authentication protocols, which send the credentials as clear text, are insecure. The ***authentication protocol*** should not set the password at all, and the authentication information should be transmitted over a secure (encrypted) channel, whenever possible.

You use authentication every day of your life. When you are at a seminar or training event, and the speaker says he is an expert on the topic of his speech you use authentication mechanisms to verify this information. You listen to the information he delivers and use it to determine if he is truly an expert. In addition, suppose someone walks up to you and says, "Hi, my name is Susan and I am tall." You would look at her and compare her height with a height you consider to be tall and authenticate whether she is truly tall or not. If she is not tall, by your standards, she will lose credibility with you.

Remember the word credentials? Consider other important "cred" words: credit, credibility, and credentials. Do you see how they are related? They all have to do with having proof of something. When you have good credit, you have proof of your trustworthiness to pay debts. When you have credibility, you have proof that you are authentic, persuasive, and dynamic. When you have credentials, you have an object or the experience that proves your skill or identity. Authentication results in the verification of credentials.

Advanced authentication is more secure than basic authentication because advanced mechanisms are used to protect the user's credentials. This usually means protecting a user name and password pair, but it can also include protecting a user/certificate combination, a user/machine combination, or any other user/object combination used to identify a specific user. In addition to the extra protection offered by advanced authentication systems, when 802.1X-based systems are used, you have the benefit of standards-based technology. This means that hardware from many different vendors is likely to support the authentication process. Sometimes driver or firmware upgrades are required, but there is often a path which can be taken to implement the authentication mechanism.

Authorization

Authorization is the process used to validate the rights of a user in relation to an action or resource. Authorization depends on authentication. If the authentication system is weak and easily exploited, the authorization system is useless. In a wireless network, authorization may be performed using role-based access control (RBAC) solutions, or proprietary vendor segregation methods. For example, many wireless hotspot solutions will allow for the creation of a captive portal. Until the user is properly authenticated, he will not be authorized to get outside of the captive portal. Once the user is authenticated, he may be granted the right to access other Internet websites.

Authorization may be a network-level activity, or it may be an application-level activity. For example, when you connect to an FTP server, you perform application-level authentication and authorization. The FTP server software determines the folders to which you may be given access, and the folders to which you may be denied access. Network-level authorization may be performed using routers or switches. Depending on the vendor, the routers and switches may support specific permissions that can be used to limit the valid activities of a connecting user.

Accounting

In order to prevent repudiation (deniability for an action taken), organizations implement accounting systems. Another word for accounting is logging. When you enable logging for network access, you can track who has accessed the network and from where the access was made. Most enterprise WLAN solutions will allow you to log network access and utilization.

Confidentiality

Confidentiality is provided through the use of encryption. The process of converting data from its normal state to an unreadable state is known as encryption. The unreadable state is known as ciphertext (or cipherdata) and the readable state is plaintext (or plaindata). The normal way to encrypt something is to pass the data through an algorithm, using a key for variable results. For example, let's say we want to protect the number 108. Here is our algorithm for protecting numeric data:

original data / crypto key + (3 x crypto key)

Using this algorithm to protect (encode or encrypt) the number 108 with a key of 3, we come up with this:

108 / 3 + (3 x 3) = 45

In order to recover the original data, you must know both the algorithm and the key. Needless to say, modern crypto algorithms are much more complex than this, and keys are much longer, but this overview gives you an idea of how things work with data encryption.

Whatever encryption solution you choose, it is essential that the encryption key store be protected. For this reason, many organizations choose to implement a public key infrastructure (PKI), which is used to securely store encryption keys in certificates, using a hierarchy of authentication and authorization, that is difficult to penetrate. Some EAP methods used in WLANs may demand a PKI, particularly if they require client certificates.

Integrity

Integrity is used to ensure data reliability. With data integrity measures in place, you can determine whether or not data has been modified during transmission. Hashing algorithms are frequently used for this purpose. A hashing algorithm is a mathematical procedure that can receive variable length input and generate a theoretically unique fixed-length output. Stated more simply, a hashing algorithm creates a number that identifies a unique chunk of data. For example, you can pass a Microsoft Word document through a hashing algorithm to generate a number that can be used to determine if the document is ever modified. If the document is modified in the slightest manner, it will cause a different number to be generated by the hashing algorithm. When the same hashing algorithm is used against two documents that are assumed to be the same, and the result is two different numbers, we know that the documents are not actually the same. Through this process, data integrity can be enforced. Examples of hashing algorithms include MD5 and SHA-1.

Availability

The final element of focus is availability. We employ various security-related techniques to ensure that systems are available when needed. For example, building redundant routes through our network can provide availability. Additionally, implementing server clustering for important servers can ensure availability. You may use a RADIUS server for authentication on your network. If the client nodes must authenticate through the RADIUS server, building redundancy into the RADIUS solution is a likely requirement. You can implement more than one RADIUS server, or you can run the RADIUS service on a multi-node cluster.

Weak Security Options

Several security options may be available in WLAN gear that should not be used, or at least should not be relied upon for effective security. These include:

- WEP
- Shared Key authentication
- SSID hiding
- MAC filtering
- Improper use of WPA (TKIP/RC4)
- Open System authentication alone, with the exception of intentional public networks
- Wi-Fi Protected Setup (WPS)

This section covers these "security" options.

> WEP is covered in detail as an illustration of security weaknesses. However, it is not tested on the CWNA-108 exam, beyond knowing that it should not be used, and has been deprecated in the 802.11 standard.

Wired Equivalent Privacy (WEP)

The original 802.11 standard specified the Wired Equivalent Privacy (WEP) protocol for the purpose of providing security that was comparable to that of wired networks. Specifically, the goal was to prevent casual eavesdropping on a WLAN. The IEEE stated that "they (pre-RSNA security standards) fail to meet their security goals." Indeed, WEP has failed as a security solution, and should not be implemented in any WLAN today. It saddens me when I see systems that still allow it, and yes, I came across a brand new one in 2017.

WEP-40 used a 40-bit key for encryption. The encryption algorithm used was RC4. WEP-104 used a 104-bit key for encryption. The encryption algorithm used was also RC4, like WEP-40. 40-bit keys are certainly considered small by today's security standards, but exportability of the encryption technologies implemented, based on the standard, was the most likely reason for limiting the key size to 40-bits initially. Vendors implemented 104-bit keys quickly, and the IEEE acknowledged them in later updates of 802.11.

If you saw a configuration interface that referred to a 64-bit or 128-bit WEP key, this was because the WEP implementation used an initialization vector that was 24-bits long, for both 40- and 104-bit WEP. Of course, 40 plus 24 is 64, and 104 plus 24 is 128. The *initialization vector* (IV) was a non-static 24-bit number that was generated for each frame. However, a 24-bit pool resulted in only 16,777,216 possible unique IVs. This limited pool required the reuse of IV values at some eventual time. The 24-bit IV was transmitted in cleartext. For this reason, the encryption was said to be 40-bit or 104-bit, and not 64-bit or 128-bit, although it is still quite common to see vendors get the terms wrong today. Some vendors even expanded WEP by allowing a 128-bit encryption key, for a total of 156-bit WEP key, when the 24-bit IV was added. This was non-standard and, if implemented, required the use of a specialized supplicant (client) that could handle the non-standard encryption key size.

WEP was only intended to protect the data payload in a frame. For this reason, the header portion of the frame was not encrypted. The header includes the source and destination MAC addresses and can easily be read using a protocol analyzer that supports the capture of 802.11 frames. One major problem with WEP, as I'll discuss in detail below, was that once you have a valid WEP key, you can decrypt all the packets that use that WEP key. This worked with all captured data packets from the capture session, and could be replayed later, when a valid WEP key was used in the protocol analyzer. A hacker could use this method to capture encrypted packets, and later, after successfully performing a brute force or dictionary attack, all the packets could be viewed in their unencrypted form.

An understanding of the basic WEP process will help you to understand the weaknesses that are covered next. The WEP process starts with the inputs to the process. These inputs include the data that should be encrypted (usually called plaintext), the secret key (40-bits or 104-bits), and the IV (24-bits). These inputs are passed through the WEP algorithms to generate the output (the ciphertext or encrypted data).

Since WEP is a Layer 2 security implementation, it doesn't matter what type of data is being transmitted, as long as it originates above Layer 2 in the OSI model. In order to encrypt the data, the RC4 algorithm is used to create a pseudorandom string of bits called a keystream. The WEP static key and the IV are used to seed the pseudorandom number generator used by the RC4 algorithm. The resulting keystream is XORed against the plaintext to generate the ciphertext. The ciphertext alone is transferred without the keystream; however, the IV is sent to the receiver. The receiver uses the IV that was transmitted and the stored static WEP key to feed the same pseudorandom number generator to regenerate the same keystream. The XOR is reversed at the receiver to recover the original plaintext from the ciphertext.

With the rapid increase in processor speeds, cracking WEP has become a very short task, though it already was more than 10 years ago. The weaknesses in WEP include the following:

- Brute Force Attacks
- Dictionary Attacks
- Weak IV Attacks
- Re-Injection Attacks
- Storage Attacks

In late 2000 and early 2001, the security weaknesses of WEP became clear. Shortly thereafter, many attack methods were developed, and tools were created, that made these attack methods simple to implement for entry-level technical individuals.

The *brute force* attack method is a key-guessing method that attempts every possible key, in order to crack the encryption. With 104-bit WEP, this is really not a feasible attack method; however, 40-bit WEP can usually be cracked in one or two days with brute force attacks, using more than 20 distributed computers. The short timeframe is accomplished using a distributed cracking tool, like the older jc-wepcrack. jc-wepcrack was actually two tools: the client and the server. You would first start the tool on the server and configure it for the WEP key size you think the WLAN uses that you are cracking and provide it with a pcap file (a capture of encrypted frames) from that network. Next, you launch the client program and configure it to connect to the server. The client program will request a portion of the keys to be guessed and will attempt to access the encrypted frames with those keys.

The *dictionary attack* method relies on the fact that humans often use words as passwords. The key then is to use a dictionary-cracking tool that understands the conversion algorithm used by a hardware vendor, to convert the typed password into the WEP key.

The *weak IV attacks* are based on the faulty implementation of RC4 in the WEP protocols. The IV is prepended to the static WEP key, to form the full WEP encryption key used by the RC4 algorithm. This means that an attacker already knows the first 24 bits of the encryption key, since the IV is sent in cleartext as part of the frame header. Additionally, Fluhrer, Mantin and Shamir identified "weak" IVs in a paper released in 2001. These weak IVs result in certain values becoming more statistically probable than others and make it easier to crack the static WEP key. The 802.11 frames that use these weak IVs have come to be known as *interesting frames*. With enough interesting frames collected, you can crack the WEP key in a matter of seconds. This reduces the total attack time down to less than 5-6 minutes on a busy WLAN.

What if the WEP-enabled network being attacked is not busy, and you cannot capture enough interesting frames in a short window of time? The answer is a *re-injection attack*. This kind of attack usually re-injects ARP packets onto the WLAN. The program aireplay can detect ARP packets based on their unique size and does not need to decrypt the packet. By re-injecting the ARP packets back onto the WLAN, it will force the other clients to reply, and cause the creation of large amounts of WLAN traffic very quickly. For 40-bit WEP cracking, you usually want around 300,000 total frames to get enough interesting frames, and for 104-bit WEP cracking you may want about 1,000,000 frames.

Storage attacks are those methods used to recover WEP or WPA keys from their storage locations. On Windows computers, for example, WEP keys have often been stored in the registry in an encrypted form. An older version of this attack method was the Lucent Registry Crack; however, it appears that the problem has not been fully removed from our modern networks. An application named *wzcook* could retrieve the stored WEP keys used by Windows' Wireless Zero Configuration. This application recovered WEP or WPA-PSK keys (since they are effectively the same, WPA just improves the way the key is managed and implemented) and came with the Aircrack-ng tools still used for cracking these keys. The application only works if you have administrator access to the local machine, but in an environment with poor physical security and poor user training, it's not difficult to find a machine that is logged on and using the WLAN for this attack.

WEP makes up the core of pre-RSNA security in 802.11 networks. I hope the reality that WEP can be cracked in less than 3-5 minutes is enough to make you realize that it was a weak security solution. In the end, businesses and organizations that have sensitive data to protect must take a stand for security, and against older technologies. This means that you should not be implementing WEP anywhere in your organization. When you have the authority of a corporation, the government, or even a non-profit oversight board, you can usually sell them on the need for better security with a short (3-5 minutes or less) demonstration of just how weak WEP is.

Shared Key Authentication

Shared Key authentication was thought to be more secure than Open System authentication at the time of their joint specification, in the 802.11-1997 standard. This was due to the fact that Shared Key authentication verified the requestor using a real authentication method, whereas Open System authentication simply authenticated the requestor, regardless of identity. However, Open System authentication leaves the door open for the use of advanced and evolving security technologies that run across the association created, using null authentication. Shared Key authentication relies on a specific set of security technologies, namely

WEP and RC4, which have proven to be insecure in their 802.11 implementation. As stated by the standard, Shared Key authentication "is only available if the WEP option is implemented." The WEP option is never implemented today, or at least it shouldn't ever be, and so Shared Key authentication should no longer be available.

Shared Key authentication uses a secret key that is shared by the requestor (the STA desiring to be authenticated) and the responder (the STA performing the authentication). The method of communicating this secret key into the two STAs in the first place is not specified by the 802.11 standard, but it is most usually implemented by manually typing the key into the client's network card configuration software interface. The standard specifies that this secret key shall not be transmitted across the WLAN and assumes that a secure channel was used for installation of the secret key on the requestors, as well as the responders.

In the traditional Shared Key system, the requestor is a WLAN client STA, and the responder is a WLAN AP. The responder may also be another WLAN client STA or any other IEEE 802.11-compliant device. Unlike Open System authentication, the Shared Key authentication process involves more than just requesting authentication, and then blindly approving it. There are four frames involved in the Shared Key authentication system. The first frame is the initial authentication request frame. Assuming the responder is configured for Shared Key authentication, the responder will respond to the request frame with challenge text that will be used to authenticate the client's possession of the secret key. The requesting client will then encrypt the challenge text with the secret key and send the challenge text back to the responder in the encrypted state. The responder decrypts the challenge text using the secret key. If the result matches the challenge text, then the requestor has been authenticated, and a successful authentication response frame is sent to the client.

While this authentication process (Shared Key) appears to be much more secure than Open System authentication (and indeed it was for a short time), its dependence on WEP for the encryption of the authentication challenge response, the transmission of the challenge in clear text, and the ongoing communications was its greatest weakness.

MAC Filtering

Vendors of wireless devices and books on wireless networking often provide a list of the "Top 5" or "Top 10" things you should do to secure your WLAN. This list, sadly, often still includes MAC filtering and SSID hiding or cloaking — particularly in consumer or SOHO gear. The reality is that neither of these provides a high level of security. MAC addresses can easily be spoofed, and valid MAC addresses can be identified in just a few moments. For example, an attacker can find out the AP in an infrastructure BSS by looking for the MAC address that sends out Beacon frames. This will always be the AP in the BSS. With this filtered out of the attacker's protocol analyzer, he has only to find other MAC addresses that are transmitting with a destination MAC address equal to that of the AP. Assuming the captured frames are data frames, the attacker now knows a valid IP address.

There is no question that MAC filtering will make it more difficult for an attacker to access your network. The attacker will have to go through the process I've just outlined (or a similar process) in order to obtain a valid MAC address to spoof. However, you are adding to your workload by implementing such MAC filtering and you have to ask, "Am I getting a good return on investment for my time?" The answer is usually no. Assuming you are using CCMP with a strong EAP method for authentication (or even pre-shared keys), this will be so much more secure than MAC filtering could ever hope to be, diminishing value with the extra effort. I recommend that you do not concern yourself with MAC filtering in an enterprise or SMB implementation. It may be useful in a SOHO implementation, but I even question its value then.

SSID Hiding

Hiding or cloaking the SSID of your WLAN falls into a similar category as MAC filtering. Both provide very little in the way of security enhancement. Changing the name of your SSID from the vendor defaults can be very helpful, as it will make dictionary attacks against PSK implementations more difficult. This is because the SSID is used in the process of creating the pairwise master key. Hiding the SSID only makes it difficult for casual eavesdroppers to find your network. Hiding the SSID also forces some of your valid clients to send out probe requests in order to connect to your WLAN. This means that, when the user turns on his or her laptop in a public place, the laptop is broadcasting your SSID out to the world. This could be considered a potential security threat, since a rogue AP of any type can be configured to the SSID that is being sent out in the probe requests.

I always recommend changing the SSID from the default, but I never recommend hiding the SSID for security purposes. Some people will hide the SSID for usability purposes. Turning off the SSID broadcast in all AP Beacon frames will prevent client computers from "seeing" the other networks to which they are not supposed to connect. This may reduce confusion, but SSID hiding should not be considered a security solution.

An argument can be made such as the following, "MAC filtering and SSID hiding are not strong security techniques, but they are security techniques because they make it more difficult to penetrate the network." Let's compare this to our physical world. For example, closing your door at night is not a strong security technique, but it is a security technique because the intruder now has to turn the door knob and push the door open to enter your house. Does that sound right to you? Me either.

Improper Use of WPA (TKIP/RC4)

The *temporal key integrity protocol (TKIP)* is an optional encryption method defined in 802.11. TKIP uses RC4 encryption like WEP; however, the weaknesses of WEP are addressed by enlarging the IV pool (it is 48-bits instead of 24-bits) and using true 128-bit static keys. TKIP also implements a stronger integrity-checking algorithm in the message integrity check (MIC) algorithm, instead of the ICV used with WEP.

TKIP is not as processor-intensive as CCMP. For this reason, many older devices were able to be upgraded through firmware patches to support TKIP. The Wi-Fi Alliance released a certification known as Wi-Fi Protected Access (WPA) before the IEEE 802.11i-2004 amendment was ratified. WPA is essentially the TKIP/RC4 implementation documented in the 802.11 standard. However, it was a transitional security solution and even though it implements the 4-way handshake, it should not be used today unless absolutely required.

The improper use of WPA is twofold: using it when you don't need it and using it with weak passphrases. If you don't need it, and the only true need today is very old hardware, don't use it. If you do need it because of old irreplaceable hardware, use very long and complex passphrases, and change them at least twice a year.

Open System Authentication — By Accident

Open System authentication has not been deprecated, since it is still used as the starting point for modern authentication and encryption implementations, such as WPA2. Open System authentication is essentially a null authentication, in that any client requesting authentication is approved for authentication, as long as the AP (or recipient STA in an IBSS) is configured for Open System authentication (the dot11AuthenticationType is set to Open System). There is no actual verification of identity, but it moves the 802.11 state machine forward in the association process.

Open System authentication includes the transfer of only two frames. Both frames are management frames and are of the subtype *authentication*. The first frame is transmitted from the authentication initiation STA to the authenticating STA (an AP in an infrastructure BSS). This frame includes an authentication transaction sequence number equal to 1. The second frame is transmitted from the authenticating STA to the authentication initiation STA and includes an authentication transaction sequence number equal to 2. This second frame will include a status code field that indicates the success or failure of the authentication. A value of 0 in the status code field, indicates that the authentication was successful.

> The 802.11 standard does specify that a "STA may ***decline to authenticate*** with another requesting STA," but it does not specify when it would be appropriate to do so. The declining of an authentication request, when using Open System authentication, is left up to the vendors implementing the 802.11 standard.

The improper implementation of Open System authentication comes when you use it alone, but do not have to and sensitive data is in play. In such cases, you should be using WPA2-Personal or WPA2-Enterprise.

Wi-Fi Protected Setup (WPS)

Wi-Fi Protected Setup (WPS) was introduced to allow for simpler setup of unique security information between a client and an AP. The idea sounded good, but several years ago, it was discovered to be vulnerable to brute force attacks. Utilities are readily available for Linux distributions that can take advantage of this vulnerability. The simple solution: don't use WPS. Many will debate ways in which it can be used securely, but the reality is that WPA2-Personal, with a strong passphrase, is secure and should be used instead.

Effective Security Mechanisms

RSNA equipment is said to be capable of creating a Robust Security Network Association, and pre-RSNA equipment is not capable of such. It is also interesting to note that the standard specifies that an RSN (Robust Security Network) can only truly be established if mutual authentication occurs. The standard does not control the type of authentication, but it does specify that EAP-MD5 would not be considered a valid solution, since it does not perform mutual authentication.

As you can see from the preceding paragraph, there are many terms that need to be understood, in order to comprehend the full functionality of the 802.11 security standards. The following definitions will act as a foundation for our further discussion:

- **Robust Security Network Association (RSNA)** — an authentication or association between two stations that includes the 4-way handshake.
- **Robust Security Network (RSN)** — a WLAN that allows for the creation of RSNAs only. To qualify as an RSN, there can be no support for associations not based on the 4-way handshake. The Beacon frame will indicate that the group cipher suite being used is not WEP.
- **4-way handshake** — an 802.11 pairwise key management protocol that confirms mutual possession of a pairwise master key (PMK) between two parties and distributes a group temporal key (GTK).

- **Pairwise Master Key (PMK)** — a key derived from an EAP method or obtained directly from a pre-shared key (PSK). The highest-level key in the 802.11 standard.

- **Group Temporal Key (GTK)** — a key used to protect multicast and broadcast traffic in WLANs.

To summarize these definitions, an RSN is a WLAN that will only allow for RSNAs. These RSNAs are established through a 4-way handshake that results in the generation of the PMK, and the provision of the GTK to the authenticating STA. The PMK is used with other materials in the 4-way handshake, to generate the unicast encryption key. Once this RSNA is set up, the STA may communicate on the WLAN with confidentiality and integrity.

WPA2 (CCMP/AES)

WPA2 or CCMP/AES is the best security solution commonly supported in modern 802.11 WLANs and should be preferred above all other deprecated solutions. TKIP, WEP and Shared Key authentication are all deprecated in the 802.11 standard.

WPA2-Personal

WPA2-Personal uses pre-shared keys, as discussed in earlier chapters of this book. The pre-shared key (PSK) is typically generated from a passphrase entered into the AP/management system and the clients. Best practices with PSK passphrases include:

- Use a longer and complex passphrase — 15-20 characters and non-word alphanumeric works well.

- Change the passphrase periodically. Some organizations change it every month but changing it every 3-6 months is certainly sufficient for most deployments.

WPA2-Enterprise

WPA2-Enterprise depends on the 802.1X port-based authentication solution and EAP methods. The remainder of this section is primarily focused on the technologies that allow for the use of WPA2-Enterprise. Remember, that WPA2-

Enterprise is a Wi-Fi Alliance certification based on CCMP/AES, as defined in the 802.11 standard.

802.1X/EAP framework

802.1X provides a generic framework that allows for the implementation of the extensible authentication protocol (EAP) over 802 networks, including 802.3 and 802.11. The benefit of using such a framework is that organizations can choose the specific implementation that works for them, without worrying about whether their authenticator devices (usually access points) will support the specific implementation of 802.1X. This is because authenticators simply forward EAP messages between the supplicant and the authentication server. The authenticator, supplicant and authentication server are all defined and explained in more detail, later in this chapter.

The 802.1X framework, though widely supported in wireless systems, was originally created and used in wired networks. It is an authentication framework that can be used in 802.3 Ethernet networks and others, including 802.11 wireless networks. 802.1X was born out of a desire to have *port-based network access control.* This means that a user's device will be unable to communicate on the network, with the exception of authentication communications, until the port has been opened by the 802.1X/EAP authentication process. This provides the added potential benefit of linking a port to a user, for security auditing and control purposes.

The benefits of using 802.1X are many. It provides a more secure authentication system for wireless networks than the default open system authentication mechanism. In addition, when used with the proper implementation of EAP, it will provide mutual authentication. Port-based control is often the primary motivator that leads to the implementation of an 802.1X/EAP solution.

Mutual authentication can be defined as the authentication of both the client/user and the authenticator/authentication server. By authenticating the client or user, the network can be protected from invalid intrusions, but this does not protect the client or user from an invalid authenticator. By authenticating the authenticator or authentication server, the user can be protected from rogue devices, such as rogue access points, but this does not protect the network from an intrusion. In order to protect both points, an EAP implementation that supports mutual authentication must be used.

When implemented in Ethernet networks, 802.1X restricts the access of a device connected to a port (Ethernet connection), until the device has been authenticated. This prevents access to sensitive information on your network and reduces management overhead. The management overhead is reduced because ports do not need to be disabled in conference rooms and other public areas. Since users cannot connect to the port and access the network unless they have a properly configured supplicant, the port can be left in an "enabled but protected" state.

Wireless connections do not have a true port to which they are connected. In an 802.11 implementation of 802.1X, the wireless link to the AP can be thought of as the port. Once the wireless connection is created, using Open System authentication, the port (connection) can be authenticated using the configured authentication solution. The port (connection) will only allow authentication communications to pass through to the network, until the authentication is complete. If the authentication fails, the client device will never be granted access to the network for anything other than authentication purposes.

802.1X/EAP Functionality

The 802.1X standard provides for port-based network access control on wired and wireless 802 LANs. In order to implement, maintain, and troubleshoot an 802.1X authentication system, you will need to understand its basic functionality, and the more specific functionality of the various EAP methods it supports. The following concepts must be understood:

- Authentication Roles
- Controlled and Uncontrolled Ports
- 802.1X Generic Authentication Flow Framework

There are three primary authentication roles in an 802.1X authentication system: the supplicant, the authenticator, and the authentication server. The *supplicant* is the device wishing to authenticate to the network. The *authenticator* is the device or service acting as a mediator between the supplicant and the authentication server. The *authentication server* is responsible for processing the applicable EAP authentication chosen for the network.

As Figure 10.1 indicates, EAPOL (extensible authentication protocol over LAN) is used to communicate between the supplicant and the authenticator and RADIUS is the most common protocol used to communicate between the authenticator and the authentication server. Depending on the EAP selected, the communications will vary.

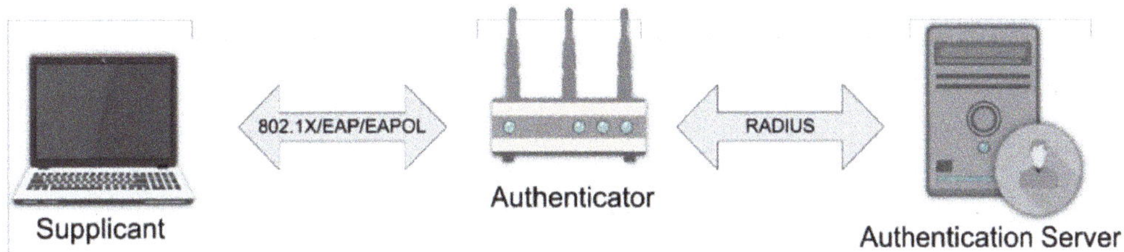

Figure 10.1: 802.1X Authentication Roles

> Remember that EAPOL is used between the WLAN clients and the APs or controllers (the authenticators), and RADIUS is used between the APS or controllers and the authentication server.

Two ports are defined by the 802.1X standard for the purpose of authenticating connected systems. These are best thought of as virtual ports existing inside the authenticator. The *uncontrolled port* is the port that allows communications to pass through for authentication and authorization only. The *controlled port* is the port that can only be used once authentication has completed. Once authentication has completed, the controlled port is switched to an authorized state, and all authorized traffic is allowed. Only the controlled port can pass standard network data and it cannot be used until authentication takes place through the uncontrolled port, switching the controlled port from unauthorized to authorized. Figure 10.2 shows an unauthorized wireless connection, and Figure 10.3 shows an authorized connection.

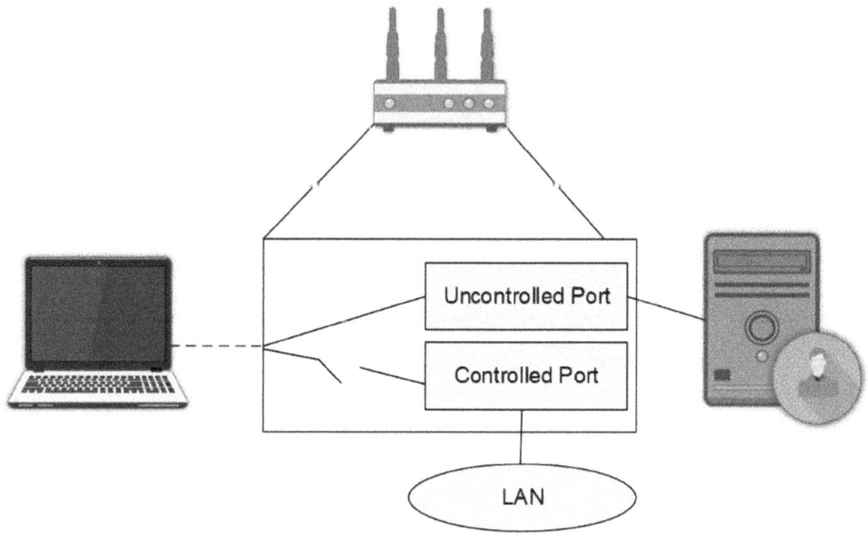

Figure 10.2: Unauthorized Connection

In a wireless implementation, the wireless side of an access point provides the connection for the client and is referenced as a port in 802.1X terms. Each wireless connection is seen as an independent controlled port and will disallow non-authentication traffic until the port is switched to authorized mode. The uncontrolled port is used for authentication traffic (EAP/RADIUS messages) only. General communications cannot occur until the authentication is complete. Figure 10.2 illustrated the unauthorized connection by showing that the controlled port is disabled and that this port provides general access to the LAN. Figure 10.3 shows the enabled controlled port in an authorized state. It is important to remember that the access point (authenticator) will have only one "true" port connecting to the LAN, but that it internally manages the virtual uncontrolled port for authentication and many controlled ports for general LAN access.

Figure 10.4 shows the generic flow of authentication information according to the 802.1X standard. While each EAP method will have a unique authentication flow, they will all use this framework for communications. In most cases, the authenticator requests identity information from the supplicant and then passes this information on to the authentication server once the identity information is provided by the supplicant. The authentication server will respond with needed authentication information requests, depending on the EAP method used. The supplicant can then provide this information and will either be validated (authenticated) by the authentication server or rejected. Assuming the authentication is successful, the authenticator switches the controlled port from unauthorized to authorized and begins permitting general LAN traffic to and from the wireless client (supplicant).

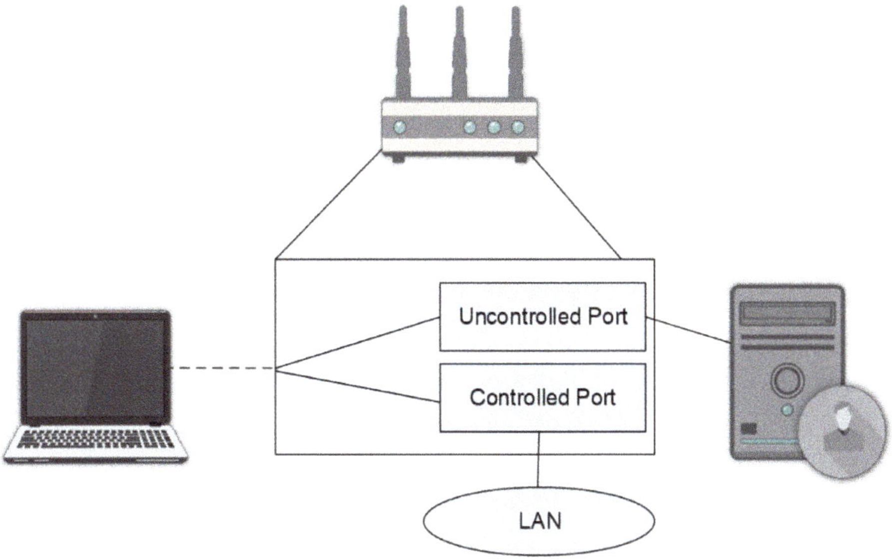

Figure 10.3: Authorized Connection

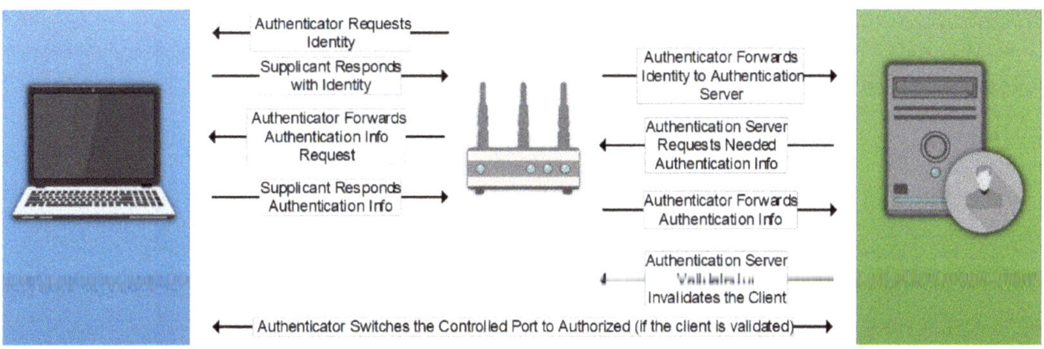

Figure 10.4: Generic 802.1X Authentication Flow

In the end, to implement 802.1X/EAP, you will need the following functional components:

- A client device that can act as an 802.1X supplicant
- Supplicant software installed on the client
- An access point that supports 802.1X authentication
- A RADIUS server to act as the authentication server, or some form of backend authentication service
- A public key infrastructure, if using an EAP method that requires client certificates on the client devices

EAP Methods

Just as there are different types of credentials used for authentication, there are different EAP methods that support these differing credentials. For example, EAP-TLS supports the use of certificates while EAP-MD5 only supports the used of passwords or known information. Depending on budgets, technical availability and security needed, you will choose among the available EAP methods. The following section provides a general discussion that should help you make this decision.

There are many factors that should be considered when selecting an EAP method for your organization. As much as possible, you will want to be consistent with your choice throughout your wireless networks. A consistent environment is usually easier to manage, but it can also be considered less secure due to the use of a single security solution. When determining which EAP method you should implement, consider the following factors:

- Mutual Authentication
- Certificate Requirements
- Dynamic Key Generation
- Costs and Management Overhead

- Industry Support

Some EAP methods support mutual authentication while others do not. As indicated in Table 10.1, EAP-MD5, which is supported by some wireless authentication systems, does not support mutual authentication and for other reasons, should never be used in production networks. EAP-TLS, EAP-TTLS and PEAP all support it.

	EAP-MD5	EAP-TLS	EAP-TTLS	PEAP
Mutual Authentication	No	Yes	Yes	Yes
Certificates Required	No	Client/Server	Server Only	Server Only
Dynamic Key Generation	No	Yes	Yes	Yes
Costs and Management Overhead	Low	High	Low/Medium	Low/Medium
Industry Support	Low	Medium	High	High

Table 10.1: EAP Selection Quick Reference for Three Common Types

When you select an EAP method, certificate requirements become an important consideration. Some EAP methods, such as EAP-TTLS and PEAP require only a server certificate while others require both a server and client certificate, such as EAP-TLS. The reason certificate requirements become important is that they will determine your need for an outside certificate authority, or an internal PKI.

EAP methods that support dynamic key generation can manage the generation and distribution of encryption keys used to protect the wireless connection.

The cost and management overhead of an EAP solution will be determined by the complexity and requirement of the solution. For example, EAP-TLS will usually require a PKI, since it demands both server and client certificates. For this reason, it is listed as a high-cost solution. The management of all the client certificates (revocation, renewal and issuance) will also be more intensive, resulting in higher management overhead. Due to the fact that EAP-TTLS and PEAP require only server certificates, they may be considered a low- to medium-cost solution. In addition to these cost factors, items such as additional hardware and software should also be considered.

It is important that the solution you choose be supported by the industry. This means that hardware and software should support the EAP you've chosen. In Table 10.1, EAP-MD5 is listed as having low industry support. This is due to the fact that many vendors are pulling support for EAP-MD5 from their products. The support is being removed because of the known vulnerabilities in the solution. For example, Microsoft has removed support for EAP-MD5 from their most recent versions of their operating systems.

In addition to the EAP methods that are supported, you should understand the legacy authentication protocols that are still in use today. These protocols may be used in conjunction with an EAP method, such as PEAP, or they may stand alone.

Password Authentication Protocol (PAP) is an older protocol that provides no protection of authentication credentials. It was originally designed for use with PPP (point-to-point protocol), as was EAP itself, and was documented in RFC 1334. PAP is only supported by EAP-TTLS and PEAP (and not in all available authentication servers) and should only be used in this context because the TLS tunnel protects the PAP authentication. Outside of an encrypted tunnel, the password is sent as clear text with the PAP authentication method.

The Challenge Handshake Authentication Protocol (CHAP) was originally defined in RFC 1994 and, like PAP, was designed for use with PPP. Many other systems have adopted CHAP, or some form of it, since that time. CHAP does not send the password as clear text, but instead sends a challenge to the client. The client responds by using MD5 to create a hash of the challenge text using the password as the key. This is then sent to the CHAP authentication server. If the authentication server gets the same results, the user is authenticated, otherwise authentication fails. Because passwords are stored, at the authentication server, in either plain text or reversible encrypted text, CHAP is not considered the most secure authentication mechanism by today's standards.

Microsoft CHAP (MS-CHAP) is a proprietary protocol created by Microsoft, though it is defined in RFC 2433. When implemented in a Microsoft environment, the password is not stored, but the hash of the password is stored in the database of the Windows server. This initial version of the protocol has been proven weak and should only be used as a last resort with older Microsoft clients such as Windows 95 — yes, I said it. In other words, don't use it. MS-CHAP uses the same three-way handshake as standard CHAP.

Because of the vulnerabilities in MS-CHAP, Microsoft created version 2 and it is usually referenced as MS-CHAPv2. RFC 2759 defines this proprietary protocol, and it was first released with Windows 2000 Professional and Server. MS-CHAPv2 improves on MS-CHAP by storing the passwords with a stronger hashing and encryption mechanism, as well as adding mutual authentication. MS-CHAPv2 authenticates both the client and the server. This protocol is commonly used as an internal authentication mechanism in the EAP method known as PEAP. However, it is not a good choice for non-tunneled authentication.

A secure system can be defined as one having an acceptable level of risk. This is because what is impenetrable today, will often be penetrated tomorrow. EAP methods are no exception, and what was once considered secure is now considered insecure. EAP-MD5 and LEAP are considered insecure EAP methods today.

EAP-MD5 was not actually created with the intention of use in production networks. It was created, and is still beneficial, for testing connection and verifying communications among the three nodes involved in the EAP framework (the supplicant, authenticator, and authentication server). In fact, EAP-MD5 was included alongside EAP itself in the original RFS 2284.

Cisco created Lightweight EAP (LEAP) for use with their equipment and systems. LEAP is supported by a variety of network operating systems and client devices, other than Cisco equipment. These systems include Windows and Linux. Due to a vulnerability discovered in the protocol, it is no longer considered secure. A tool called ASLEAP can be used to hack the LEAP authentication process, due to LEAP's close ties with MS-CHAPv2.

While EAP-MD5 and LEAP have shown their security weaknesses, other EAP methods have proven to be moderately to very strong at this point. This section provides a brief overview of the basic EAP methods that are still considered to be strong, while providing details related to their authentication flows and benefits.

When most people speak of PEAP (Protected EAP), they are referring to what is now called PEAPv0. PEAP addresses the vulnerabilities of an EAP-like EAP-MD5, by encapsulating the authentication process in a TLS (transport layer security) tunnel. Microsoft supports two implementations of PEAP: PEAP/EAP-MS-CHAPv2 and PEAP/EAP-TLS. When MS-CHAPv2 is used, either passwords or certificates may be used to authenticate the clients. Note that MS-CHAPv2 only supports the use of certificates for authentication, when used by PEAP. PEAP/EAP-TLS requires the use of certificates for the authentication of each client.

One of the advantages of PEAP, when used with MS-CHAPv2, is that it only requires a server-side certificate. This can reduce management overhead and total cost of ownership. Figure 10.5 illustrates the authentication flow of the PEAP process. Notice the stage called "EAP in EAP Authentication." This is named so because the encrypted tunnel that is established allows for EAP method authentication to initiate at that point. Because of this behavior, PEAP is often called a two-stage protocol. In stage one, a secure tunnel is established, and in stage two, the authentication process takes place.

Cisco created PEAPv1, or PEAPv1/EAP-GTC (Generic Token Card), to be used instead of PEAPv0. EAP-GTC was defined in the original RFC 2284 EAP specification and allows for the use of simple username/password pairs. Cisco, who originally worked with Microsoft to create and support PEAP, parted ways with them when Microsoft opted against supporting EAP-GTC across the PEAP tunnel. Due to the strength of Microsoft and Cisco, PEAPv0 and PEAPv1 are the most common EAP methods implemented today.

Unlike PEAP, EAP-TLS does require the use of certificates on the clients, as well as the authentication server. This usually requires the implementation of a PKI system and added costs. However, certificate-based authentication is usually considered more secure than password-based authentication due to the fact that you can't "guess" a certificate. On the other hand, if the certificate store is not secure, it can easily be stolen. For this reason, a certificate-based authentication system is only as secure as its certificate store. EAP-TLS is defined in RFC 2716. Since TLS is the successor to SSL, it is generally considered very secure. In addition, TLS tunnels are used by both PEAP and EAP-TTLS. The authentication flow is detailed in Figure 10.6.

Like PEAP, EAP-TTLS uses a TLS tunnel to protect less-secure authentication mechanisms. These may include PAP, CHAP, MS-CHAPv2, EAP-MD5 and more. EAP-TTLS is still useful for organizations who want to use older authentication systems on the backend, like Windows NT, and have the protection of the TLS tunnel. Though EAP-TTLS was released to market before PEAP, the strength of Microsoft and Cisco seems to have overshadowed any benefit of being first to market, as PEAP has a large install base today. Figure 10.7 shows the authentication flow of EAP-TTLS.

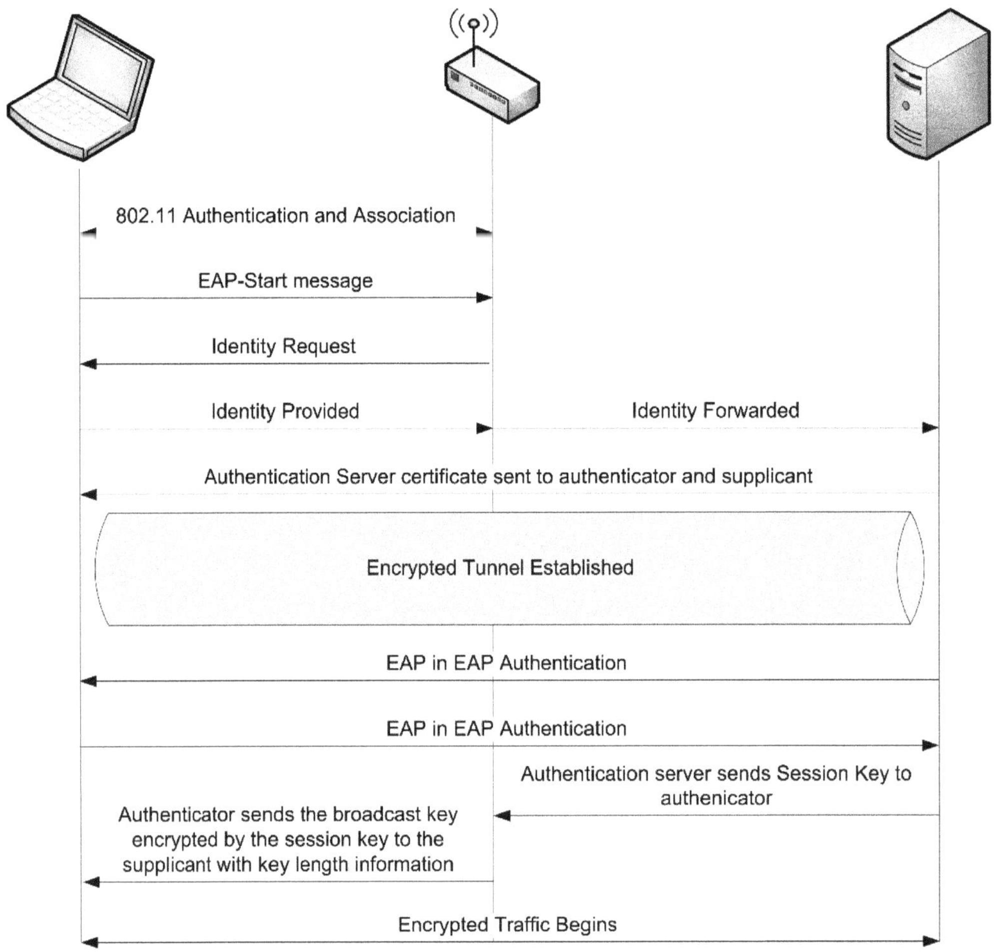

Figure 10.5 PEAP Authentication Flow

EAP-FAST (Flexible Authentication via Secure Tunneling) is another Cisco-developed EAP method. It was proposed and developed to address the weaknesses of the LEAP protocol and is supported by Cisco hardware and some client supplicants. Even server certificates are optional with EAP-FAST, so it provides a lower total cost of implementation than PEAP or EAP-TTLS. Like Kerberos, EAP-FAST, uses a PAC to establish the TLS tunnel for protection of credential transfer.

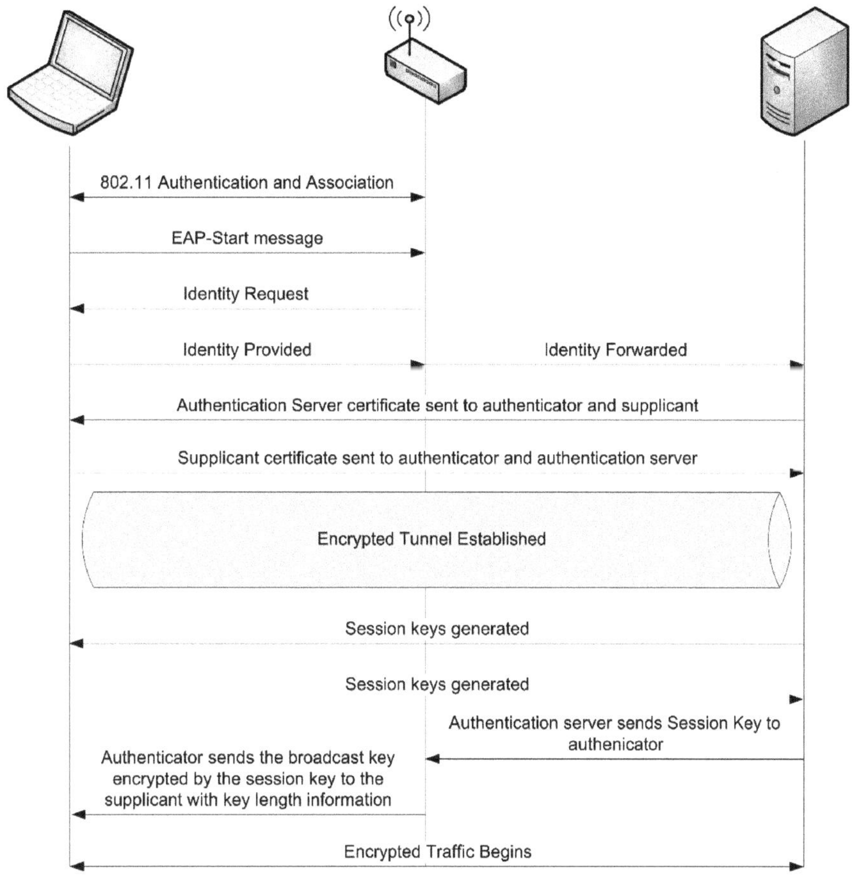

Figure 10.6: EAP-TLS Authentication Flow

EAP-FAST can be configured in one of two ways: server-side certificates or PAC usage. When using server-side certificates, EAP-FAST is thought to be as secure as PEAP. When using PACs for the creation of the secure tunnel, it is as easy to implement as LEAP. You will want to remember that EAP-FAST cannot be both easy to implement (no certificate required), and as secure as PEAP, at the same time. This is a decision you will have to make. Is greater security more important, or faster and easier implementation?

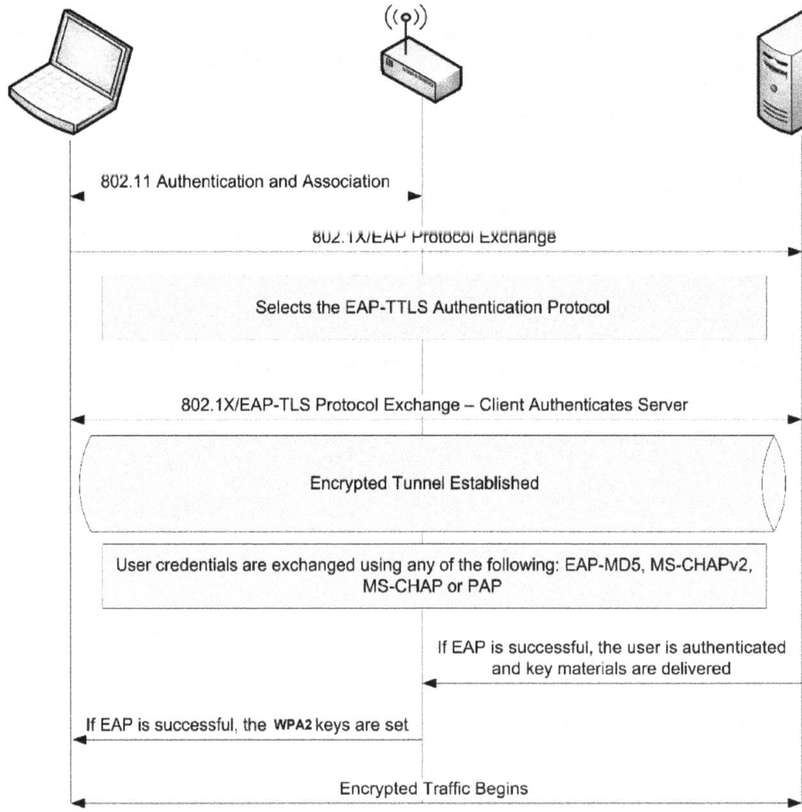

Figure 10.7: EAP-TTLS Authentication Flow

EAP-FAST uses MS-CHAPv2 for the internal authentication mechanism. However, any normal vulnerabilities of the MS-CHAPv2 protocol, such as dictionary attacks, become non-issues with the use of EAP-FAST. This is because the MS-CHAPv2 exchange is encrypted inside the EAP-FAST TLS tunnel.

The EAP method or type used is defined in a special packet known as an EAP-Method packet. The first field of this packet is a Type field and it defines the EAP method requested for authentication.

Per-User PSK (PPSK)

A proprietary solution is provided by some vendors that allows for a PSK to be used, but the PSK is unique to each user. PPSK allows for this implementation, and provides for fast roaming, because 802.1X/EAP is not required. The most well-known implementer of PPSK is Aerohive.

Security Enhancements and Tools

Several additional tools may be used in addition to WPA2-Personal and WPA2-Enterprise to provide enhanced security on WLANs. These tools may work above Layer 2 of the OSI model. Examples include captive portals, BYOD solutions, and guest networks. Layer 2 solutions include protected management frames and fast secure roaming methods.

Captive portals

A captive portal is implemented when all the traffic coming through an AP is initially directed to an access control device on the wired LAN. The access control device is used to authenticate the user and provide access to resources on the wired LAN, which may include Internet access. If you've connected to a WLAN at a hotel or hotspot, that first routed you to a logon screen which required you to agree with the terms of use, you've likely experienced the concept of a captive portal. When you connected, though your homepage may have been Google.com, you were redirected to the captive portal page before you could navigate to your normal homepage. After authenticating (which can be as simple as click a button that reads, "I agree," or as complicated as providing a code and your contact information) you can communicate with other websites as you normally would.

A captive portal is usually implemented using a WLAN controller. These captive portals may support more than just logging you onto the network. They may be able to provide VPN tunnel endpoints, or other security mechanisms, that protect the data transfers that occur after the authentication, as well as the initial authentication itself.

It is important to note that all captive portals are not created equal. Many devices or services only reroute HTTP (TCP port 80) to the web server used for authentication and authorization. If the client computer uses some other protocol, most commonly ICMP or DNS is used, the device or service that normally reroutes the client to the authentication server will allow the packets through to the Internet. All an attacker (in this case a freeloader or someone who wants to use the

Internet access for free) must do is setup a service on an Internet-connected machine to which they can connect using ICMP or DNS. This concept is sometimes called ICMP tunneling or DNS tunneling. Basically, the normal HTTP information is tunneled through the ICMP or DNS connection to the attacker's Internet-connected machine (you can call this the tunnel server). From there, the Internet-connected machine routes the HTTP information back out to the Internet, and then tunnels responses from the Internet back to the attacker through the ICMP or DNS tunnel. There are video and text tutorials floating around the Internet that teach attackers how to perform this penetration.

BYOD and Guest Networks

BYOD and guest networks have their own security concerns. BYOD introduces the concern for data leakage and the introduction of viruses to the network. These can be addressed with MDM and NAC solutions respectively.

Guest networks cause several questions to arise related to security:

- Will you leave the network open or require authentication?
- Will you implement a captive portal?
- Will you use encryption in the network?
- Will all guests be tunneled to the DMZ, for example, in a GRE tunnel?
- Will you require guest registration?

Answering these questions is very important when implementing a guest network, to ensure that you implement the needed security for your specific use case.

Protected Management Frames

802.11w-2009 (now part of 802.11-2016) introduced management frame protection. However, this does not encrypt the MAC headers of frames and it only applies to certain management frames. The only frames protected are de-authentication, disassociation and robust action frames. Data frames do not include this protection.

Fast Secure Roaming methods

As you've learned in previous chapters, the ability to roam while maintaining a network connection is one of the primary benefits of WLANs. However, this roaming may come at cost. Depending on the network infrastructure chosen and the roaming technologies supported, you may be unable to support some applications while roaming. Early wireless networks had little support for applications that required, what you might call, a near persistent connection. The main reason for this limit, was the time it took for a client to disconnect from one AP and connect to another AP, as the client moved around in the coverage areas. The movement from one AP to another could take 300 milliseconds or more. Additionally, the Layer 3 (IP) connection was often lost and, since most modern applications rely on TCP/IP, this meant that any transfers or connections had to be reinitiated. While you could speed up the move from one AP to another with some technologies, the added layer of security was often still a deal breaker.

Fast Secure Roaming (FSR) is the solution to the problem. FSR has been implemented in proprietary solutions for several years now and a standardized solution is also now available, thanks to the 802.11i, 802.11k and 802.11r amendments. Thankfully, the standardized solutions take advantage of existing hardware, so that most vendors can implement the new standardized FSR features with simple firmware or software upgrades.

In many cases, traditional WLAN roaming is sufficient for user needs. For example, if users need to have network access in their offices and in conference rooms, but not while moving from their offices to the conference rooms, Layer 3 roaming that breaks the IP connection is perfectly sufficient. On the other hand, if the users are making VoIP calls on voice over WLAN phones, and they want to have a conversation while walking from one building on campus to another, fast secure roaming is important.

VoIP communications are very sensitive to delays. In fact, the voice engineer must focus on reducing latency (delay), reducing jitter (variance in delay) and reducing dropped packets to achieve a well-performing voice network. This requirement is true of wired networks, and it is true of wireless networks. In early WLANs, VoIP was often taken off the table, due to roaming problems when security was used on the WLAN. However, 802.11i and 802.11r have provided mechanisms to allow for security and fast roaming, and many vendors are focused on implementing these solutions.

The following list includes the most common networked applications requiring fast BSS transition (FT) roaming technologies:

- Voice over IP
- Video over IP
- Large file transfers with non-resuming services
- Any IP-dependent application sessions

The following list represents examples of common networked applications not requiring FT roaming technologies (a change in IP address is acceptable):

- Web browsing
- Local network file access
- Local printing

Of course, any application, even something simple like local printing, must maintain the connection until the action is completed; however, long term connections (more than a few seconds) are seldom maintained for services like web browsing, file access and local printing.

> Remember that voice over WLAN (sometimes called VoWLAN) is a primary service requiring fast secure roaming or fast BSS transition within an ESS. Video over IP can benefit from it as well.

Roaming works in one of three primary ways:

- Layer 2 roaming across APs within a single controller
- Layer 2 roaming across APs connected to separate controllers

- Layer 3 roaming

When Layer 2 roaming occurs, the IP configuration is not lost. With the same IP address and roaming times of typically less than 20 to 30 milliseconds, Layer 2 roaming can support streaming technologies like VoIP. Layer 2 roaming across APs within a single controller is called *intracontroller* roaming. Layer 2 roaming across APs connected to separate controllers is called *intercontroller* roaming. Vendors handled the actions that take place within or between the controllers, according to their proprietary algorithms. The 802.11 standards, as amended, define only what should take place as a client STA roams from one AP to another and they do not specify exactly how the communications must occur within the infrastructure. This flexibility allows the vendors to provide competitive features in this area, and with the recent flood of VoIP activity, this infrastructure solution can indeed be a deal breaker if the vendor's roaming solutions are inefficient. The good news is that the major vendors all have intracontroller and intercontroller Layer 2 roaming solutions that can accomplish the roaming speeds required for wireless voice over IP.

Layer 3 roaming occurs when the client STA roams to an AP that cannot provide the same IP configuration because the AP is located on a different wired network. In such a roaming scenario, the IP address must be reallocated from the DHCP server, and the client STA is placed on a new subnet of the network. The problem with this action is that the client's Layer 3 connections will be lost. If a file was in the process of copying from the client to a server, the file copy process will most likely have to be started again from the beginning. The same is true in the reverse scenario, where the file is copying from the server to the client. While this situation is painful for the users, it can't begin to compare with frustrations of dropped calls, due to Layer 3 roaming on voice over WLAN phones. To solve this problem, FT roaming must be implemented and the APs across which users would roam must somehow be part of the same wired network. They may be part of the same wired network through tunneling solutions within the infrastructure, or they may simply be connected to the same controller, but they must somehow allow the client STA to maintain its IP address.

Implementing Layer 2 roaming without IP configuration loss is not actually very difficult at all in open wireless networks. It has been available for many years. The problem is that, previous to 802.11i and 802.11r, you faced difficulties in implementing a standards-based secure wireless network that offered very fast roaming. For secure wireless networks to support fast roaming, the 802.1X authentication process has to be accommodated in rapid fashion, so that the user can roam without requiring an 802.1X re-authentication exchange to occur. This is typically accomplished using some form of key.

Finally, in order for roaming to work in a seamless manner, the coverage cells of the APs must overlap. If there is no point of overlap, the client STA will always lose network connectivity for a brief time, as the user moves the STA across the non-covered area. Vendors recommend cell overlaps ranging from 15 to 30 percent. However, indicating 6-16 dB overlap, or visibility of multiple APs, is a better option. Check with your vendor to determine the best overlap configuration for your network.

> When a user moves his or her laptop around within the coverage area of a single AP, roaming is not required. The user has mobility, but the connection to a single AP is maintained, and no roaming occurs.

To understand how 802.11i impacts roaming, you must first understand the different keys used in 802.11i networks. The keys are used to secure communications on the wireless link. The following keys are used:

- **Pairwise Master Key (PMK):** The is the top-level key used in the standard. The PMK is derived from a key generated by EAP or from a pre-shared key (PSK) in smaller implementations.
- **Pairwise Transient Key (PTK):** The key derived from the PMK, Authenticator (access point) address (AA), supplicant (client) address (SPA), Authenticator NONCE (number used once) (ANONCE), and supplicant nonce (SNONCE). A pseudo-random function is used to generate up to five keys. The five keys are the EAPOL-Key confirmation

key, the EPOL-Key encryption key, the temporal encryption key, and two temporal message integrity code keys.

- **Group Master Key (GMK):** A supporting key that may be used to generate a group temporal key. The GMK may be regenerated within the AP periodically, to reduce the exposure of the group temporal key.
- **Group Temporal Key (GTK):** The key used to protect broadcast or multicast MPDUs on a wireless link.

The term pairwise simply refers to two devices associated with each other. A pairwise master key, for example, is a key used between an AP and a client STA to secure the communications.

Once a STA is associated and authenticated to the wireless network, with 802.11i, a PMK secure association (PMKSA) exists between the authentication server and the STA. A PTK secure association (PTKSA) exists between the AP and the STA once the 4-way handshake is completed. The problem, when discussing roaming, is that the accomplishment of such a PMKSA and PTKSA takes time. It can take too much time for real-time applications, if some additional mechanism is not in play.

The 802.11i solution to the delay cause by establishing a PMKSA was PMK caching. With PMK caching, the authenticator (the AP) and the STA can cache PMKSAs, so that regeneration of the PMKSA is not required at the time of roaming. Instead, the first step in the four-way handshake is that the authenticator specifies the identifier of the PMK in the first message (Message 1) of the handshake. This functionality means that the process of PMKSA establishment is removed, and only the PTKSA must be established. Always remember this rule: If you can remove steps from a process, you will typically reduce the time required to complete the process. By removing the step of PMKSA establishment at roaming time, we speed up the process or reassociation with the new AP. The PMKSA must still be established, however it is cached ahead of time so that, at roam time, the effort is not required.

The 802.11r amendment was the first 802.11 attempt to truly define fast secure roaming in any level of detail. It was ratified in 2008 and is not part of the 802.11

standard as amended. The 802.11r amendment assumes the 802.11i amendment — as would be expected, since 802.11i was ratified in 2004. You must always remember, when studying IEEE standards, that an amendment ratified today is based upon the original standard and all amendments ratified before today. If you don't keep this in mind, the standard will become very confusing to you very quickly.

Let's begin our discussion of 802.11r with a few key definitions from the standard:

- **Fast Basic Service Set (BSS) Transition:** A station (STA) movement that is from one BSS in one extended service set (ESS) to another BSS within the same ESS, and that minimizes the amount of time that data connectivity is lost between the STA and the distribution system (DS).
- **Fast Basic Service Set (BSS) Transition (FT) 4-Way Handshake:** A pairwise key management protocol used during FT initial mobility domain association. This handshake confirms mutual possession of a pairwise master key, the PMK-R1, by two parties and distributes a group temporal key (GTK).
- **Fast Basic Service Set (BSS) Transition (FT) Initial Mobility Domain association:** The first association or first reassociation procedure within a mobility domain, during which a station (STA) indicates its intention to use the FT procedures.
- **Mobility Domain:** A set of basic service sets (BSSs), within the same extended service set (ESS), that support fast BSS transitions between themselves and that are identified by the set's mobility domain identifier (MDID).
- **Over-the-Air Fast Basic Service Set (BSS) Transition (FT):** An FT method in which the station (STA) communicates over a direct IEEE 802.11 link to the target access point (AP).
- **Over-the-DS (distribution system) fast Basic Service Set (BSS) Transition (FT):** An FT method in which the station (STA) communicates with the target access point (AP) via the current AP.

As painful as it may be, memorizing the preceding list of definitions — at least in your own words — is a key part of understanding FT. However, it is not a key part of the CWNA-108 exam, so you need not worry about memorizing them for that. In addition to these terms, you need to understand that we no longer deal with a single PMK, such as was introduced in 802.11i. Instead, we must deal with a FT key hierarchy. The following definitions will help you understand this hierarchy:

- **PMK-R0:** The first level (or top-level) PMK. The PMK-R0 is derived from the master session key (MSK) when 802.1X/RADIUS is used or from the pre-shared key (PSK) when personal implementations are used.
- **PMK-R1:** The second level PMK. The PMK-R1 keys are derived from the PMK-R0 key.

Remember this hierarchy. The first level is not PMK-R1, but it is PMK-R0.

The core of what 802.11r is all about is allowing a non-AP STA to pre-authenticate with an AP, to which it may roam at a later time. During the pre-authentication process, in an FT implementation, the PTK is derived from the PMK-R1. It's important to remember that the PTK is not derived directly from the PMK-R0, but that it is derived from the PMK-R1.

Pre-authentication is optional. If it is to be used, it must be available and enabled on the APs and the client devices. Remember that it is not required of an 802.11-compliant device; however, it will be very useful for wireless networks that must carry voice or other real-time traffic and provide for roaming ability.

One example of a proprietary roaming solution is opportunistic key caching (OKC). OKC is implemented in Microsoft wireless clients and some WLAN infrastructure vendors have support for it. With OKC, the client performs a standard 802.1X/EAP authentication on initial connection to the network. This authentication generates a PMKSA called PMKSA1. The PMKSA is generated on both the AP and the client. The PMKSA has a PMKID called PMKID1. When the STA desires to roam to another AP, an opportunistic PMKID called PMKID2 is generated. When the client requests a reassociation with the alternate AP, it sends the PMKID2 in the reassociation request. Assuming the second AP has a matching

PMKSA, the four-way handshake begins immediately and PMK generation is avoided during the roaming process.

This OKC process if very similar to that ratified in 802.11r and we are likely to see it fall by the side, as the standards-based roaming solutions are implemented in enterprise-class equipment. However, many organizations have existing infrastructure in place based on 802.11g and have no intentions to upgrade before 2012 or later. In these organizations, continued support for OKC is important and, thankfully, most wireless clients can support it through either built-in software or add-on supplicants.

Wireless Intrusion Prevention System (WIPS)

Wireless networks are vulnerable to the same intrusion attacks as wired networks. Additionally, the RF medium of wireless networks makes them susceptible to additional intrusion methods to which wired networks are not exposed. For example, a wireless network can be accessed from distances far greater than the 100 meters typical of wired Ethernet cables. For this reason, you must understand intrusion monitoring solutions that protect both the wireless access medium and the network layer.

Intrusion monitoring systems come in two basic forms. The first form is the intrusion detection system (IDS) and the second is the intrusion prevention system (IPS). From the perspective of detection, both systems are the same. The difference between the two lies in the way they respond to a detected potential intrusion. An IDS solution will usually detect the intrusion and notify an administrator, or simply log the intrusion; however, it will not take actions to stop or mitigate the attack. An IPS solution will also detect the intrusion and notify an administrator, but it will also take actions to stop or mitigate the attack.

When it comes to the detection procedures, both IDS and IPS solutions use similar algorithms. Three basic detection methods are used:

- Anomaly-based detection
- Signature-based detection
- Behavior-based detection

An anomaly-based detection system works in two phases. In the first phase, the system establishes a baseline of normal network activity. This baseline may be established through monitoring, through administrative configuration, or both. Once the baseline is established, the second phase of anomaly detection begins. In this phase, actions falling outside the boundaries of the baseline are flagged as potential intrusion attempts. Based on features like point systems or decision trees, the intrusion monitoring system may prioritize the detected event and log the action, report the action, or even make network adjustments to prevent further action.

Signature-based systems look for event sequences that usually indicate an attack. As a simple example, a user attempting to crack passwords may attempt to logon with more than three accounts from the same computer in less than two or three minutes. If the intrusion monitoring systems sees this activity, it may report it as an intrusion attempt. While this example is an oversimplification of signature-based detection, it illustrates the basic concept.

Behavior-based systems are similar to signature-based systems, with the exception that they are usually configured to watch for specific actions that the administrator does not want to allow. For example, the administrator may configure the system to watch for users who are uploading large numbers of MP3 files or MKV video files. Since the corporate policy indicates that such files should not be placed on the network, this behavior would affect a security breach and qualify as an intrusion event.

> Remember that an IDS detects and logs possible intrusions, and an IPS reacts to the intrusion as it occurs in order to prevent it from progressing. This difference is important to understand.

Intrusion monitoring solutions come in two implementations from a monitoring point, as well. You can implement network-based IDS/IPS or host-based IDS/IPS. Network-based systems are installed at ingress/egress points on the network, and possibly with sensors installed throughout the network. Network-based systems monitor and protect the entire network. Host-based systems are installed on each node on the network and protect only that node. Host-based systems are very useful for laptop wireless stations because they continue to protect the laptop when it is connected to public hotspots and other unprotected networks.

Wireless IPS (WIPS) solutions are specifically designed to prevent wireless attacks. Most WIPS solutions can detect and log or prevent the following types of attacks:

- Detection of MAC spoofing
- Detection of frame injection
- Detection of rogue APs
- Detection of rogue clients

In addition, wireless DoS attacks can usually be detected, and may be prevented through dynamic channel reassignment. For example, if a WIPS solution detects a DoS on channel 11 in the 2.4 GHz spectrum in one area of a building, it may be able to reconfigure the network so that channel 1 is used in that area instead. This behavior can prevent single-channel DoS attacks. However, a skilled attacker will likely attack all three non-overlapping 2.4 GHz channels at the same time, so the best the WIPS can hope to do is notify the administrator, who may then choose to use laptop-based intrusion analysis to locate the offender.

Protocol and Spectrum Analyzers

In many smaller networks, administrators will choose to perform manual analysis. This decision is often based on smaller budgets; however, larger networks may also do spot checks in addition to running enterprise-class IDS or IPS solutions. Manual analysis is performed with spectrum and protocol analyzers. An example of a spectrum analyzer is the AirMagnet Spectrum XT device and software, or Metageek Wi-Spy DBx. AirMagnet Spectrum XT and Wi-Spy DBx are USB form factor protocol analyzers for use with your laptops (or desktops, if you have a need for stationary analysis). The features of the AirMagnet Spectrum XT include:

- USB form factor for use with practically any modern computer.

- Combining spectrum analysis with traffic analysis (traffic analysis required a compatible Wi-Fi adapter).
- Automatic identification of WLAN and non-WLAN interference sources.
- Real-time RF spectrum and WLAN graphs.
- Integration capabilities with other AirMagnet solutions, such as Survey PRO and Wi-Fi Analyzer PRO.
- Recording and playback of spectrum analysis sessions.
- Support for both the 2.4 GHz and 5 GHz bands.

When using a laptop-based analyzer for intrusion analysis, you are usually looking for the following types of intrusions:

- Rogue APs
- Unauthorized clients
- Denial of Service attacks
- MAC layer wireless attacks

For example, to locate a physical DoS attack source, you can use the laptop analysis software to find the location where the signal is strongest. Once you've found this location, you can simply look around until you find the RF generating source. Tools like AirMagnet Spectrum XT and Metageek Chanalyzer often include device locator tools with graphical elements.

In most cases, laptop-based spectrum analysis products are used for auditing purposes. Audits can expose security risks and accidental interference sources. Devices like AirMagnet Spectrum XT can locate and identify non-WLAN interference sources, such as:

- Baby monitors
- Cordless phones
- Microwave ovens

- Bluetooth devices
- Wireless cameras
- Game controllers
- Digital video devices

Of course, intentional security attacks can also be detected during the audit. You can locate RF jammers using AirMagnet Spectrum XT, and you can locate laptops being used to flood WLANs with "junk" frames (frames that do not contain meaningful data), thanks to the ability to perform parallel protocol analysis as well as the spectrum analysis.

When AirMagnet Spectrum XT locates an interference source, it can provide detailed information about the source including:

- Peak and average output power
- Center channel frequency
- Channels impacted by the interference
- First and last times the source was detected
- Number of times the source was detected

When used with a directional antenna, spectrum analyzers can often pinpoint the location of the interference source down to a few meters. While I've used the AirMagnet Spectrum XT device as an example of a laptop-based analyzer in this section and feel that it is one of the best devices for the money, it is just one example of the laptop-based spectrum analysis solutions available. Similar features may be found in devices from vendors such as Cisco and MetaGeek.

Secure Management Solutions

It is not only the users' connections that must be secured, but the management connections must be secure as well. In this section, I will focus on two key elements of WLAN security: secure management and rogue AP detection.

SNMPv3 / HTTPS / SSH2 / VPNs

If you manage the APs in your WLAN independently (meaning they are not lightweight APs), you should be sure to use a secure method of management. While you can connect to many APs using standard HTTP by default, this is not a practice you want to follow. All HTTP traffic is transmitted as clear text. Figure 10.8 demonstrates this. In this case, I've blocked out the identifying information to protect the site owners, but you can clearly see the logon is "swettmarden" and the password is "drow1ssap1." This is because the web server does not use HTTPS for the logon process and the credentials are passed in the clear. Of course, this scenario was created completely for this document, but this scenario occurs every day, thousands (if not millions) of times around the world.

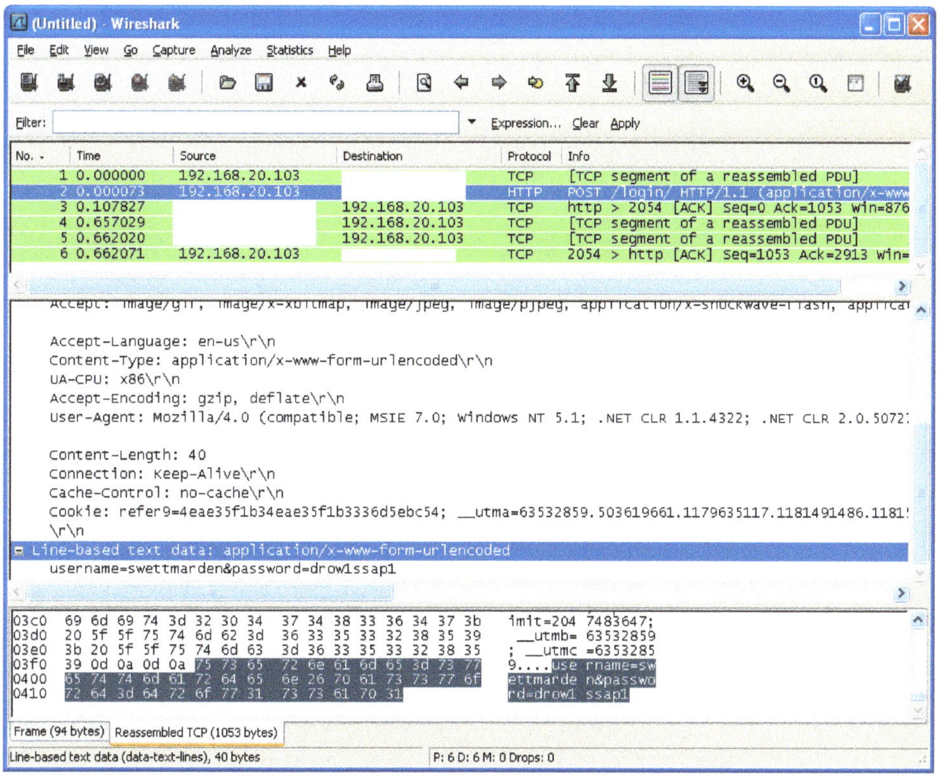

Figure 10.8: Screen Capture of a Logon Captured in WireShark

For this reason, HTTPS should always be used when a web-based interface is used to manage your APs. If the AP does not support HTTPS, it is best not to use HTTP to manage the device. HTTPS actually uses SSL and requires that a certificate be made available to the server. APs that support HTTPS have a certificate installed in the AP already. SSL is a Layer 7 encryption technology.

Another Layer 7 encryption solution is SSH. The first version of SSH has known vulnerabilities and should be avoided, but SSH2 is considered secure at this time. SSH2 is usually used to provide command line interface (CLI) access to the managed device. SSH2 provides the following benefits in a secure networking application:

- Public and private key authentication or username and password authentication.
- Data signing through the use of public and private key pairs.
- Private key passphrase association.
- Multiple encryption algorithms are supported such as AES, 3DES and DES.
- Encryption key rotation.
- Data integrity enforced through hashing algorithms.
- Data compression may be supported.

Because of these strong security features, SSH2 can help to mitigate against eavesdropping on management communications between you and the managed device. It can also help prevent man-in-the-middle attacks and replay attacks. The most common use of SSH2 is to implement a secure command shell or CLI across the network instead of having to connect to the console (serial port) of the managed device. Remember that Telnet is just as insecure as HTTP by default, because they both send their data packets as clear text that is easily readable by network protocol analyzers like WireShark.

The *Simple Network Management Protocol* (SNMP) is a standard solution for centrally monitoring and managing network devices. SNMP was plagued by security vulnerabilities early on, and these weaknesses have been addressed in SNMP v3. Version 3 has added authentication and privacy controls to help protect the management information passed on your network. You should ensure that any device you will manage with SNMP uses version 3 or higher of this protocol. Of course, as is true with any technology, you must be proactive and continually be on the lookout for new vulnerabilities that would impact your network. That which is secure today may be vulnerable tomorrow.

Virtual private network (VPN) solutions still have a role related to WLANs. The most common use today is that of running a VPN connection when using an open network, like a public hotspot. By initiating a VPN link after connecting to the hotspot, you ensure that all communications are encrypted.

WPA3

WPA3 is a Wi-Fi Alliance certification that is based on the 802.11 specified use of Simultaneous Authentication of Equals (SAE) and 802.1X authentication procedures. SAE uses the Dragonfly Key Exchange instead of the typical 4-way handshake, which is considered more secure. Opportunistic Wireless Encryption (OWE) is not a requirement of WPA3, but it is required for 6 GHz operations with 802.11ax. 6 GHz operations of 802.11 networks do not allow for Open Authentication networks without additional security. Instead, they use OWE for "open" networks.

WPA3-Personal (Mandatory in all 802.11ax devices), also known as WPA3-SAE comes in two (2) different modes, one being WPA3-SAE Transition Mode. This mode is configured on the access point for the purpose of supporting WPA2-PSK and WPA3-SAE on the same SSID. This allows for the same passphrase to be used on both WPA2-PSK and WPA3-SAE connections. The main difference for the WPA3 user is that the passphrase will grant a hacker access to the network, but not grant them the ability to decrypt traffic on any of the WPA3 sessions. Protected Management Frames are optional in this mode and not mandatory. The network administrator can turn this mode on or off if needed, until all WPA2 client devices are updated.

The second mode in WPA3-Personal operates strictly in WPA3-SAE mode with no connectivity for WPA2 certified devices. In WPA3-SAE mode, Protected Management Frames (802.11w) is required, which helps to prevent spoofed management frames, and connections are more secure through a unique cryptographic exchange process.

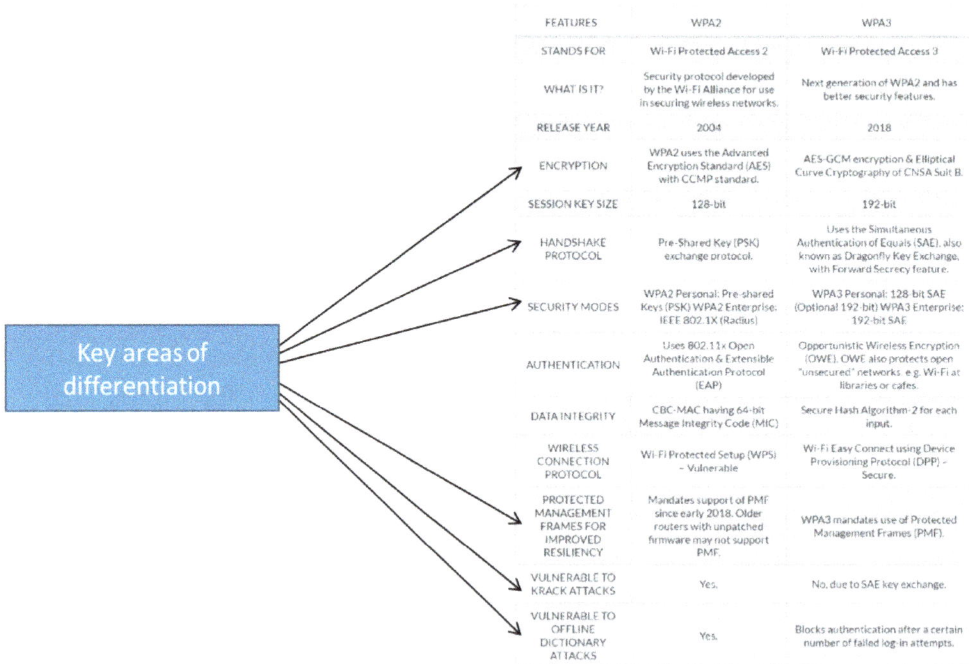

FEATURES	WPA2	WPA3
STANDS FOR	Wi-Fi Protected Access 2	Wi-Fi Protected Access 3
WHAT IS IT?	Security protocol developed by the Wi-Fi Alliance for use in securing wireless networks.	Next generation of WPA2 and has better security features.
RELEASE YEAR	2004	2018
ENCRYPTION	WPA2 uses the Advanced Encryption Standard (AES) with CCMP standard.	AES-GCM encryption & Elliptical Curve Cryptography of CNSA Suit B.
SESSION KEY SIZE	128-bit	192-bit
HANDSHAKE PROTOCOL	Pre-Shared Key (PSK) exchange protocol.	Uses the Simultaneous Authentication of Equals (SAE), also known as Dragonfly Key Exchange, with Forward Secrecy feature.
SECURITY MODES	WPA2 Personal: Pre-shared Keys (PSK) WPA2 Enterprise: IEEE 802.1X (Radius)	WPA3 Personal: 128-bit SAE (Optional 192-bit) WPA3 Enterprise: 192-bit SAE
AUTHENTICATION	Uses 802.11x Open Authentication & Extensible Authentication Protocol (EAP)	Opportunistic Wireless Encryption (OWE). OWE also protects open "unsecured" networks. e.g. Wi-Fi at libraries or cafes.
DATA INTEGRITY	CBC-MAC having 64-bit Message Integrity Code (MIC)	Secure Hash Algorithm-2 for each input.
WIRELESS CONNECTION PROTOCOL	Wi-Fi Protected Setup (WPS) - Vulnerable	Wi-Fi Easy Connect using Device Provisioning Protocol (DPP) - Secure.
PROTECTED MANAGEMENT FRAMES FOR IMPROVED RESILIENCY	Mandates support of PMF since early 2018. Older routers with unpatched firmware may not support PMF.	WPA3 mandates use of Protected Management Frames (PMF).
VULNERABLE TO KRACK ATTACKS	Yes.	No, due to SAE key exchange.
VULNERABLE TO OFFLINE DICTIONARY ATTACKS	Yes.	Blocks authentication after a certain number of failed log-in attempts.

Figure 10.9: WPA2 and WPA3 Compared

With the use of a Diffie-Hellman key exchange and the NIST elliptical curve cryptography (ECC), an attacker can know the password and still not be able to decrypt traffic because it isn't used as a credential in the authentication protocol. The password is only used to index a secret point on an elliptic curve and that point on the curve becomes the generator for use in the cryptographic exchange known as the Dragonfly Key Exchange (see slide). The result is a 32-byte PMK, unknowable to the attacker. If an attacker were to be passively observing this exchange, knowing the password, he/she wouldn't be able to discover or calculate the session's PMK, leaving the encryption keys unknown and unavailable to the attacker.

WPA3 refuses authentication after a certain number of attempted log-ins. This added security helps mitigate Brute-Force-Attacks. This attack is mitigated in WPA3-Personal by using tokens to limit the number of connection attempts. As noted in the slide, WPA3-Personal authentication has four (4) frames instead of two (2) found in WPA2-Personal. The last two (2) WPA3-Personal authentication frames contain a confirmation token. When the access point gets too many SAE requests, it uses the tokens to limit how many simultaneous connections can be attempted, providing protection against Brute-Force-Attacks.

OWE

Opportunistic Wireless Encryption (OWE) uses exchanges similar to those in an HTTPS website to provide secure, encrypted connections without require the client to be provisioned with passphrases or certificates.

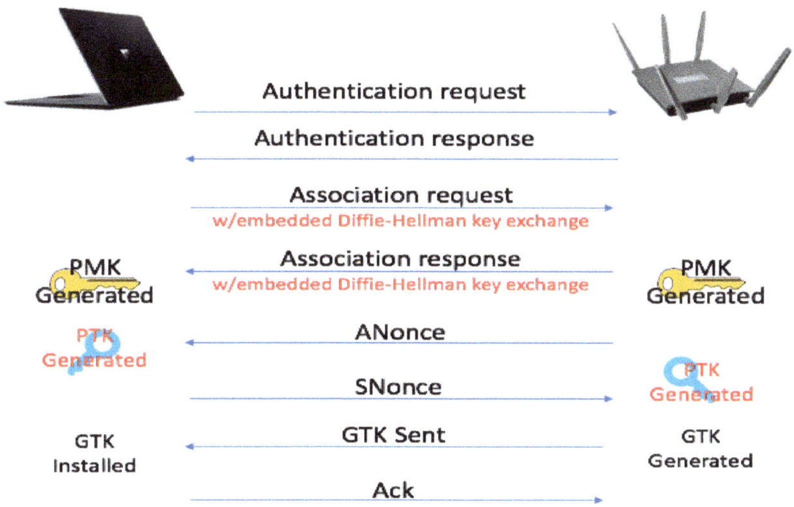

Figure 10.10: OWE Process

Let's first look at how our devices become OWE capable. The good news is that we do not have to buy all new equipment or devices. OWE can be implemented through minor software changes and can even run on legacy equipment. Because OWE is optional and not a mandate in WPA3, open system authentication will function both ways if you want. In the open authentication process, there is an added information element signaling OWE client devices to associate to a hidden

BSS. If a client device isn't OWE capable, then it simply ignores the OWE information element in the open system authentication process and the four (4) frames of open authentication is all that is performed. If a client device is OWE capable, it will recognize the OWE information element and the client will be sent to a hidden BSS that performs the OWE process. Here, the client and the access point will initiate the Diffie-Hellman key exchange during open authentication, then utilize the generated PMK to start the 4-way handshake. Once the 4-way handshake is complete, encryption keys are generated on both the supplicant and the authenticator and all traffic during this session gets encrypted. There is also the option to have your access point(s) function in OWE only mode, not allowing non-OWE clients to associate.

Finally, the new 6 GHz band has specific security requirements, including:

- No Open Authentication without encryption
- Use of OWE for open networks
- No WPA or WPA2 for pre-shared key
- Use of SAE for passphrase-based networks
- Protected management frames will be used

So only WPA3 and OWE are supported in the 6 GHz band. The old days of continue to support older security solution are gone. At least, until the next PHY is released or the next security specification is provided in the 6 GHz band.

Chapter Summary

In this chapter, you learned about security in 802.11 WLANs. You explored weak security solutions that should not be used, and then effective solutions that should be used. You also evaluated some enhanced security solutions and secure management protocols.

Points to Remember

Remember the following important points:

- WEP, Shared Key Authentication, and TKIP are all deprecated security standards.
- SSID hiding and MAC filtering provide no protection against attackers.
- WPA2 is the proper security solution to use today, and it is effectively 802.11 CCMP/AES.
- WPA2-Personal uses pre-shared keys (PSKs) based on passphrases.
- WPA2-Enterprise uses 802.1X/EAP.
- 802.1X implements virtual ports: controlled and uncontrolled. The uncontrolled port is always open, but only allows EAP authentication. The controlled port is closed until authentication is complete.
- When 802.X is used and authentication is incomplete, it is called an unauthorized connection, or said to be in the unauthorized state.
- EAP-TLS, EAP-TTLS, and PEAP all provide mutual authentication.
- Captive portals work by redirecting DNS requests to the captive portal address.
- Management frame protection protects de-authentication, disassociation and robust action frames.
- Voice and video are typical examples of WLAN applications that require fast secure roaming when 802.1X/EAP is used.

- The PTK contains the TK, which is the encryption key used to encrypt unicast traffic.
- A WIPS can both detect an intrusion and take actions to try and prevent it.
- Spectrum and protocol analyzers are often used in smaller organizations to detect security breaches, or for spot checks in large organizations.

Review Questions

1. Which one of the following is not deprecated in the 802.11 standard?
    a. CCMP
    b. TKIP
    c. WEP
    d. Shared Key Authentication

2. Why is MAC filtering not secure?
    a. The encryption used is not strong enough
    b. The password is sent as clear text
    c. Attackers can easily identify allowed MAC addresses
    d. It doesn't work after the AP has been running for three or more hours

3. What term is defined as an association between two stations that includes the 4-way handshake?
    a. RSNA
    b. Pre-RSNA
    c. WEP
    d. WIDS

4. What is a common security recommendation when using WPA2-Personal?
    a. Use a passphrase of at least 40 characters
    b. Only implement it through WPS
    c. Reset the passphrase periodically
    d. None of these

5. What phrase is used to describe when the client and server are both authenticated?
    a. Mutual authentication
    b. Bi-directional authentication
    c. Full duplex authentication
    d. Total authentication

6. What role does the AP or controller play in an 802.1X implementation?
    a. Supplicant
    b. Authentication Server
    c. Access Server
    d. Authenticator

7. What protocol is used between WLAN clients and Aps or controllers when EAP is implemented?
    a. GRE
    b. CAPWAP
    c. LDAP
    d. EAPOL

8. Which virtual port is used to access LAN resources like files and printers in an 802.1X port-based authentication implementation?
    a. Uncontrolled
    b. Controlled
    c. Auth_Port
    d. No-Auth_Port

9. What do EAP methods used with WLANs need to provide?
    a. Dynamic key generation
    b. Anomaly detection
    c. Hashing
    d. Repudiation

10. Which one of the following frame types is not protected by management frame protection?
    a. De-authentication
    b. Disassociation
    c. Probe Requests
    d. Robust Action

Review Answers

1. **A is correct.** Only CCMP, of those listed, is not deprecated in the current standard.

2. **C is correct.** Attackers can see MAC addresses traversing the medium and those addresses are valid.

3. **A is correct.** An RSNA is an association between two stations that includes the 4-way handshake.

4. **C is correct.** Any PSK solution should be reset periodically, including WPA2-Personal.

5. **A is correct.** Mutual authentication indicates that the client and the server are authenticated.

6. **D is correct.** The AP or controller acts as the authenticator in an 802.1X implementation.

7. **D is correct.** EAP over LAN (EAPOL) is used between wireless clients and the authenticator.

8. **B is correct.** The controlled port is used to access LAN resources, but it cannot be used until authentication is successful across the uncontrolled port.

9. **A is correct.** They need to offer dynamic key generation as this material is used to start the 4-way handshake after authentication.

10. **C is correct.** Only de-authentication, disassociation and robust action frames are protected by management frame protection.

Chapter 12 – Site Surveys, Network Design and Validation

The concept of a site survey is beyond the scope of the CWNA-108 exam from a pre-deployment perspective. However, a post-implementation validation is important and the CWNA candidate should understand the process involved. This chapter covers post-implementation validation and the tools used to perform it. Before covering the testable information, however, I want to spend a few pages explaining the general concept of a site survey and the importance of network design. For detailed coverage of these topics, see the "CWDP Official Study Guide."

Site Surveys

Site surveys can be placed into two major categories. The first is the physical site survey, and the second is the RF site survey. Some resources differentiate between the physical site survey and the RF site survey; however, many overlaps exist between these two types.

The physical site survey is an examination of the physical environment in which the WLAN or wireless links will operate. The environment is inclusive of the physical premises owned by the organization operating the WLAN, and possibly physical locations leased for antenna placement or cable runs. It may also include an analysis of the physical space at the install locations of two or more wireless bridges in a point-to-point or point-to-multipoint link implementation. The primary objectives of a physical site survey are to ensure that the location can accommodate a WLAN.

The RF site survey is the process of examining the current RF activity in the physical space where the WLAN must operate and determining RF behaviors in the space. It involves evaluating how your WLAN will function within that physical space as existing RF may or may not be generated by 802.11 radios. The RF site survey should answer the following key questions:

- Is the current RF utilization low enough to allow for the implementation of my new WLAN in the desired channels/frequencies?
- How must I implement the WLAN in order to provide the needed RF coverage within the designated service areas?

- Will I need to negotiate with neighboring WLAN administrators for such demands as reduction in output power on their WLANs, or even channel adjustments on the WLANs?
- Should I implement 5 GHz or 2.4 GHz, or both, for my WLAN?
- Will I have sufficient unlicensed channel space for 802.11n/ac (HT) channel bonding?

While it may seem simple to think of having to answer only four or five questions, many other questions must be answered in order to fully answer these five high-level questions. For example, the current RF frequency utilization is both a factor of frequency or channel usage, and the signal strength of the frequency usage within the area. You might be able to detect a WLAN on the 2.4 GHz channel 6, for example, but is it strong enough to prevent you from implementing a WLAN on that same channel? Additionally, you might discover that channels 1 and 11 are utilized by networks near you that are showing strong signals; however, you could implement a BSS on channel 6 and not be seriously impacted by the neighboring WLANs. You may still have to negotiate a reduction in output power on the neighboring WLANs.

WLAN Design

The importance of design for WLANs cannot be overstated. The best way to grasp the importance of good design is to consider the problems that occur because of bad design, cookie cutter designs or no design at all. This section provides an overview of these issues

The WLAN design process includes defining the network, designing the network, implementing the network based on the design, and then validating that the network design performs as you desire it to perform.

The define phase of a WLAN design and installation project includes important tasks like requirements analysis, information gathering, and pre-site survey checklists.

The design phase of the project includes site surveying and the design of the WLAN. The site survey may be predictive, AP-on-a-stick, or a mixture of these (hybrid). Predictive site surveys are performed using software that simulates RF propagation, based on blueprints and proper data entry about materials. AP-on-a-stick involves placing Aps at locations that are likely to provide the needed coverage, and then testing to ensure that they do

Deployment or implementation of the WLAN is the next logical step. Once the design plans are approved, the network can be configured and installed.

The final phase of the WLAN implementation project is the validation phase. In this phase, the network is evaluated to ensure that it meets the requirements determined in phase one. Therefore, the validation phase is a cyclical procedure. You must validate and tune (or adjust) the network components as many times as required to accomplish the demands of the original requirements. This phase may take a single afternoon, for smaller networks, and several days or weeks for large-scale deployments.

This design process is explored in detail in the "CWDP Official Study Guide."

Post-Implementation Validation

Verify Design Requirements

Metrics are the data values gathered and used to validate the WLAN. These are used to define minimum expectations for data rates and other WLAN parameters. The following metrics should be understood:

- **Connectivity** – The ability to connect to the network and maintain a link. To achieve connectivity, the proper signal strength must be provided when compared to the noise floor.
- **RSSI** – Received Signal Strength Indicator (RSSI) is a measurement of the signal strength at a given location. Devices have received sensitivity ratings and must have a signal at a particular level to receive transmissions at a particular data rate. RSSI is a relative number, as each vendor calculates it using a different range. But higher numbers mean a better signal, up to some maximum, and lower numbers mean a worse signal, down to some minimum.
- **Noise Floor** – Noise is effectively any signal or RF energy, other than the signal being monitored. The noise floor is the level of energy in the

environment without the introduction of local intentional signals, typically measured in dB. For example, -94 dB is a common noise floor reading.

- **SNR** — The signal-to-noise (SNR) ratio is not so much a ratio, as simply the difference between the noise floor and the desired signal's strength. For example, if the noise floor is -94 dB and the signal is -53 dB, the SNR is 41 dB. The ability to properly interpret (demodulate) signals is directly related to the SNR. When SNR values are higher, more complex modulations may be used that result in higher data rates. The reality is that SNR is the key factor in getting a data rate, assuming CCI is not a problem. RSSI is relative, SNR is actual and real. (NOTE: Technically, if the signal is greater than the noise, then the SNR is a ratio higher than 1:1, but we measure it in dB for practical usability in WLANs.)
- **Interference Levels** — Interference levels can be higher in some channels than in others. For example, video cameras and other devices may be operating on channel 6 in an area, but not on channel 11. The result is that the interference levels are higher on channel 6. Effectively, the interference level can be thought of as the noise floor on that channel, as it is a measurement of signals on the channel and the strength and duty cycle of those signals.
- **Cell Coverage** — Cell coverage is a reference to the size of a single AP cell. That is, how much physical space should an AP cover? This is typically defined as a signal strength metric, and then the actual size is simply accepted, given some output power level determined for the Aps. Such as 10 mW for high density, 25 mW for standard density, or even 50 mW for low density.
- **Cell Overlap** — Cell overlap is the measurement of overlap among cells, which allows for effective client roaming. Vendors sometimes encourage 25 percent or more overlap, but of course, this is not possible to measure. Instead, the goal should be two or more Aps (depending on density) at each measurement location. This suggestion is particularly true for 5 GHz WLANs, though a bit harder to achieve for 2.4 GHz WLANS, where accepting that two Aps should be visible to a client for an acceptable distance to allow roaming while mobile, for example, for 20-40 feet, is more realistic.

- **Data Rates** — The speed at which bits can be sent across the RF medium is the data rate. The data rate is determined based on several factors, including, signal strength, channel width, guard intervals, modulation and coding methods.

- **Throughput** — A measurement of the usable data passed through the network. For example, the speed at which a file is transferred, as opposed to the data rate. On WLANs, the data rate is always significantly higher than the throughput. To ensure application functionality, throughput, and not simply data rates, should be used to test for achieved capacity.

- **Latency** — Latency, also called *delay*, is a measurement of the time it takes to move data from one point to another. That is to say, how long does it take for data to get from a VoWLAN phone to the other VoWLAN phone in the conversation. This value should typically be less than 150 ms unidirectional for VoIP implementations.

- **Jitter** — Jitter is the variance in delay (latency). If one packet takes 130 ms to get from point A to point B and the next packet takes 80 ms, it can cause problems in VoIP communications. Jitter buffers are typically used to circumvent this problem.

- **Loss** — Loss is a generic term used to reference packets that do not reach their destination for any reason. Packet loss is typically measured in percentages.

- **Retries** — With connection-oriented protocols and the 802.11MAC, when frames are lost (or packets with TCP), they are resent. This resending is called a retry. High retry rates indicate a problem in a WLAN.

- **Capacity** — The capability of the WLAN to provide services for the required number of users and the applications they use. Capacity is second only to coverage as a WLAN requirement.

- **Roaming** — Roaming is required by many WLAN application use cases, and the implemented network should be tested to verify that roaming occurs within a tolerable amount of time (measured in milliseconds).

- **Aesthetics** — Not usually considered a metric, but it is a measurement of success in WLAN implementations. If the organization desires specific implementation methods (mounting, placement, etc.) for aesthetic reasons, compliance with these requirements should be assured in post implementation.

The primary purpose of the post-implementation survey is to validate that the important metrics to your organization from this list have been achieved, based on requirements. The final step would be documentation.

The tools used to perform the post validation are discussed later in this chapter.

Document WLAN Implementation Results

After validation, you should document the results. Items to document include the metrics from the preceding section, and you will likely create network diagrams, configuration documents (baselines, change management, etc.) and more. Site survey software, if used for the post implementation validation, will typically provide a reporting option to automatically generate much of the documentation required.

Locate and Identify Sources of WLAN Interference

One important skill to use after WLAN installation, is that of interference management. You may have problem spots in the installed WLAN, strictly because of interference. The following types of interference should be understood:

- **WLAN Devices:** WLAN devices may cause two kinds of interference. One is based on contention and normal WLAN operations. The other is usually caused by poor design or improper implementation.
    - **Co-Channel Interference (CCI):** CCI is based on contention. When multiple devices share the same area and channel, but are in different BSSs, they must wait on each other's communications. This is CCI.
    - **Adjacent Channel Interference (ACI):** When an adjacent non-overlapping channel transmits with a strong signal, it may cause interference because of the sideband energy generated. This is ACI. An additional kind of ACI is adjacent overlapping channels, like channel 1 and channel 2 in 2.4 GHz.

- **Non-Wi-Fi Devices:** If a non-Wi-Fi device transmits in the same bands used by WLAN devices, it can cause interference.
    - **Airtime Utilization:** When non-Wi-Fi interference is detected, it is useful to know the utilization level of the interferer. If it is communicating only 5% of the time, then it should degrade your WLAN no more than that. If it is communicating 70% of the time, you have a serious problem.
    - **Frequencies Used:** The other important factor related to non-Wi-Fi interferers, is the frequencies they use. This should be discovered as well.

Spectrum Analysis

A spectrum analyzer can be used to identify non-Wi-Fi interferers and locate them. It can also show you the airtime utilization and frequencies used. Figure 12.1 shows AirMagnet Spectrum XT, and Figure 12.2 shows CommView for Wi-Fi (a protocol analyzer) with a Wi-Spy DBx adapter connected to the laptop.

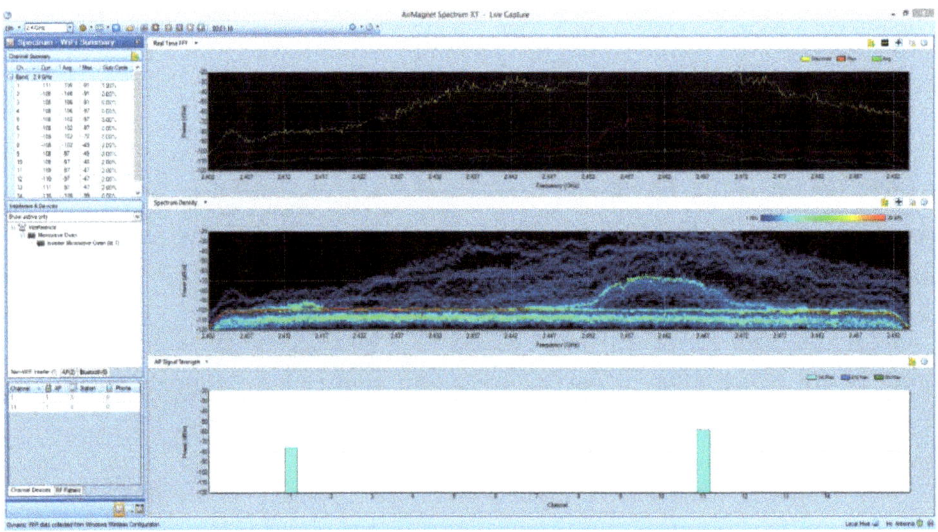

Figure 12.1: AirMagnet Spectrum XT

Interference Solutions

Depending on the kind of interference, solutions may vary. Consider the following recommendations:

- **CCI**: Reconfigure AP settings, including output power and location. Move more STAs to the 5 GHz band to reduce excessive CCI in 2.4 GHz.
- **ACI**: Configure the Aps properly. Do not use excessive output power. Do not use adjacent overlapping channels in 2.4 GHz.
- **Non-Wi-Fi**: Locate the interferer and either remove it or change the local BSS to a different channel that is not on the same frequencies.

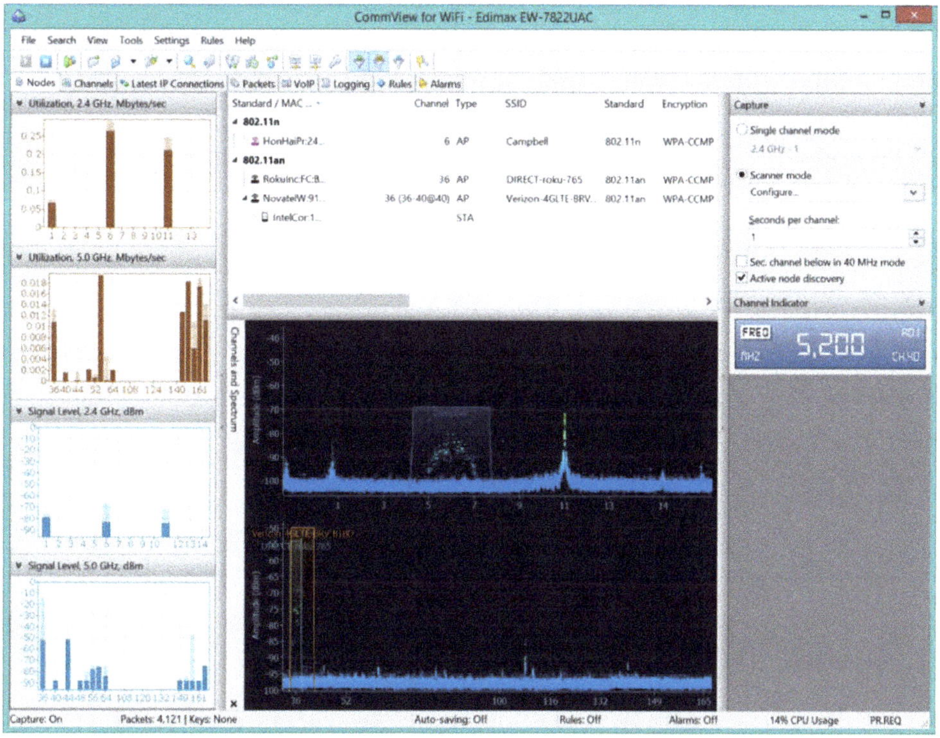

Figure 12.2: CommView for Wi-Fi with Spectrum View

Application Testing

Application testing is performed to ensure that applications work as expected. The tests that may be required include network and service availability, VoIP testing, real-time application testing, throughput testing, and load testing.

Network and service availability should be tested, and the focus here is on the user's ability to access the network and services. In most cases, this testing is simple. Using tools like PING and TRACERT/TRACEROUTE, you can verify basic network connectivity. Services that should be reachable by the users include:

- DNS
- DHCP
- Email
- File servers
- Printers
- Databases
- Application servers
- Internet locations

VoIP testing should be performed if the WLAN must support VoIP handsets. This testing can be performed by simply connecting to the WLAN with a handset and validating the functionality. Three basic tests must be performed to ensure effective VoIP operation:

- **Placing a call:** If you cannot make a call, nothing else matters. Start here.
- **Roaming:** Once the call is connected, move around so that the handset roams to different Aps and ensure that call quality does not suffer, and the call is not dropped.
- **Loading:** Placing a call when you are the only person using the WLAN doesn't prove much. Do it while a load is on the network and, if possible, test multiple concurrent calls on the WLAN.

Real-time application testing may also be required. Such applications include streaming video, video conferencing, and push-to-talk devices. The best way to test such applications is to use them and verify an acceptable user experience. You can also capture the communications in a protocol analyzer to ensure proper QoS markings throughout the network.

Throughput testing can also be performed. It is not a real-world test, but it tells you what the network can do when a single client is connected. However, if you perform a throughput test the first day the users begin using the network and do it again periodically, you can detect changes in network performance. For example, as more uses use the network, performance may suffer. Using this method, you can predict when upgrades may be required.

Load testing is the most challenging test type. You are attempting to simulate a real network load. Special applications are available to assist with this process. WAN killer by SolarWinds is such a tool.

Validation Tools

Finally, you should be aware of the validation tools available to you for WLAN post-installation validation. These include throughput testers, wireless design software, protocol analyzers, and spectrum analyzers.

Throughput Testers

Throughput testers, like TamoSoft Throughput Test and iPerf, send data between a client and server to evaluate the throughput capabilities of a link Figure 12. 3 shows TamoSoft Throughput Test. This tool can test TCP and UDP or only TCP communications. You can tag the frames with QoS markings to see the impact it has on the throughput. This is a free tool that can be downloaded from TamoSoft. You run one copy on the server and the other on the client.

> When using throughput testers, the term server does not reference a physical server, but rather any computer running the server side of the throughput test solution. It can be a laptop, desktop, or an actual server.

TamoSoft Throughput Test comes in both Windows and macOS versions.

iPerf is an open source testing tool available for free download for Windows, Linux, and macOS. The same command line utility is used on the server and the client. You will use different command line parameters to setup a server than those used to connect to the server with a client. Figure 12.4 shows the iPerf command running in Windows.

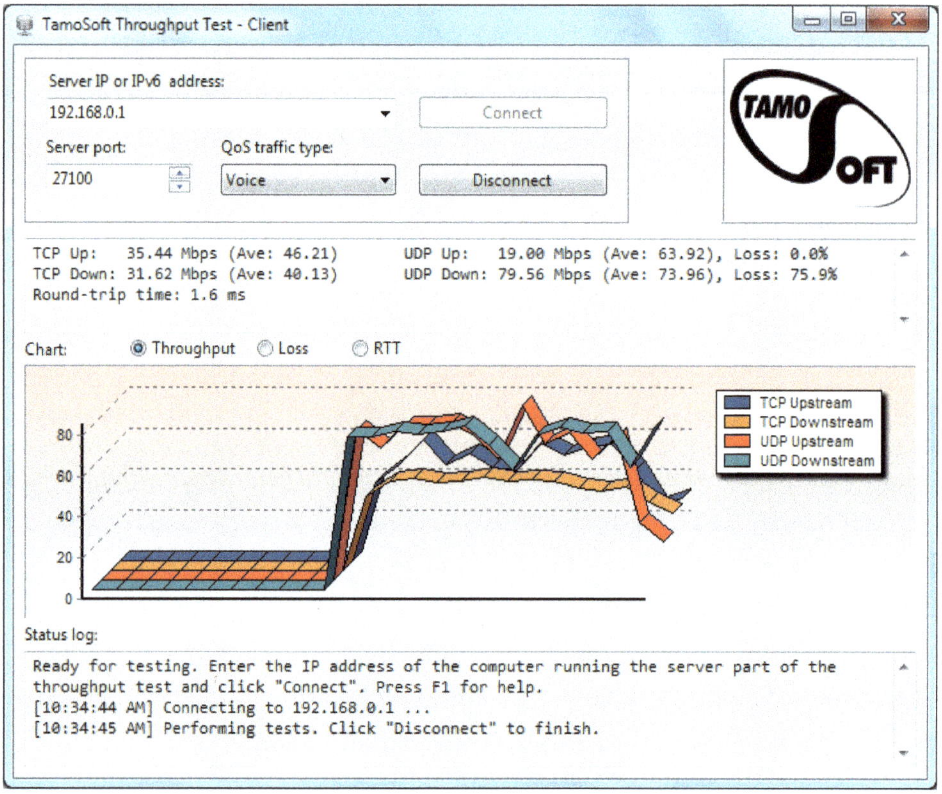

Figure 12.3: TamoSoft Throughput Test

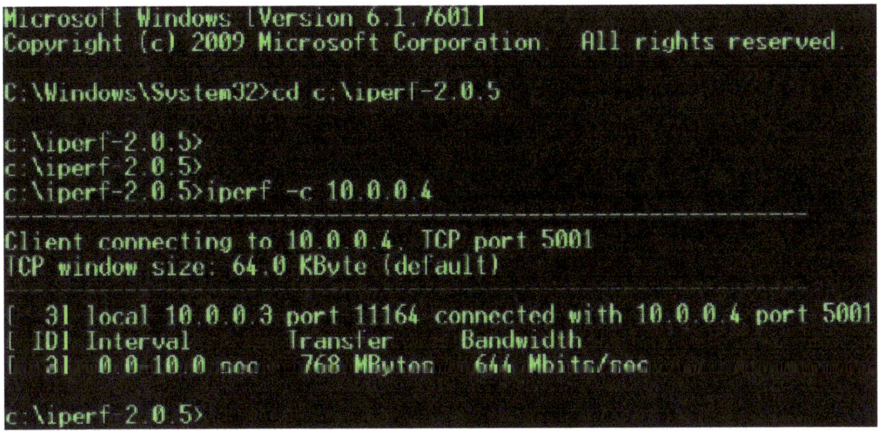

Figure 12.4: iperf on Windows

Wireless Design Software

Wireless design software is available from several vendors including:

- Ekahau: Ekahau Site Survey
- iBwave: iBwave Wi-Fi
- NETSCOUT: AirMagnet Survey Pro
- TamoSoft: TamoGraph Site Survey

Additionally, WLAN hardware vendors often include wireless design software in their controllers or as downloadable software. For example, Aruba Networks provides Aruba RFPLan and Aerohive has design capabilities built into the Hive Manager, to name two.

To be effective for post-implementation validation, the software must be able to gather actual metrics from the WLAN. Most wireless design software solutions can do this. You will simply run the software, load a floor plan and then walk through the facility, periodically pausing and clicking on the floor plan map where you are located. Each time you click, the software will gather live metrics from the network.

Additionally, most of the solutions allow you to perform active surveys. In this case, you actively send and receive data across the network, and this performance is measured as well.

Figurc 12.5 shows TamoGraph Site Survey and Figure 12.6 shows Ekahau Site Survey.

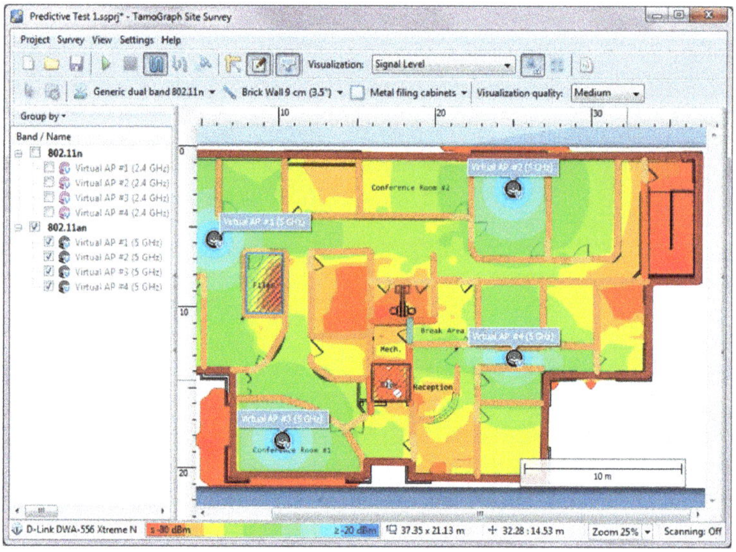

Figure 12.5: TamoGraph Site Survey

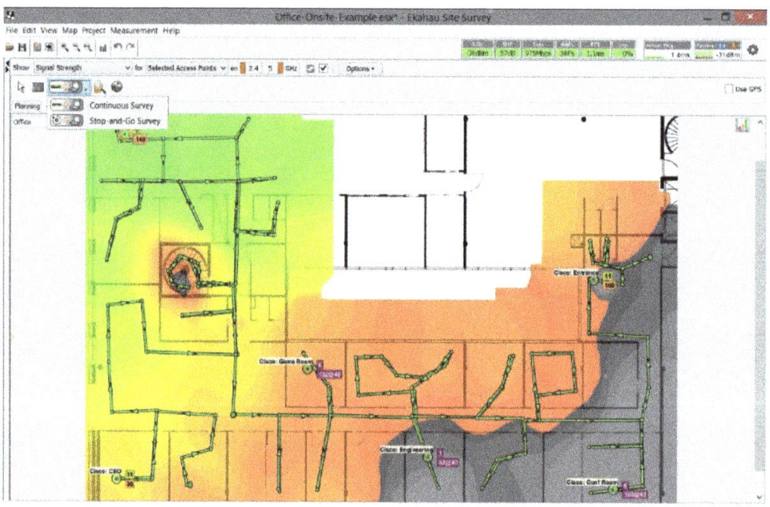

Figure 12.6: Ekahau Site Survey

Protocol Analyzers

Protocol analyzers can also be used in post-implementation validation surveys. For example, the screenshot in Figure 12.7 shows the dashboard in OmniPeek, which can reveal significant information about network activity on the WLAN.

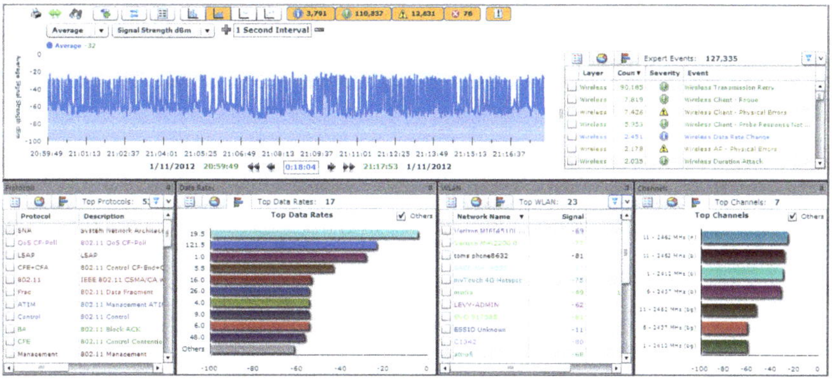

Figure 12.7: OmniPeek Dashboard

Figure 12.8 shows the statistics for all frame types in a channel, as well as measured signal levels in CommView for Wi-Fi.

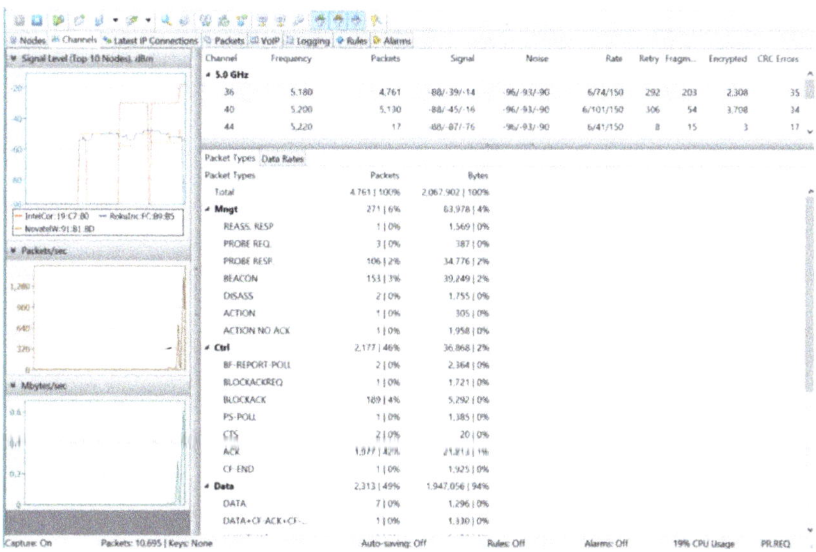

Figure 12.8: CommView for Wi-Fi

Spectrum Analyzers

Spectrum analyzers are most useful in identifying interferers. Figure 12.9 shows the Spectrum XT adapter and the Wi-Spy DBx adapter.

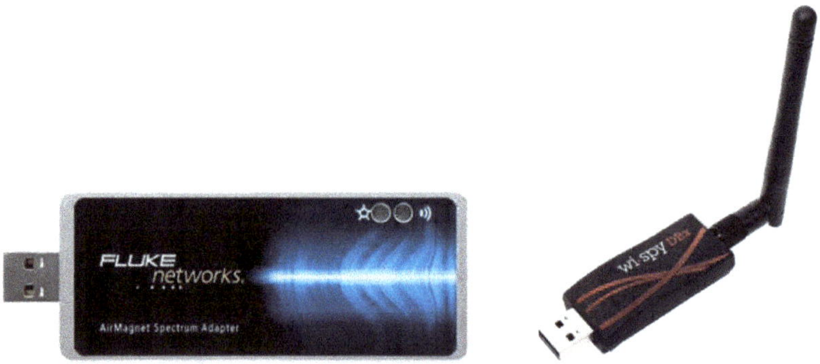

Figure 12.9: Spectrum Analyzers

Figure 12.10 shows the Spectrum XT software interface, which is a rich full-featured spectrum analysis solution for WLANs.

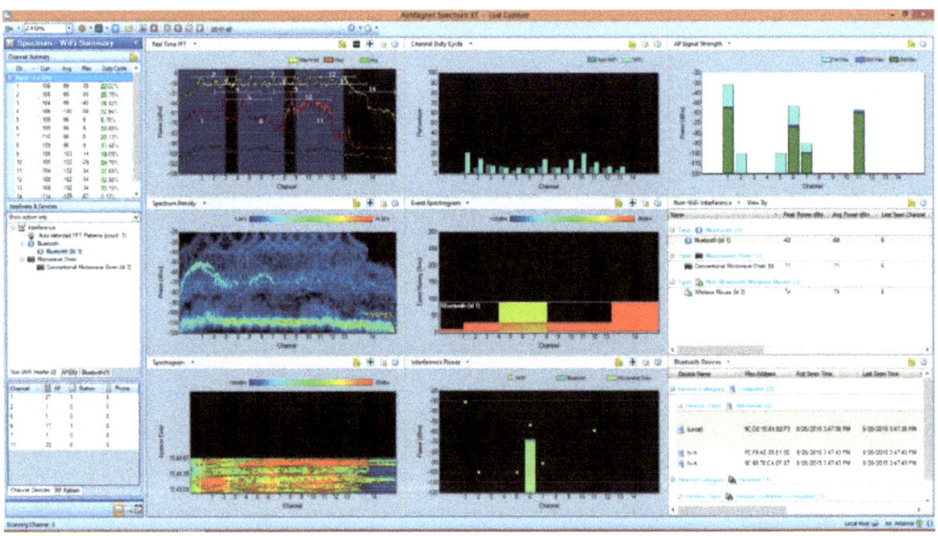

Figure 12.10: Spectrum XT Software

Figure 12.11 shows the Metageek Chanalyzer software interface. The software does not offer all of the views and features of Spectrum XT, but it is significantly less expensive. The question you'll have to answer when selecting between them (if they are your choices), is whether you want the advanced features or the lower price.

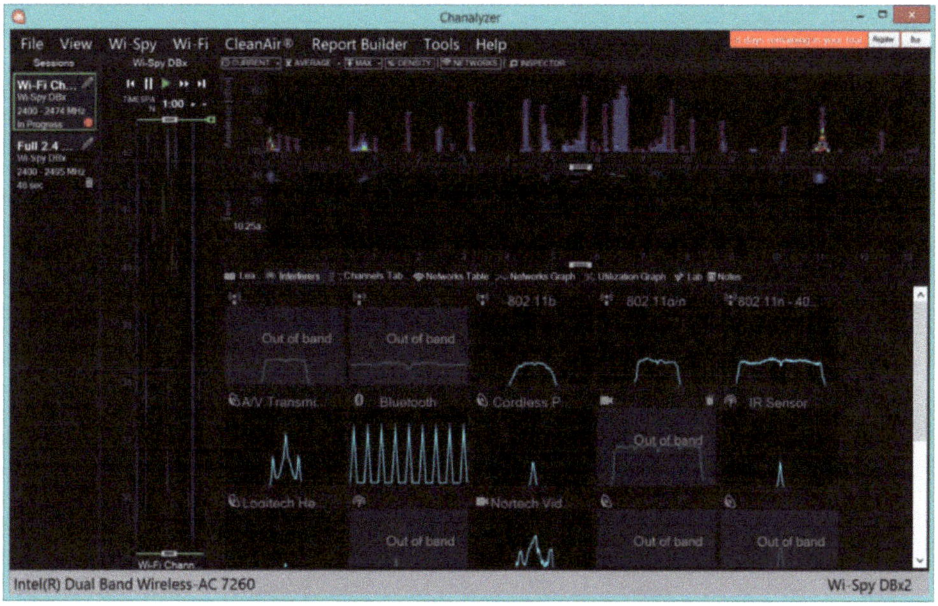

Figure 12.11: Chanalyzer Software

To use a spectrum analyzer, like Wi-Spy DBx, follow these steps:

1. Insert the Wi-Spy DBx adapter into an available USB port.
2. Launch the Chanalyzer software.

3. Select Wi-Spy > Full 2.4 GHz Band from the menu.

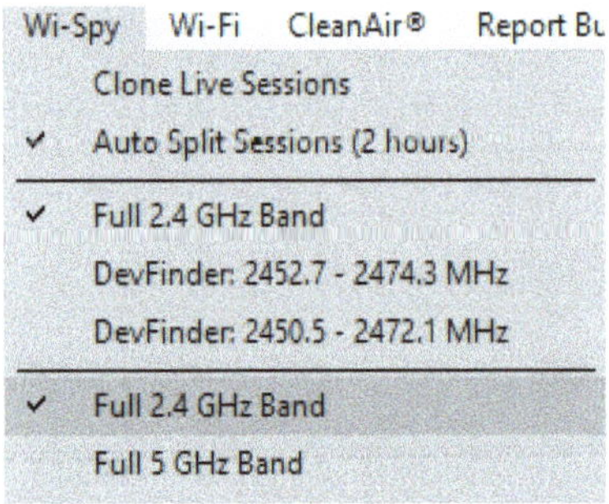

4. Allow the spectrum analyzer to run for a minute or two, to gather spectrum data.

5. Select Wi-Fi > *Your Adapter* to enable Wi-Fi integration.

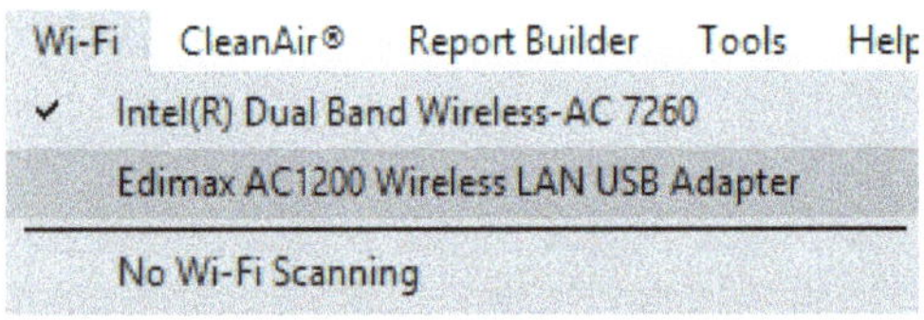

6. Choose the Networks Table in the lower right pane of Chanalyzer.

7. Select (check) the networks you want to see in overlay in the density view.

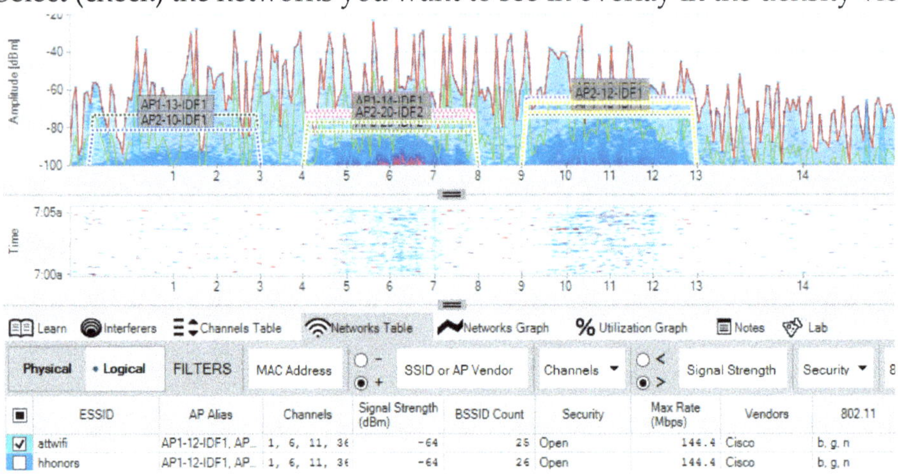

8. Above the density graph, enable the INSPECTOR feature.

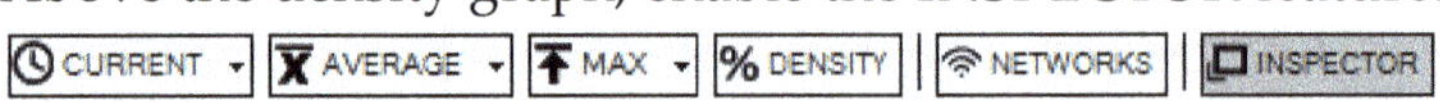

9. Hover over an area of the density graph and notice the spectrum data it reveals with INSPECTOR enabled.

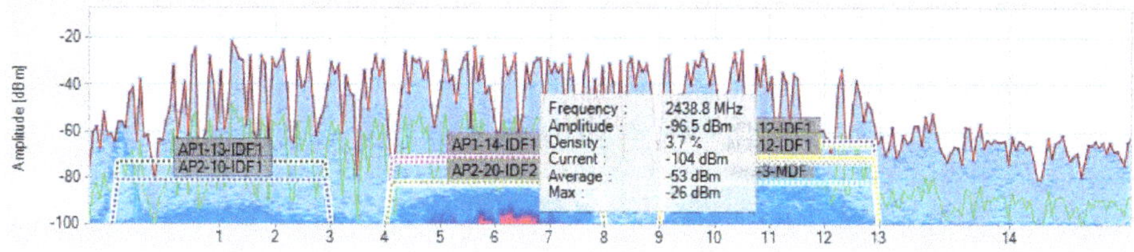

10. Change to the Network Graph tab in the lower-right pane. View the signal over time for the various networks.

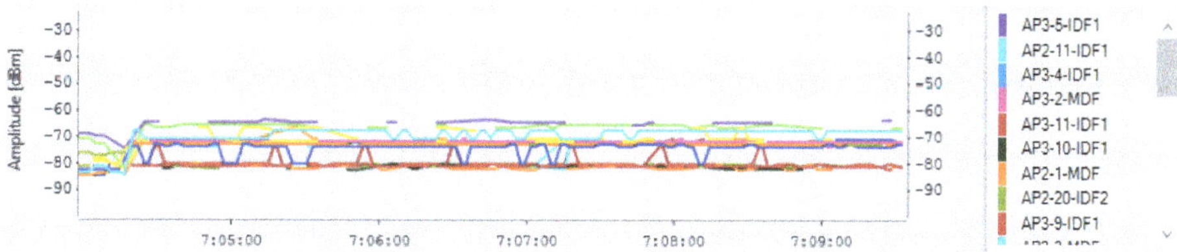

11. Change to the Utilization Graph and view the utilization. Notice you can change the signal strength at which to measure utilization (-90 dBm is shown).

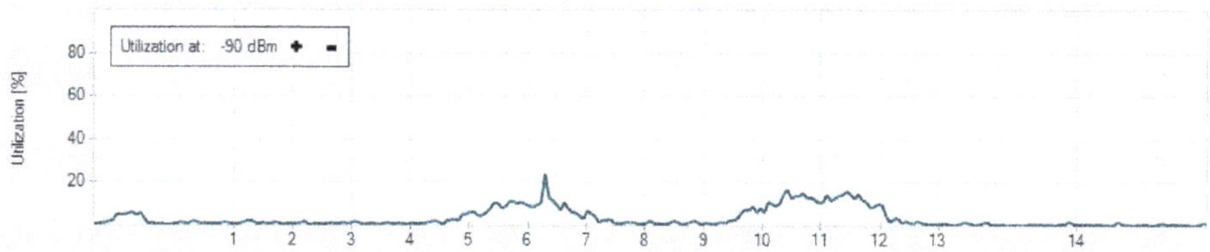

12. Select the Channels Table and note the information that can be gathered there. Grade is a measurement of interference impact versus a "perfect" channel. Higher grades are better.

Channel	Grade	Utilization	Average (dBm)	Current (dBm)	Max (dBm)	Noise Floor (dBm)	Access Points
11	90.4	8.8%	-58.0	-93.5	-31.5	-100.0	15
12	91.4	8.1%	-59.0	-84.0	-32.5	-100.5	0
10	91.4	8.3%	-57.5	-93.5	-31.0	-100.5	1
13	94.2	5.3%	-62.5	-84.0	-36.0	-101.0	0
9	94.2	5.9%	-57.0	-94.5	-30.5	-100.5	0
8	96.6	4.9%	-57.0	-92.5	-30.5	-100.5	0
6	97.1	4.7%	-55.5	-78.0	-29.5	-99.5	13
7	97.2	4.7%	-56.0	-78.0	-29.5	-100.0	0
5	97.3	4.7%	-56.0	-78.0	-29.5	-100.0	0
4	97.6	3.1%	-56.0	-78.0	-30.0	-101.0	0
1	97.7	1.6%	-60.0	-93.5	-33.5	-102.0	9
[illegible]	97.0	1.6%	-57.0	-95.5	-31.0	-101.5	0
2	98.1	1.1%	-68.5	-95.5	-32.0	-102.0	0
14	98.6	0.9%	-87.5	-84.0	-62.5	-102.5	0

13. Select Wi-Spy > Full 5 GHz Band to switch to 5 GHz mode.

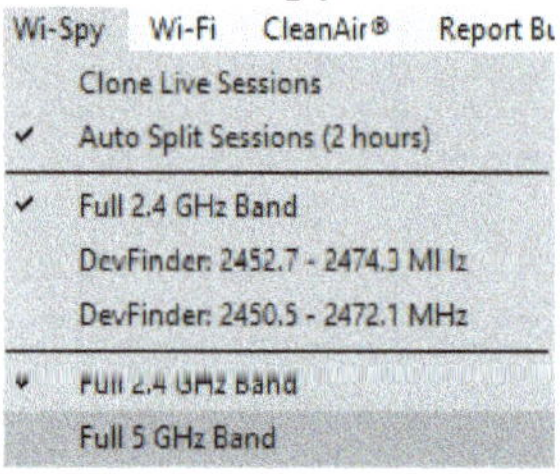

14. Use the same features previously used in the 2.4 GHz band to gather information about the 5 GHz band.

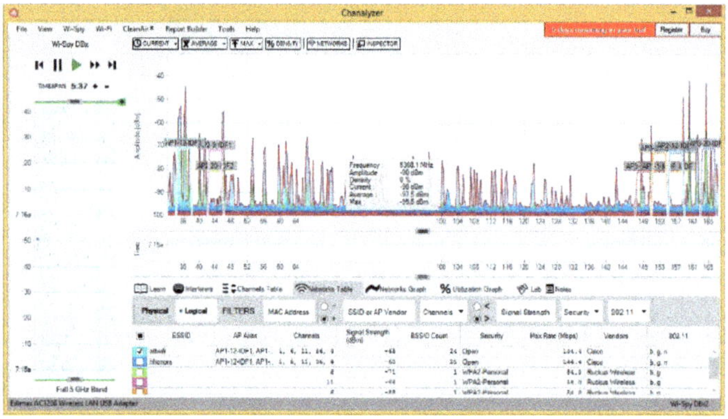

Chapter Summary

In this chapter, you learned about post-installation validation surveys. You discovered the importance of WLAN design and validation process and then considered the different metrics used to validate the WLAN. Finally, you reviewed the various tools that may be used for WLAN validation.

Points to Remember

Remember the following important points:

- You should target two or more detectable Aps in each location to provide overlap, as percentage-based overlap metrics are not reasonable.
- Delay or latency is a key metric in WLANs, and for VoIP should be at 150 ms or less.
- High retry rates indicate a problem in the WLAN.
- You can reduce CCI in 2.4 GHz by moving more STAs to the 5 GHz frequency band.
- When non-Wi-Fi interferers have low airtime utilization, they will not have as significant an impact on the WLAN.
- Simple network availability testing can be performed with tools like PING and TRATEROUTE/TRACERT.
- You should test VoIP on WLANs by placing a call, roaming across Aps, and loading the network while testing, when possible.
- iPerf is a commonly used open source throughput tester.
- Wireless design software that can perform active and passive surveys can also be very useful for post-implementation validation.

Review Questions

1. What is always required to achieve connectivity to the WLAN?
   a. The proper signal strength when compared to the noise floor
   b. The built-in macOS supplicant
   c. External antennas
   d. No interference

2. What is the key factor in achieving a particular data rate?
   a. The transmit power at the AP
   b. The transmit power at the client
   c. That antenna gain on the AP
   d. SNR

3. What is the best way to ensure you have cell overlap?
   a. Measure the percentage of overlap with a ruler
   b. Ensure multiple Aps are visible at cell boundaries
   c. Disable RRM
   d. Enable ARM

4. What is the maximum acceptable unidirectional delay in milliseconds for most VoIP implementations?
   a. 300
   b. 200
   c. 150
   d. 100

5. You have detected interference cause by one AP on channel 1 and another on channel 2. What kind of interference is this?
   a. Non-Wi-Fi
   b. CCI
   c. Adjacent non-overlapping channel
   d. Adjacent overlapping channel

6. Which one of the following actions may resolve non-Wi-Fi interference problems?
   a. Move the local BSS to a different channel
   b. Decrease the output power on the AP
   c. Use external antennas on the client
   d. None of these

7. What kind of interference may occur if an AP on channel 6 is configured with 400 mW of output power and another AP is a few feet away on channel 1?
   a. CCI
   b. Non-Wi-Fi
   c. ACI
   d. Full-time interference

8. What example tool can be used for load testing?
   a. WAN killer
   b. Spectrum analyzer
   c. Protocol analyzer
   d. Wi-Fi scanner

9. What simple networking tool can verify network connectivity?
   a. PING
   b. Protocol analyzer
   c. Spectrum analyzer
   d. IPCONFIG

10. What is an example of an application that can test throughput, use only UDP, and simulate Voice QoS traffic?
    a. PING
    b. WAN killer
    c. TamoSoft Throughput Test
    d. TRACERT

Review Answers

1. **A is correct.** In order to connect you must be able to receive the proper signal strength at the location compared to the noise floor.
2. **D is correct.** The signal strength alone is not as important as the SNR (which is the signal strength compared to the noise floor).
3. **B is correct.** By ensuring multiple Aps are available at the cell boundaries, you ensure overlap. Overlap in percentages cannot really be measured.
4. **C is correct.** Most VoIP implementations will require latency or delay measurements of less than 150 ms unidirectional.
5. **D is correct.** The scenario described is ACI, but it is specifically adjacent overlapping interference.
6. **A is correct.** By moving the local BSS to a different channel, you may be able to move away from the interferer.
7. **C is correct.** Because of the high output power and the close proximity, ACI is likely to occur.
8. **A is correct.** WAN killer from SolarWinds is a load testing tool.
9. **A is correct.** PING and TRACEROUTE/TRACERT are simply network connectivity testing tools.
10. **C is correct.** TamoSoft Throughput test can perform all of these functions.

Chapter 13 — WLAN Troubleshooting

Troubleshooting WLANs is an art and a science. It is a science, because you can learn the skills. It is an art, because even after learning the skills, it takes time to fine tune your abilities to narrow symptoms to a problem. In this final chapter of the CWNA study guide, you will explore troubleshooting process, tools and specific scenarios. In the end, you will have learned a lot of the science, but you must put the science into practice to master the art.

Troubleshooting Processes

Troubleshooting processes can be defined in methodologies. Vendors, and the IT industry in general, define various methodologies. For example, Cisco defines a specific troubleshooting model that looks like this:

1. Define a clear problem statement with symptoms and potential causes.
2. Gather the facts to help isolate the possible causes.
3. Consider possible problems, based on the facts discovered.
4. Create an action plan, based on the remaining potential problems and the most likely cause.
5. Implement the action plan.
6. As changes are made, gather results.
7. Analyze the results and determine whether the problem has been resolved.
8. If the problem is not resolved, create a new action plan based on the next most likely cause, and proceed with steps 5-8. Repeat until resolved or escalated.

Microsoft's troubleshooting process looks like this:

- **Phase 1:** Discovery. Gather information about the problem.
- **Phase 2:** Planning. Create a plan of action.
- **Phase 3:** Problem Reproduction. Reproduce the problem or determine that you cannot reproduce it. If you cannot reproduce the problem, then you might not have enough information to confirm that there is a problem.

- **Phase 4:** Problem Isolation. Isolate the variables that relate directly to the problem.
- **Phase 5:** Analysis. Analyze your findings to determine the cause of the problem.

CompTIA, in their Network+ objectives, defines the process like this:

1. Identify the problem.
2. Establish a theory of probable cause.
3. Test the theory to determine cause.
4. Establish a plan of action to resolve the problem and identify potential effects.
5. Implement the solution or escalate as necessary.
6. Verify full system functionality and, if applicable, implement preventative measures.
7. Document findings, actions and outcomes.

Regardless of the source, one thing is consistent: start by defining the problem. It is amazing how often technicians (not me, of course) start troubleshooting a problem without verifying that they understand the actual problem itself. So, troubleshooting starts with a problem, and you start troubleshooting by defining the problem.

The CWNP Troubleshooting Methodology

Because CWNP exams are focused on WLANs and the CWNA exam is focused on WLAN administration, the CWNP methodology includes the steps and actions that should be performed in such an environment. It is based on industry experience and feedback and will aid the WLAN professional in resolving network issues quickly and effectively.

You will be tested on the CWNP methodology. Make sure you understand this sequence of steps and be prepared to answer questions that test your knowledge of this troubleshooting flow.

The CWNP methodology includes the following steps:

1. Identify the problem.
2. Discover the scale of the problem.
3. Define the possible causes of the problem.
4. Narrow to the most likely cause.
5. Create a plan of action or escalate the problem.
6. Perform corrective actions.
7. Verify the solution.
8. Document the results.

The first step is to *identify the problem,* which is shared by nearly all troubleshooting methodologies. The worst mistake a troubleshooter can make is to assume the specifics of a given problem. Think of identifying the problem as defining the objective. When you define objectives for a WLAN design, for example, you lay the foundation on which the entire design and implementation is built. Without this foundation, the design is sure to fail. The same is true in troubleshooting. Many hours can be wasted by troubleshooting an assumed problem. Assumptions can come from faulty communications with the users experiencing the problem. The problem must always be verified. Ask questions like the following to identify the problem:

- Do you see any error messages?
- Specifically, what results are you experiencing that make you feel the network is down?
- Has this happened before and, if so, how often?

- Where are you located?
- Have you moved since your initial connection to the wireless network?
- What device are you using?
- What software are you using?
- Does any other software work on the network?
- Is the problem related to time of day?

As you can see from these questions, you are narrowing the problem to the location, the device and the application. These questions, and others like them, can reveal the true problem.

The second step is to *discover the scale of the problem*. This step is very important, as it can reveal a local network outage that impacts all users, as opposed to a single user problem. If you are receiving reports from multiple users in a coverage area, it is likely a network problem or application problem, and not an issue with individual user device configuration. If you are addressing the first report of a problem, ask the user if other users in his or her area are experiencing the same or similar problem.

Remember that application problems can be larger in scale than a single individual, as well. For example, if users use a PC-based softphone for VoIP on their laptops, and the first user calls from another phone to inform you that the network is down. The reality may be that the call manager is down for that segment, and only the VoIP application is experiencing problems. In this case, it is not an actual network problem, but an application problem with scale impact.

The third step is to *define the possible causes of the problem*. A single problem can occur because of many different potential causes. The troubleshooter must narrow the pool of potential causes to the most likely for a given scenario, but first the common causes must be identified. For example, if a user cannot connect to the WLAN, many issues could cause this problem, including:

- The client is configured improperly.

- The AP is down.
- The controller is down.
- The DHCP pool is depleted.
- The DHCP server is down.
- The DNS server is down.
- The switch or router is experiencing problems.
- The Internet connection is down.
- The application server is down or overloaded.
- The client hardware is failing.
- The switch for the wireless adapter is turned off on their laptop.

The point is simple: all of these potential causes, and more, tell the user that they cannot connect to the WLAN. In reality, with many of these causes the device is, in fact, connected to the WLAN but something else is wrong. This truth is why step one is so important. The real problem must be identified. If it is, the cause list will shrink dramatically for this third step.

In these first three steps, you will also use technical methods to define the problem and causes. For example, you may use the OSI model troubleshooting methods common to the industry. You may use networking tools to identify possible causes, such as spectrum analyzers, protocol analyzers and operating system commands like PING, IPCONFIG, TRACEROUTE and NETSH.

The fourth step is to *narrow to the most likely cause*. One cause is more likely than the others for a given problem in a given environment. Stated differently, each production environment includes a set of devices and standard configurations. A specific environment will experience common problem causes that another environment may not experience as frequently. For this reason, step four is experiential. Over time, you will learn the most likely cause or causes for a given problem in the environments you support.

For example, when using Aruba Networks WLAN solution, you will have access to configuration options that do not even exist in a Cisco solution (and vice versa). Therefore, you will experience configuration-related problems in one network that you would not experience in another. After having experience with a solution in your environment, you will develop the experiential expertise that allows for faster troubleshooting. This reality is why step eight is so important. The documentation will allow you to determine the most common causes of problems over time and, therefore, make you a better troubleshooter.

The fifth step is to *create a plan of action or escalate the problem*. In the real world of network support, you will not always have the required access to resolve an issue. In such scenarios, you must escalate the problem to the appropriate individual or group. For example, if you determine that your WLAN users are experiencing problems only with VoIP and that it is likely the call manager that is causing the problem, you may not have the appropriate administration permissions to do anything about it. This issue should be escalated to the call manager administrator with all of the details that you have gathered. When you can resolve the issue yourself (assuming you've identified the appropriate cause), you should create a plan of action.

The plan of action may or may not be documented, but you should know what you're going to do and the results that you expect. For example, the plan of action may be to reinstall the device drivers for the WLAN adapter on a client device. You expect that this will result in the repair of corrupted driver files and allow for connectivity to the WLAN. Given a system that supports recoverability features, the following action plan may be in mind:

1. Create a backup of the current configuration.
2. Uninstall the drivers completely from the device.
3. Reinstall the drivers.
4. Attempt to connect to the WLAN.

The sixth step is to *perform corrective* actions. If the previous plan of action results in a working system, you have resolved the issue and are ready for step seven, *verify the solution*. The reality is that you may cycle through steps four through seven, many times before finding the solution. In cases where you have altered configuration settings and the problem is not resolved, it is often best to reconfigure the system back to the original settings before moving on to the next possible cause. Otherwise, the system may experience different problems related to the unneeded changes, and you can lose track of where you are in the process.

The eighth and final step is to *document the results*. I would argue that this is equal in importance to the first step, *identify the problem*. If you do not document the results, you do not learn from the experience as you should. Additionally, if you have shared documentation within the organization, others can benefit from your knowledge, as well. I call this OPK (other people's knowledge). It is for this reason that, immediately after identifying the problem and its scale, you should research your own documentation, and possibly online resources, to see if others have experienced the same problem and found a solution.

Today, with the global scale of the Internet, it is very unlikely that you are the first one to experience a given problem. Do some research to help focus your *step three* process of defining possible causes. In many scenarios, this research can save you dozens of hours of effort. Use OPK to enhance your troubleshooting abilities. Many WLAN professionals blog, participate in forums, and write other online content that will help you. Additionally, vendors often have troubleshooting guides that provide insightful information for their specific solutions. Take advantage of these resources and of your internal documentation to reduce your troubleshooting time and to become a better WLAN analyst.

In the end, the primary benefit of a troubleshooting methodology is that it ensures the right problem is solved, and time is not wasted. In other words, it brings focus to the troubleshooting process.

Troubleshooting Tools

Tools define a professional. You would be shocked if an electrician come to your house to repair an electrical problem and all she had was a hammer. Or imagine a plumber without pipe wrenches, or an automobile mechanic without ratchets. It just wouldn't make sense. In the same, it doesn't make sense to try to administer and troubleshoot a WLAN without the right tools. Some of these tools will be free, and others will be very costly; however, the time they save will be well worth the cost in the long run.

Networking tools are used to analyze and troubleshoot network connection and throughput issues, and include throughput testers, protocol analyzers and spectrum analyzers. These tools are not included as native software in operating systems, and therefore exist in their own category, as they must be installed before use.

Throughput testers are used to evaluate the useful data bits that can pass through a network. They typically test at Layer 4 but may be able to test at higher layers as well. At Layer 4, the Network layer, they are testing TCP and UDP traffic. TCP is used for standard data communications and UDP is used for real-time communications. Figure 13.1 shows the help output for the Windows iperf command (specifically iperf3, available at: bit.ly/1Ut2fs7). Figure 13.2 shows the output of an executed command.

Throughput testers typically work on a client/server model. That is, one machine will act as the server and another as the client. GUI-based throughput testers provide a graphical interface used to configure the server and the client and to execute the testing. Command-based throughput testers work at the Command Prompt in Windows, or at the shell in Linux environments. They use commands with switches to configure the server and to execute the test on the client.

The default behavior of iperf is to test the throughput from the client to the server. Therefore, when testing a wireless client, to test the downlink, the wireless client should be configured as the iperf server. To test the uplink, the wireless client should be configured as the iperf client. Some versions of iperf allow for bidirectional testing, so that this concern no longer exists. You will find, when working with wireless links, that downlink traffic often performs better than uplink traffic

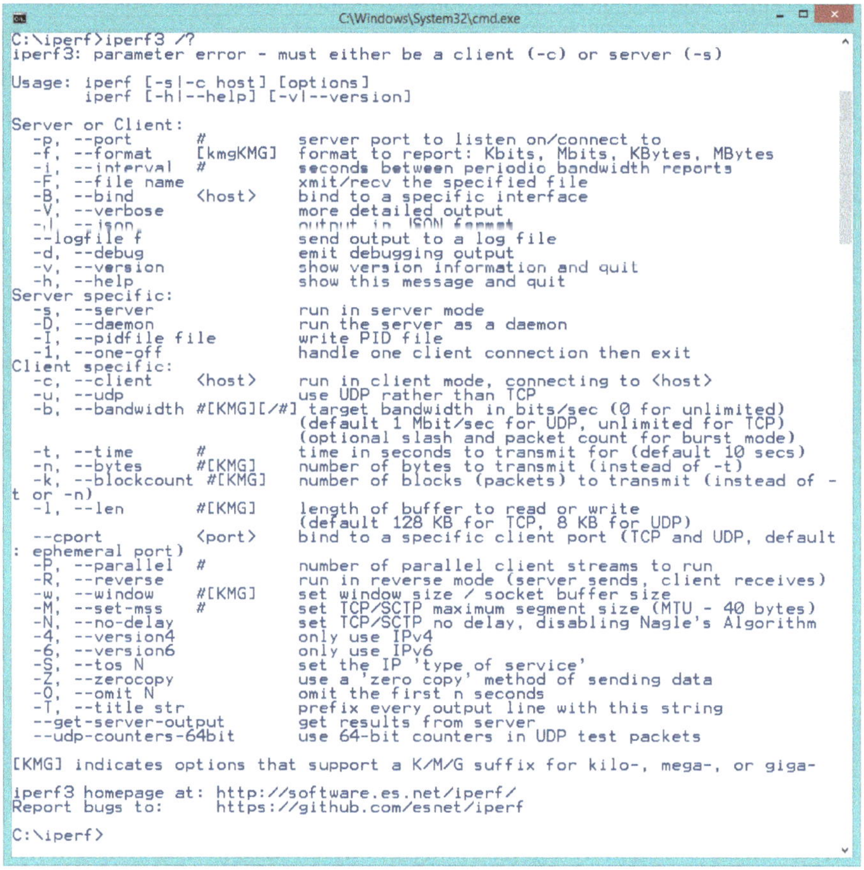

```
C:\iperf>iperf3 /?
iperf3: parameter error - must either be a client (-c) or server (-s)

Usage: iperf [-s|-c host] [options]
       iperf [-h|--help] [-v|--version]

Server or Client:
  -p, --port      #         server port to listen on/connect to
  -f, --format    [kmgKMG]  format to report: Kbits, Mbits, KBytes, MBytes
  -i, --interval  #         seconds between periodic bandwidth reports
  -F, --file name           xmit/recv the specified file
  -B, --bind      <host>    bind to a specific interface
  -V, --verbose             more detailed output
  -J, --json                output in JSON format
  --logfile f               send output to a log file
  -d, --debug               emit debugging output
  -v, --version             show version information and quit
  -h, --help                show this message and quit
Server specific:
  -s, --server              run in server mode
  -D, --daemon              run the server as a daemon
  -I, --pidfile file        write PID file
  -1, --one-off             handle one client connection then exit
Client specific:
  -c, --client    <host>    run in client mode, connecting to <host>
  -u, --udp                 use UDP rather than TCP
  -b, --bandwidth #[KMG][/#] target bandwidth in bits/sec (0 for unlimited)
                            (default 1 Mbit/sec for UDP, unlimited for TCP)
                            (optional slash and packet count for burst mode)
  -t, --time      #         time in seconds to transmit for (default 10 secs)
  -n, --bytes     #[KMG]    number of bytes to transmit (instead of -t)
  -k, --blockcount #[KMG]   number of blocks (packets) to transmit (instead of -
t or -n)
  -l, --len       #[KMG]    length of buffer to read or write
                            (default 128 KB for TCP, 8 KB for UDP)
  --cport         <port>    bind to a specific client port (TCP and UDP, default
: ephemeral port)
  -P, --parallel  #         number of parallel client streams to run
  -R, --reverse             run in reverse mode (server sends, client receives)
  -w, --window    #[KMG]    set window size / socket buffer size
  -M, --set-mss   #         set TCP/SCTP maximum segment size (MTU - 40 bytes)
  -N, --no-delay            set TCP/SCTP no delay, disabling Nagle's Algorithm
  -4, --version4            only use IPv4
  -6, --version6            only use IPv6
  -S, --tos N               set the IP 'type of service'
  -Z, --zerocopy            use a 'zero copy' method of sending data
  -O, --omit N              omit the first n seconds
  -T, --title str           prefix every output line with this string
  --get-server-output       get results from server
  --udp-counters-64bit      use 64-bit counters in UDP test packets

[KMG] indicates options that support a K/M/G suffix for kilo-, mega-, or giga-

iperf3 homepage at: http://software.es.net/iperf/
Report bugs to:     https://github.com/esnet/iperf

C:\iperf>
```

Figure 13.1: Help for the Windows-based iPerf Command

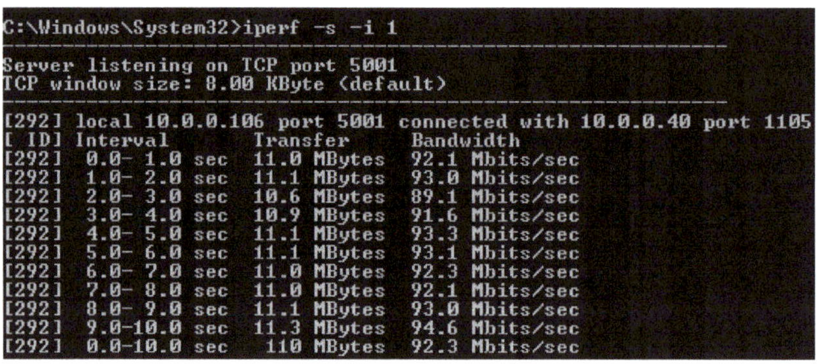

```
C:\Windows\System32>iperf -s -i 1
------------------------------------------------------------
Server listening on TCP port 5001
TCP window size: 8.00 KByte (default)
------------------------------------------------------------
[292] local 10.0.0.106 port 5001 connected with 10.0.0.40 port 1105
[ ID] Interval       Transfer     Bandwidth
[292]  0.0- 1.0 sec  11.0 MBytes  92.1 Mbits/sec
[292]  1.0- 2.0 sec  11.1 MBytes  93.0 Mbits/sec
[292]  2.0- 3.0 sec  10.6 MBytes  89.1 Mbits/sec
[292]  3.0- 4.0 sec  10.9 MBytes  91.6 Mbits/sec
[292]  4.0- 5.0 sec  11.1 MBytes  93.3 Mbits/sec
[292]  5.0- 6.0 sec  11.1 MBytes  93.1 Mbits/sec
[292]  6.0- 7.0 sec  11.0 MBytes  92.3 Mbits/sec
[292]  7.0- 8.0 sec  11.0 MBytes  92.1 Mbits/sec
[292]  8.0- 9.0 sec  11.1 MBytes  93.0 Mbits/sec
[292]  9.0-10.0 sec  11.3 MBytes  94.6 Mbits/sec
[292]  0.0-10.0 sec   110 MBytes  92.3 Mbits/sec
```

Figure 13.2: Output from the Windows iPerf Command

An example of a GUI-based throughput tester is TamoSoft's Throughput Tester, shown in Figure 13.3. This tool can test both TCP and UDP traffic and supports reporting on packet loss with visual graphs showing moment-by-moment throughput performance. The tool is available for both Windows and macOS.

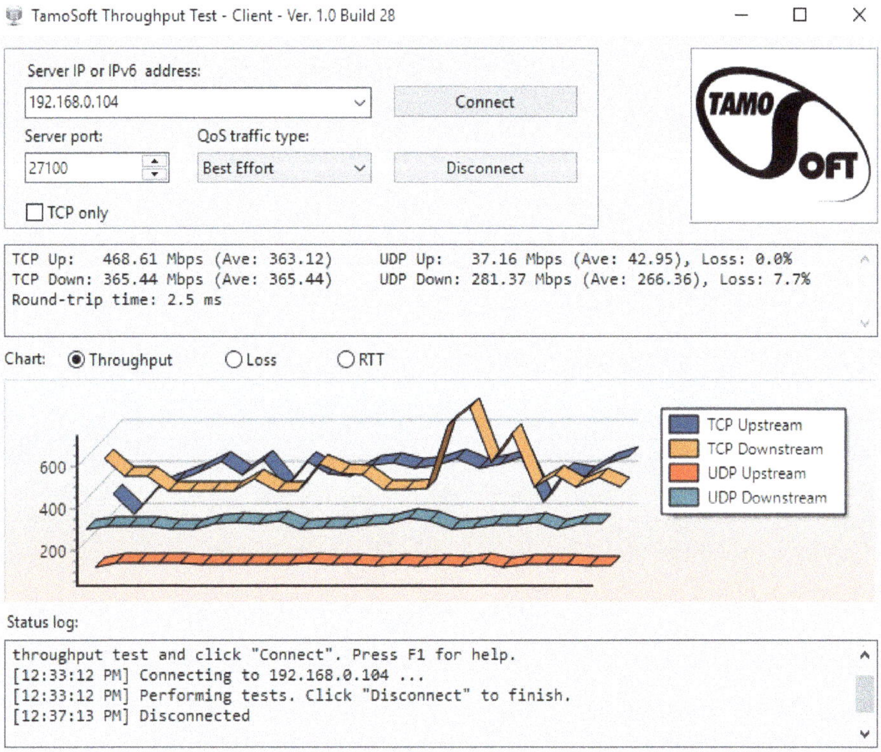

Figure 13.3: TamoSoft Throughput Tester

When testing throughput, it is important to remember that you are not testing the data rate. The data rate is the rate at which bits can be sent across the wireless medium and is entirely dependent on signal quality, and the modulation and coding used. Higher data rates use more sophisticated modulation and coding schemes and require better signal conditions than lower data rates. The data rate is a significant factor in determining network throughput for a user, but it does not stand alone. In addition, the contention for the wireless medium must be considered.

For example, if a single client has a data rate of 866.7 Mbps with an 802.11ac connection to the AP using the VHT PHY, this does not mean that client will achieve performance values, as if it were the only client connected. Other clients may be connected to the same AP at 54 and 48 Mbps. Those clients will gain access to the medium as well, and the super-fast 802.11ac client will simply have to wait its turn. This impacts Layer 4 throughput significantly, and it impacts it even more on busier WLANs with more varied clients and more activity from those clients. The point is that throughput is not a simple factor of data rate, and this will be discussed more as you continue through the book.

Throughput testers are useful to the WLAN analyst for the following:

- Verifying application performance problems
- Locating intermittent performance issues
- Validating the performance of a new WLAN
- Proactively locating problem areas of the WLAN
- Ensuring continued and consistent performance

The next networking tool is the protocol analyzer. Protocol analyzers have existed for more than two decades. They are tools that allow you to capture and decode networking frames and packets. Wired protocol analyzers are very easy to use, as they work with practically any network adapter. Wireless protocol analyzers are different, as they require specifically compatible adapters. Figure 13.4 shows Wireshark with special color coding for Association Request and Association Response frames (green), Authentication frames (red), and Probe Request and Response frames (yellow).

Know that protocol analyzers are useful to the WLAN administrator for the following:

- Analyzing network settings
- Gathering details about unsupported networks
- Checking for frame corruption and retransmissions

- Locating the source of authentication and other communication problems
- Identifying overloaded service sets or channels
- Identifying devices on the network
- Validating compliance with requirements
- Discovering supported features and behaviors of wireless devices

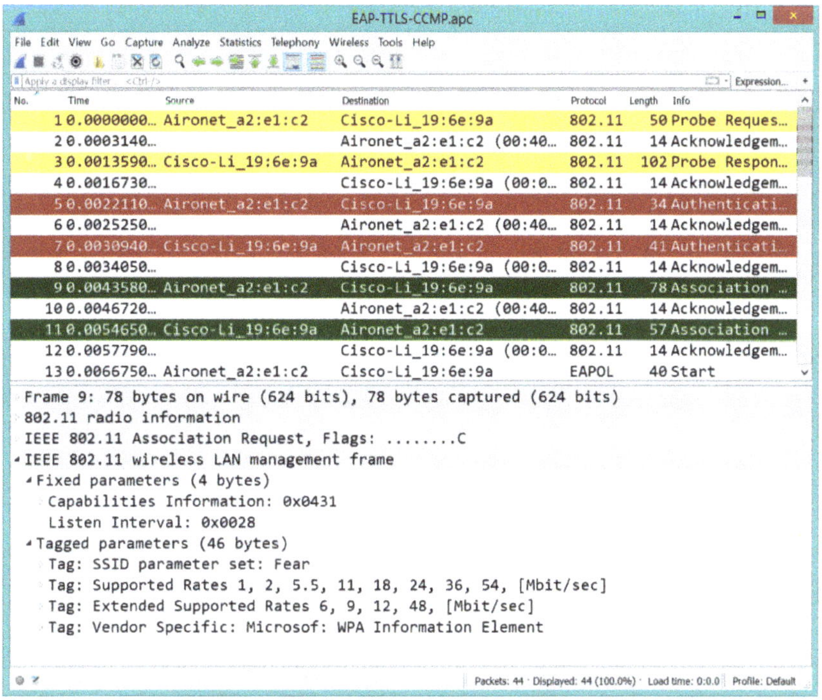

Figure 13.4: Wireshark with Color Coding for 802.11 Frames

Spectrum analyzers are used to monitor and analyze the RF activity in an area. They show all RF activity, and not just WLAN activity like a protocol analyzer does. For example, non-W-Fi devices like microwave ovens, phones, wireless peripherals and more will show up as long as they are operating in the monitored frequency.

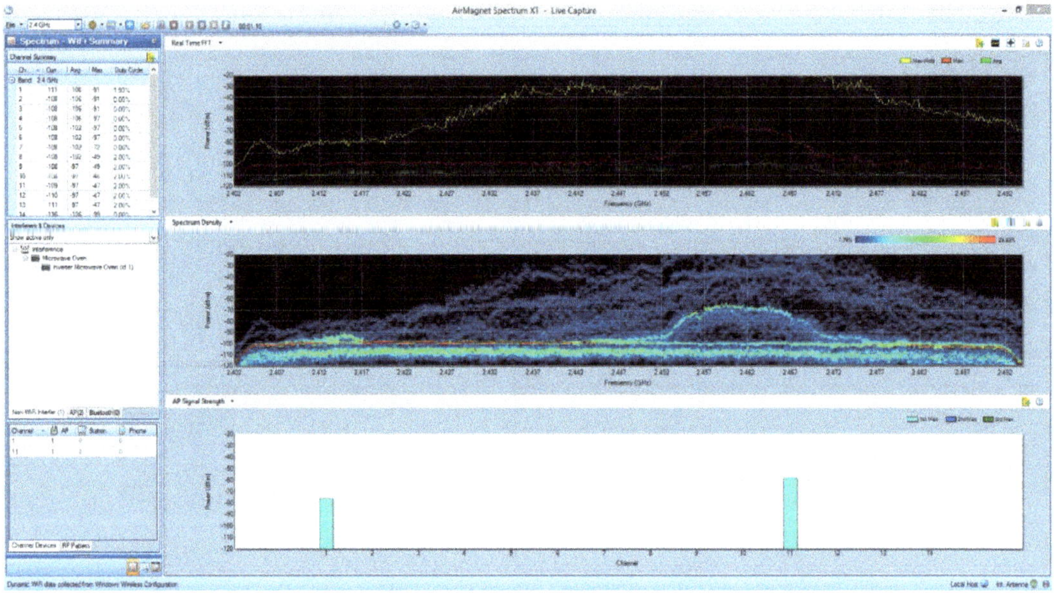

Figure 13.5: AirMagnet Spectrum XT USB-Based Spectrum Analyzer

Know that spectrum analyzers are useful to the WLAN administrator for:

- Locating sources of interference
- Determining channel utilization for Wi-Fi and non-Wi-Fi devices
- Detecting poorly constructed hardware with improper spectral masks or inconsistent spectral masks
- Discovering the presence of non-Wi-Fi activity, including incidental activity
- Viewing signal strength in important coverage areas
- Selecting the least busy channel for a new BSA (Basic Service Area)

Operating System (OS) tools come with the OS and help in the troubleshooting process. These tools are also used to analyze connection issues and view client device parameters, settings and capabilities. These include PING, TRACEROUTE, PATHPING, NSLOOKUP, NETSTAT and NETSH (in Windows).

The PING command is available in most Oses and even in many embedded Oses, such as those in switches and routers. The command is used to attempt an Internet Control Message Protocol (ICMP) communication with a remote host, based on the IP address. While a DNS host name may be used, the name is simply resolved to the IP address, and the IP address is the actual target of the ICMP PING request. The sender (the machine on which the PING command is executed) sends an ECHO ICMP message (a TYPE 8 ICMP message) to the target IP address. If the target IP address both receives the request and is configured to allow responses, it will send back an ECHO REPLY ICMP message (a TYPE 0 ICMP message – see RFC 792 for more detail).

When using this command, the size of the PING response packet is based on the size of the data field in the ECHO message. The ECHO REPLY message simply sends back the same data sent in the ECHO message. This behavior is defined in the RFC and can be validated in a simple protocol capture of a PING process, as shown in Figure 13.6. Most PING commands provide a switch to change the size of the ECHO message, like the -l switch in Windows.

In Windows, PING supports the parameters shown in Figure 13.7. Two important parameters for testing are –t and –l. The –t parameters are used to specify that the PING operation should run until interrupted (with a CTRL + C keystroke). This function is useful when testing for intermittent connectivity problems. Simply run the command, like PING 192.168.10.7 -t and then watch for lost ECHO REPLY messages during the process.

The –l parameter is used to change the data size in the ECHO message (the sent message) and therefore in the ECHO REPLY message. This function is useful when you wish to force more data through the network, which can reveal problems that a small 32-byte message (the Windows default size) will not reveal.

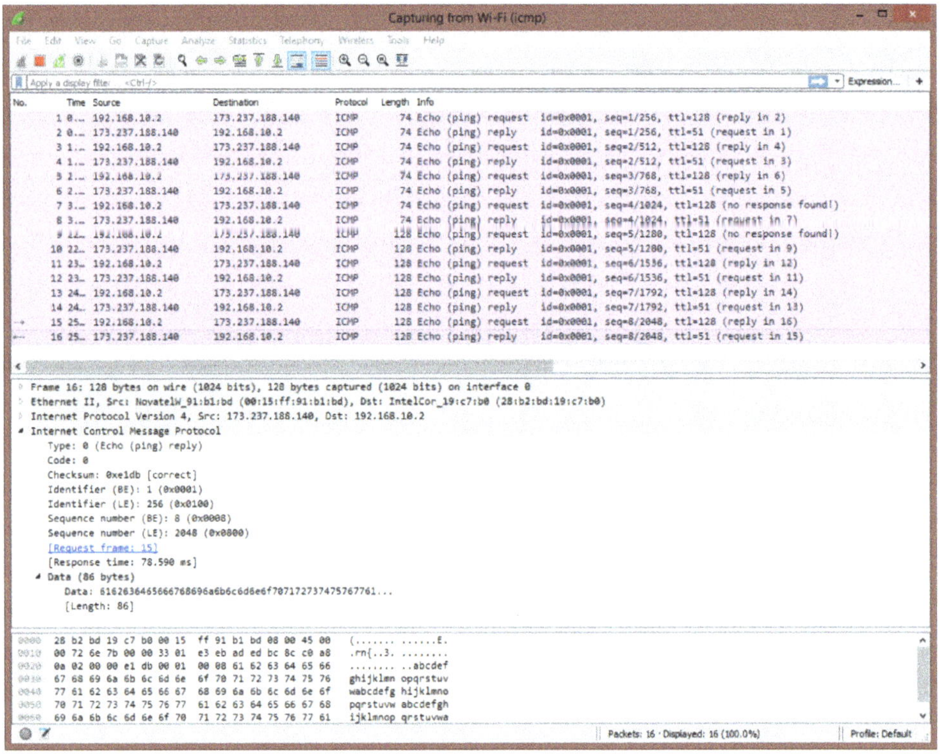

Figure 13.6: PING Captured in a Protocol Analyzer

The TRACEROUTE command differs from the PING command in that it sends ICMP ECHO messages to each node along the path to a destination. This function is accomplished with creative use of the time-to-live (TTL) field in the IP packet. First, the command sends three ICMP ECHO messages to the PING target with a TTL of 1. Therefore, when the first router receives it, it sends back a TTL Timeout message and, of course, this means the TRACEROUTE command now knows that router's address. Next, the command sends three more ICMP ECHO messages with a TTL of 2. The result, as you might imagine, is that the next router in the path receives the packets, but the TTL will be 0 and it therefore responds with a TTL Timeout message. The TRACEROUTE command now knows that IP address. This process continues until the PING target is reached.

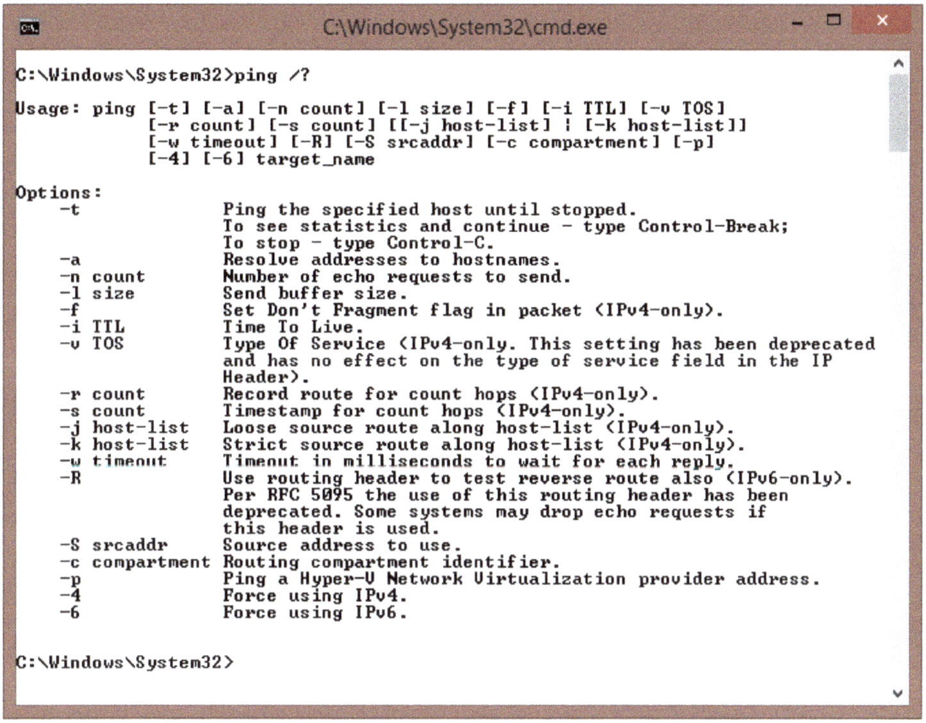

```
C:\Windows\System32\cmd.exe

C:\Windows\System32>ping /?

Usage: ping [-t] [-a] [-n count] [-l size] [-f] [-i TTL] [-v TOS]
            [-r count] [-s count] [[-j host-list] | [-k host-list]]
            [-w timeout] [-R] [-S srcaddr] [-c compartment] [-p]
            [-4] [-6] target_name

Options:
    -t             Ping the specified host until stopped.
                   To see statistics and continue - type Control-Break;
                   To stop - type Control-C.
    -a             Resolve addresses to hostnames.
    -n count       Number of echo requests to send.
    -l size        Send buffer size.
    -f             Set Don't Fragment flag in packet (IPv4-only).
    -i TTL         Time To Live.
    -v TOS         Type Of Service (IPv4-only. This setting has been deprecated
                   and has no effect on the type of service field in the IP
                   Header).
    -r count       Record route for count hops (IPv4-only).
    -s count       Timestamp for count hops (IPv4-only).
    -j host-list   Loose source route along host-list (IPv4-only).
    -k host-list   Strict source route along host-list (IPv4-only).
    -w timeout     Timeout in milliseconds to wait for each reply.
    -R             Use routing header to test reverse route also (IPv6-only).
                   Per RFC 5095 the use of this routing header has been
                   deprecated. Some systems may drop echo requests if
                   this header is used.
    -S srcaddr     Source address to use.
    -c compartment Routing compartment identifier.
    -p             Ping a Hyper-V Network Virtualization provider address.
    -4             Force using IPv4.
    -6             Force using IPv6.

C:\Windows\System32>
```

Figure 13.7: PING Command Parameters

The benefit of the TRACEROUTE/TRACERT command is that it checks each device along the path. On your internal network, assuming all routers are configured to respond to ICMP ECHO messages with ICMP ECHO REPLY messages, the TRACEROUTE command will help you ensure availability of all routers along the path. On the Internet, it is not uncommon to see request timeout errors from some nodes along the path. Some organizations disable ICMP ECHO REPLY messages on Internet-facing devices for performance and security reasons. Figure 13.8 shows a protocol analyzer capture of the ICMP messages sent and received by a TRACEROUTE command. Remember, when using TRACERT and other IP tools, all communications with private addresses (10.x.x.x, 192.168.x.x and 172.16.x.x-172.31.x.x) stay within your network under normal conditions.

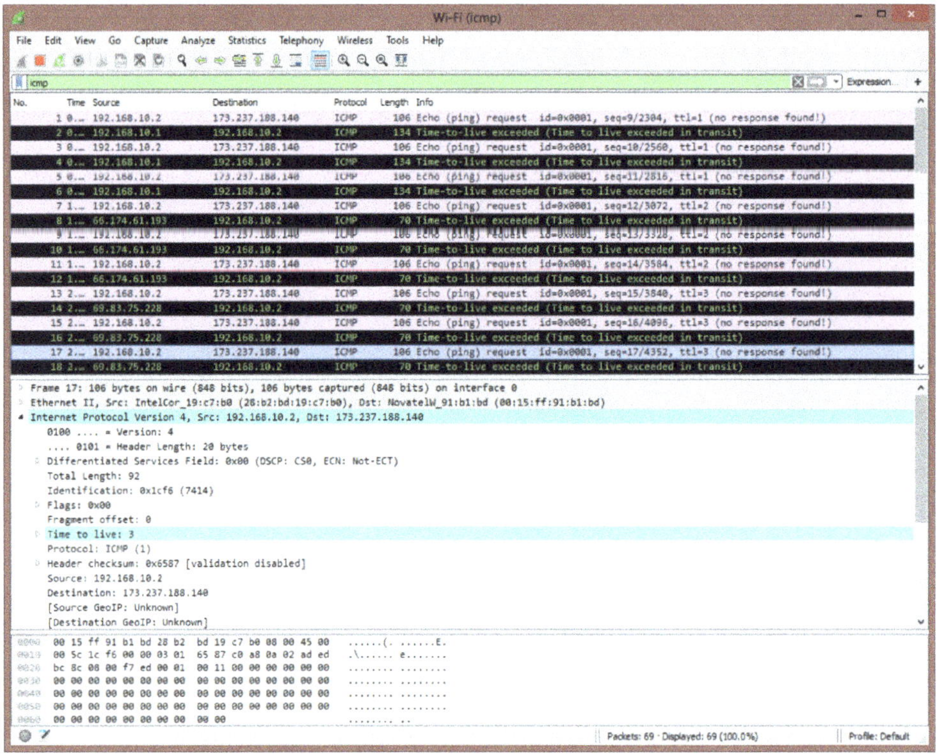

Figure 13.8: TRACEROUTE Process Captured in WireShark

The PATHPING command is a somewhat enhanced implementation of TRACEROUTE in Windows. It not only determines the route taken, but also responds with useful statistics about the performance along the path. The PATHPING command sends ICMP ECHO messages to each hop in the same manner as TRACEROUTE and then sends multiple ICMP ECHO messages to each hop to calculate performance over time for each hop. Figure 13.9 shows sample output from the PATHPING command.

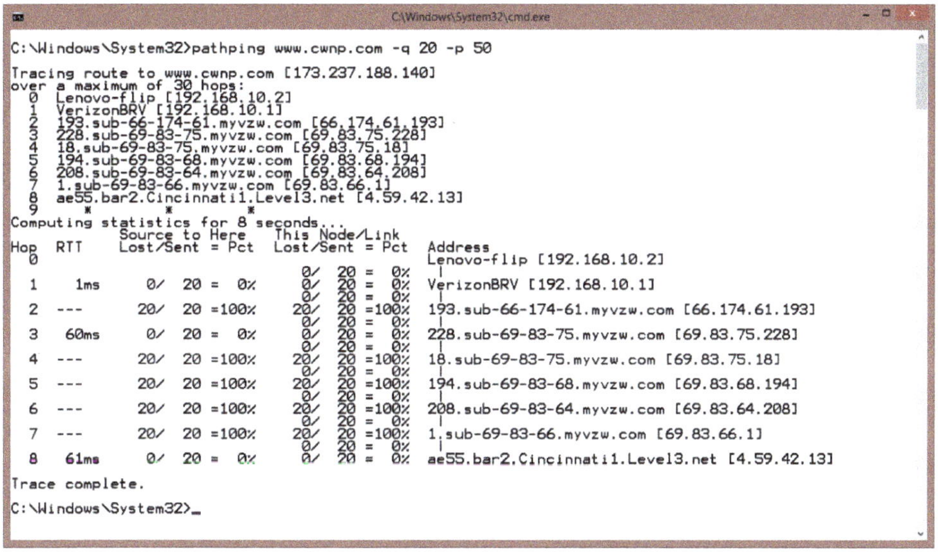

Figure 13.9: PATHPING Command Output

NSLOOKUP is used to query DNS servers. It is a useful command to use when clients cannot resolve host names to IP addresses, or when a lightweight AP is unable to locate its controller and DNS is intended to be used for such location services.

NETSTAT is used to show statistics for network connections. Simply running NETSTAT with an interval in seconds, like 10, will show active connections and, if you leave it running, it will show new connections you create. This can be useful to analyze targets for TCP sessions on the network.

The final command, unique to Windows systems, is the network shell (NETSH) command. This command reveals many things about network connections and configurations on the Windows computer. It provides extensive information about the wireless adapter and connection, when in WLAN mode. Unlike many other Command Prompt commands, the NETSH command has different modes with difference commands in those modes. For example, you can execute many commands specific to WLANs when in the WLAN mode, accomplished with the NETSH command, followed by the embedded WLAN command. Next execute the ? command to view options.

Important NETSH WLAN commands include:

- SHOW INTERFACES
- SHOW NETWORKS
- SHOW DRIVERS
- SHOW PROFILES

You should take some time to explore the difference NETSH WLAN commands available and the output they generate. These commands are useful for troubleshooting WLAN configuration issues. Specifically, familiarize yourself with the output of the SHOW INTERFACES, SHOW NETWORKS, SHOW DRIVERS and SHOW PROFILES commands.

Additional NETSH commands of interest include:

- NETSH WLAN SHOW ALL
- NETSH INTERFACES IPV4 SHOW ADDRESSES
- NETSH INTERFACES IPV4 SHOW IPSTATS
- NETSH INTERFACES IPV4 SHOW CONFIG
- NETSH INTERFACES IPV4 SHOW ICMPSTATS
- NETSH INTERFACES IPV4 SHOW TCPSTATS
- NETSH INTERFACES IPV4 SHOW TCPCONNECTIONS

The NETSH shell is a powerful interface for viewing and configuring network settings and statistics and is very useful to the network troubleshooter.

The NETSH WLAN SHOW DRIVERS command reveals the driver files used. Additionally, it reveals the security methods provided by the adapters, the radio

PHYs supported, and other features of importance, like Management Frame Protection (MFP) and driver versions.

The NETSH WLAN SHOW PROFILES command is useful for evaluating the profiles installed and configured on the local machine. These profiles include pre-shared key (PSK) passphrases, when WPA or WPA2-Personal is used in the profiles. When the name of a specific profile is provided, such as NETSH WLAN SHOW PROFILES NAME="OFFICE24", the output will reveal additional information about the specified profiles; however, PSK passphrases are not shown in the output. If you want to see the stored key, you can add the KEY=clear parameter to the command.

The NETSH WLAN SHOW INTERFACES command reveals the current profiles operation, including the authentication and key management (AKM) protocol (listed as Authentication), the encryption method (listed as Cipher and CCMP, which means AES is used), the channel, the signal strength and data rates (including transmit and receive rates, which may vary, and this is a useful measurement). Since this is a WLAN client, the transmit data rate would be the uplink rate and the receive data rate would be the downlink rate.

The NETSH WLAN SHOW NETWORKS command provides information about visible networks that the client STA (station) can see. To get more or alternate information about a network, use the NETSH WLAN SHOW NETWORKS MODE=BSSID command.

Centralize management consoles can also be useful for troubleshooting WLAN problems. They often report statistical information about dropped calls, MOS scores, average data rates, number of clients per AP, busiest Aps, signal strength information and more.

Additionally, third-party WLAN monitoring solutions are also available, like those from 7SIGNAL. These solutions also provide a central console for reporting and troubleshooting. In Figure 13.10, notice the connectivity problems that started between 2 and 4 a.m.

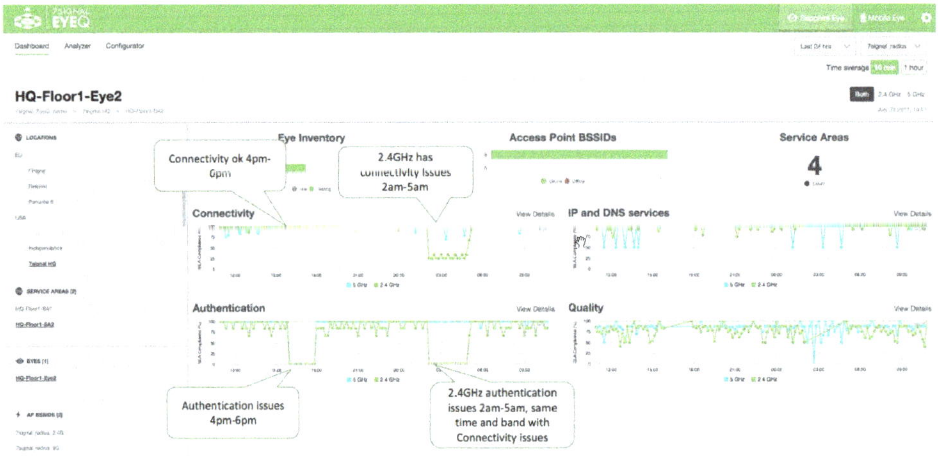

Figure 13.10: Connectivity Problems

Next, we look at the historic spectrum analysis in Figure 13.11 to notice significant RF activity in that time window.

Tools like these 7SIGNAL views can be extremely helpful in the troubleshooting process. The fact that the 7SIGNAL system is monitoring and archiving data all the time, and you can go back and look at the historic data, makes troubleshooting much, much easier.

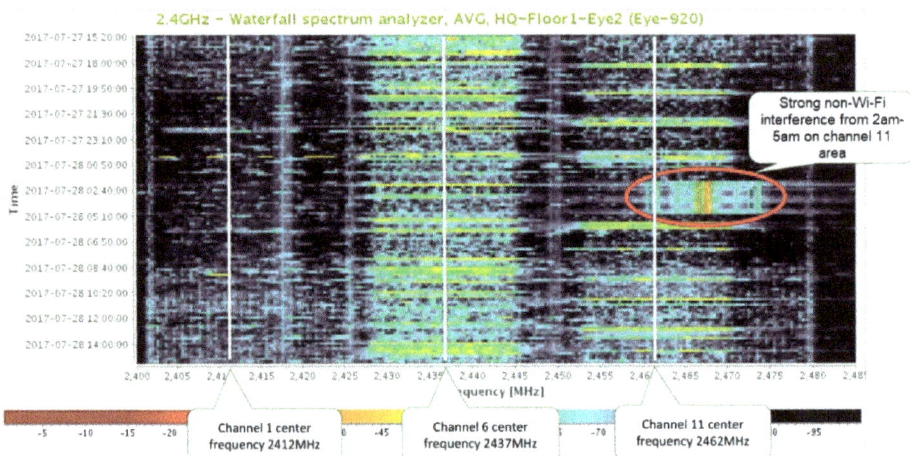

Figure 13.11: Spectrum Activity

Implementation Challenges

Implementation challenges come in many forms. Poor throughput, interference, no connectivity, and more. This section addresses some of these WLAN implementation challenges.

System Throughput

Installing a WLAN that provides access to users is only a partial solution. The access provided must be sufficient for the users' needs. This usually means providing adequate throughput or capacity for the network clients to use the applications they require. One might suggest that there is a difference between throughput and capacity. Capacity is a linkage between throughput, and the number of users that require a certain throughput in a cell. That is, as more users join the cell, at some point, overall throughput is diminished. Management of capacity is simultaneous management of both overall throughput and controlling the number of stations communicating in a channel. Manny different factors can affect the available throughput in a WLAN, including the chosen PHY, wired-side limitations and more.

Consider the following possible solutions when attempting to resolve throughput or capacity issues:

- Upgrade or replace older PHY devices.
- Install more Aps and use lower output power settings on each AP.
- Enable features like band steering and AirTime fairness where appropriate.
- Never join any device that is stationary and has access to a wired connection to the WLAN.
- Reduce the number of SSIDs on the Aps, if there are more than three.
- Evaluate the channel plan in an attempt to reduce CCI/ACI (discussed next).

CCI and ACI

Co-channel interference has been defined in previous chapters. It occurs when two cells are near enough each other and operating on the same channel, so that some clients, and often even the Aps, will be required to contend with each other for medium access. As was previously noted, for this reason, many are now calling it co-channel contention (CCC), as it is a more accurate reference to what happens.

The mitigating solution to CCI is cell sizing. By adjusting cell sizes (and shapes with directional antennas) you can reduce CCI in 2.4 GHz, and all but remove it in many 5 GHz deployments. To understand how this works, you should understand two boundaries of a cell: the association boundary and the CCI boundary. Additionally, you should be aware that every client associated to a cell (an AP) will have its own CCI boundaries as well.

The association boundary is the edge of the cell where the minimum supported data rate can be achieved. If the lowest data rates of 1 and 2 Mbps have been disabled, for example, this will result in a smaller associate cell (reduced association boundary), than when they are enabled. However, just because clients can no longer associate beyond that point does not mean that CCI cannot exist beyond that point.

The CCI boundary exists where the listening stations that are not part of the cell must be silent, because they have detected a signal at a minimum threshold. This behavior is part of the CSMA/CA protocol used in 802.11 WLANs. Even when lower data rates are disabled, such as 1, 2 and 5.5 Mbps, 802.11 standard devices must still process and wait for 1, 2 and 5.5 Mbps signals that are heard in the 2.4 GHz band. The same is true in 5 GHz, if low data rates like 6 and 12 Mbps are disabled. Therefore, CCI can occur based on low data rate signals, even though those signals are not enabled in the cell.

Additionally, just because the signal cannot be fully processed does not mean that the signal cannot be detected, and initiate a backoff based on the duration for a STA.

When considering association boundaries (also called data rate boundaries) and CCI boundaries, the AP perspective is almost always considered, and it should be.

Additionally, the client perspective must be considered. The AP perspective is simplest and is represented in Figure 13.12. The Aps in the image have low data rates of 1, 2 and 5.5 Mbps disabled. Therefore, the association boundary is smaller than the CCI boundary by a significant amount. The image shows only a 2.4 GHz representation, but 5 GHz boundaries work in a similar manner. In Figure 13.12, the CCI boundary does not reach all the way into the space of the other AP, but this does not mean it will not cause problems. The client perspective will show that problems still occur in such scenarios.

The client perspective of the association boundary is also called the data rate boundary and the CCI boundary is much different. Like Aps, the clients transmit, typically, in an omni-directional manner. Therefore, they both see the boundaries of other nearby Aps and create their own. That is, they see cells and they also create a cell of RF energy of their own. Figure 13.13 illustrates this.

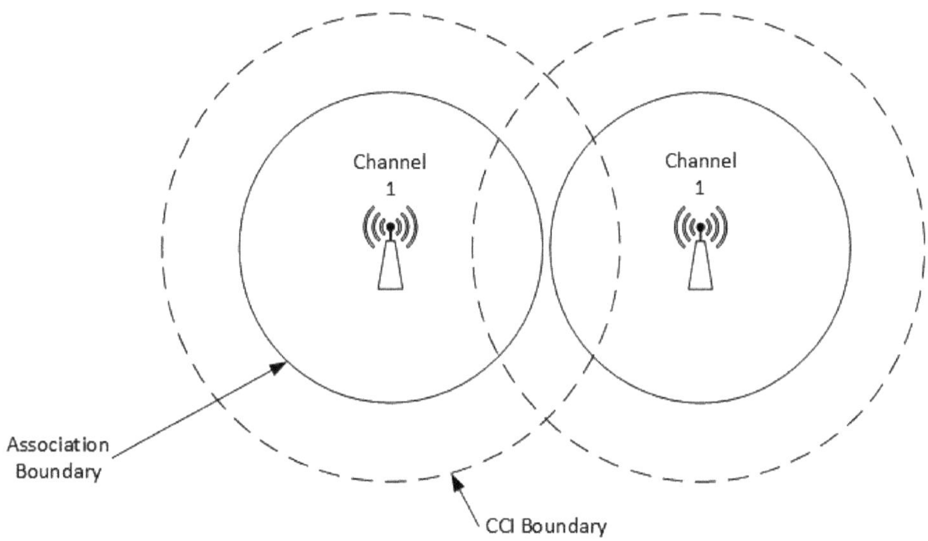

Figure 13.12: Association Boundaries for Aps

Notice in Figure 13.13 that the signals from the client station, which is associated with the AP on the left, are reaching the AP on the right. The client is causing CCI for the AP on the right and the AP on the right is causing CCI for the client even though the Aps are not causing significant CCI for each other directly. Therefore, the cells created by the two Aps are causing CCI for each other. Again, remember, that CCI is normal contention processes defined in CSMA/CA. The client station in Figure 13.13 must be silent for the duration of any frames it sees from the AP on the right or any stations associated with the AP on the right that it can see. Additionally, the AP on the right must be silent for the duration of any frames it sees from the client or any other clients it can see that are associated with the AP on the left. Therefore, CCI is not an AP-only issue, but it is an issue impacting clients, and created by clients.

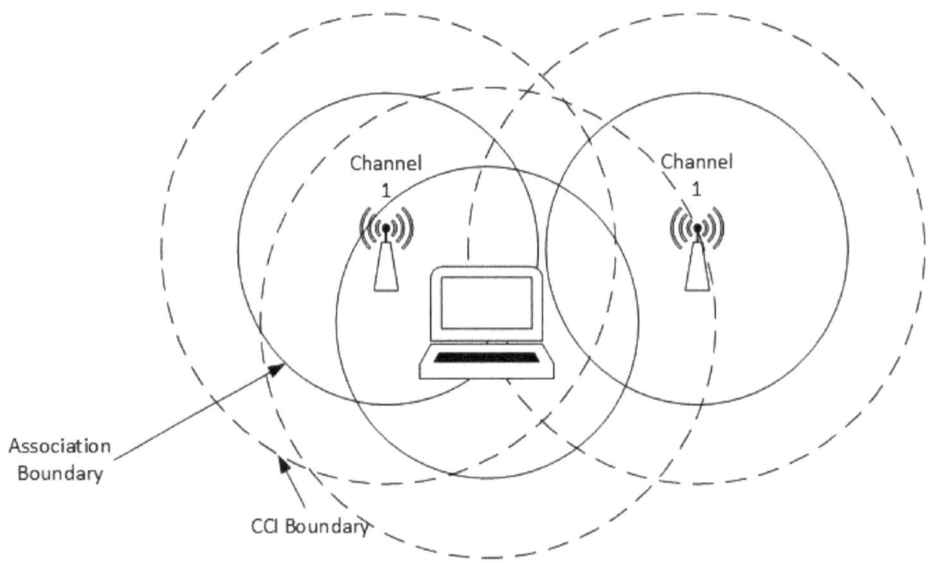

Figure 13.13: Association Boundaries for Clients

Adjacent channel interference is caused when channels overlap intentionally, like 2.4 GHz channels, or when non-overlapping channels cause side-band interference, like 5 GHz channels. 2.4 GHz channels are 5 MHz apart and 20 or 22 MHz wide, therefore it is a certainty that adjacent channels will overlap and cause interference. For example, Figure 13.14 shows the overlap of channel 1 and channel 2 in the 2.4 GHz band.

In Figure 13.14, the spectral mask for an OFDM channel is shown with an overlay of the center frequency for channel 1 and channel 2. Given that the spectral mask would be that of channel 1, it is obvious that channel 2 would overlap with it if its spectral mask were in view. When adjacent channels overlap they cause ACI, which can manifest as corrupt frames or simply prevent communications due to energy detect in 2.4 GHz. In 2.4 GHz, ACI can be prevented within your controlled networks by using channels 1, 6 and 11. However, in areas where neighbor WLANs can be seen, if they are using channels 2-5, 7-10 or 11-14, ACI may exist, regardless of your network settings. This can be mitigated by carefully planning channel selections nearest to these neighbor WLANs.

In 5 GHz, ACI is caused by side-bands when the adjacent channel is either implemented very close to the AP, or with high output power, or a client associated with the other channel AP is near the interfered AP/cell. 5 GHz ACI is typically manifest as corrupt frames. In 5 GHz it can be prevented with proper channel plans that do not place adjacent channels near each other.

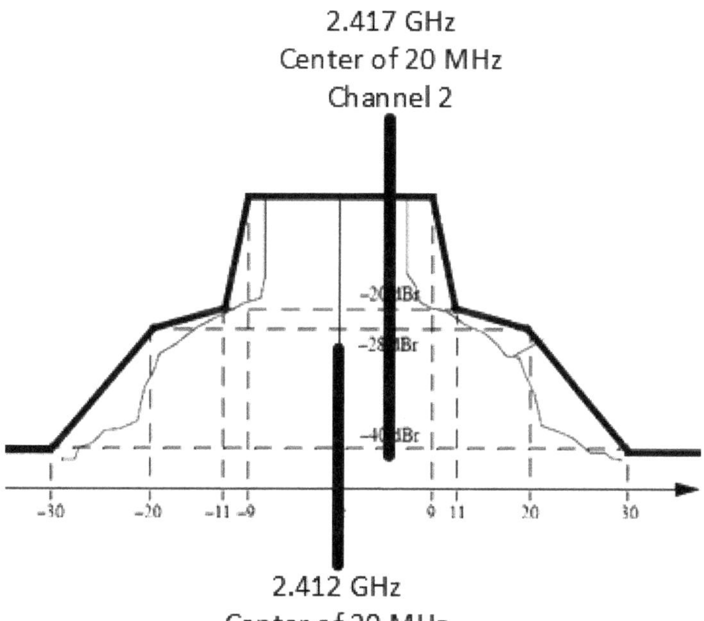

Figure 13.14: Channel Overlap in 2.4 GHz

RF Noise and Noise Floor

RF noise may be defined as RF energy or signals generated by RF systems, other than those systems with which the detecting system intends to communicate. For example, a WLAN STA configured to listen to an AP on channel 11 may consider RF signals transmitted from an AP on channel 9 at high power levels to be RF noise. This RF noise may cause corruption of frames. Interestingly, what is RF noise to one device may be the RF signal to another.

The noise floor is defined as the background level of RF noise, and the signal-to-noise ratio is the difference between the strength of the signal for which a device is monitoring, and the strength of the noise floor.

RF Interference

Narrowband and wideband interference can cause corruption of data in WLANs. You can often detect that interference exists by looking at the frames in a WLAN analyzer, which may report CRC errors or corruption. When CRC errors are reported, it indicates that the signal strength was great enough to receive the RF signal, but that noise joined with the signal and corrupted the data, as the signal arrived at the receiver. This results in retransmissions and, therefore, reduced throughput.

WLAN administrators can deal with these retransmissions in different ways. One way is to reduce the data rate, which provides for more fault tolerance in the data transfer, and the ability to handle more interference without losing data. Another way of dealing with the retransmissions is to fragment the WLAN frames. Smaller frames get on and off of the medium faster, and fewer of the frames will become corrupted. The fragmentation threshold can be used to control the point at which fragmentation is utilized. A lower fragmentation threshold value should be tested when intermittent interference is suspected. If the problem is not resolved by lowering the threshold, you should immediately raise the threshold again.

If you determine that RF noise or interference is a problem in your environment, take these steps to diminish RF noise as much as possible:

- Remove or replace all RF devices that communicate on the same channels as the WLAN.

- Reduce the output power to the minimum possible, to create acceptable links for all non-Wi-Fi devices.
- Replace leaky microwaves with better sealed units.
- Replace 2.4 GHz and 5 GHz phones with WLAN VoIP handsets.
- Strategically plan the channel selections in your environment to work around RF noise.

Hidden Nodes

Hidden nodes are STAs that can be seen by the AP and that can see the AP, but they cannot see one or more other STAs, and one or more other STAs cannot see the hidden nodes. Because of this scenario, the hidden nodes cannot hear at least one of the other clients communicating, and so may attempt to communicate while the other nodes or nodes are active. Hidden nodes usually occur because of some large obstacle like a solid wall that's between the STAs, or because of insufficient transmit power. For example, the AP may be placed on top of a thick block or brick wall, and clients that are lower and on either side of the wall can see the AP, but they cannot see each other. The result of the hidden node paradigm will be collisions that cannot be avoided, without the implementation of some function to clear the channel, such as RTS/CTS.

Commonly recommended solutions to hidden node problems include:

- Use RTS/CTS
- Increase power output at the client STAs
- Remove obstacles
- Move the client STAs

Insufficient PoE Power

When an AP will not power up or powers up with limited features, insufficient PoE power may be the problem. Line tester tools, like the LinkSprinter, can be used to test for the presence of PoE power (as well as other wired line tests). Figure 13.15 shows the LinkSprinter 300. This line tester provides a wireless connection to your smartphone so that you can view detailed information about test results.

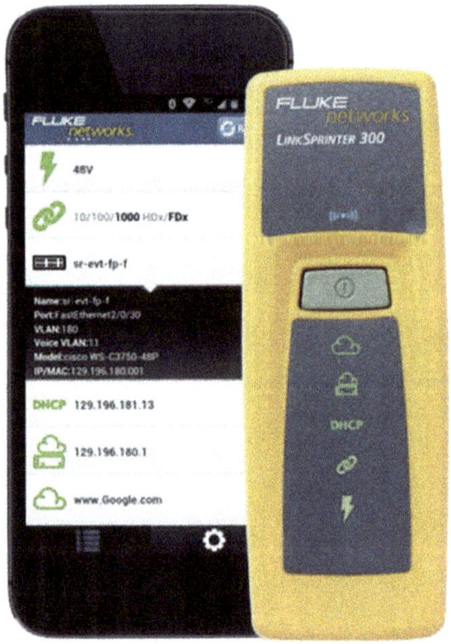

Figure 13.15: LinkSprinter 300

If it turns out that insufficient PoE power is the problem, check the following:

- Is the PoE port enabled in the switch, if a switch is used as the injector?
- Is the cable run too long (more than 100 meters)?
- Is the PoE port configured properly in the switch?
- Does the injector have enough power left in the budget?

Lack of Coverage / No Signal or Weak Signal

Lack of coverage, indicated by a weak signal or no signal, is a serious problem. I would rather have a slow car than no car. Of course, I would rather have a fast car than a slow car too, but the point is some WLAN access is often better than none. When users cannot connect at all due to lack of coverage, they are sure to let you know. When they do, consider the following potential solutions:

- **Install more Aps:** this is the simplest solution and often the one that works best when it does not result in significantly increased CCI. By adding a new AP in the general area of lacking coverage, you can typically resolve the issue.

- **Increase the output power on the Aps:** this may be a good solution if the Aps nearby are simply configured with the output power setting too low; however, if the Aps already match the clients closely, turning up the output power is not likely to be a good solution.

- **Use RRM coverage hole detection:** RRM coverage hole detection can attempt to algorithmically adjust the AP setting for Aps in the area, so that coverage is provided.

- **Install directional antennas to get coverage to the area:** directional antennas may be a good solution if you have Aps supporting external antennas. For example, a patch antenna aimed toward the non-coverage area is likely to accomplish a better signal in that area than an omni antenna originating from the same location.

Connectivity Problems

Some problems prevent connectivity, either to the WLAN or to services. Many times users will say, "The Wi-Fi is down." In actuality, the Wi-Fi may be up and working perfectly, but the Internet is down. To users, in many cases, the Internet is Wi-Fi, and Wi-Fi is the Internet. In this section we will discuss how to deal with common connectivity problems.

Common Connectivity Issues

Some issues are very common and should be considered early on in the troubleshooting process, if signal strength analysis indicates sufficient coverage in the area:

- **Security Configuration Mismatch:** If PSK is used, make sure the passphrase was entered properly in both the client and AP. Of course, if other clients are connecting to the AP, you need only validate the client. If 802.1X/EAP is used, make sure the entered credentials are correct.

- **Improper AP Configuration:** Check settings like disabled data rates, band steering (sometimes it breaks and doesn't let a 2.4 GHz client connect at all), PHY support, etc.

- **Improper Client Configuration:** Verify that band preference or band force settings are not preventing the client from connecting. Test the simple things. For example, is the user connecting to the right SSID.
- **Faulty Drivers/Firmware:** Check the drivers on the client and the firmware on the AP. In the past, some AP firmware updates have broken entire networks. Client drivers can do everything from disabling a band to simply not connecting.
- **Hardware Failure:** Of course, this is always a possibility. If you can swap out the wireless adapter, try it. Remember, if you suspect hardware failure in the AP, but 50 clients are connecting and three cannot, it's probably not AP hardware failure.

That last point about 50 clients connecting brought something important to mind. If many clients are connected and suddenly a few cannot, check the maximum clients setting in your AP. One time, I spent half a day to troubleshoot a wireless connection problem only to realize that someone had configured the AP to allow only 20 clients to connect. Needless to say, my problem client was number 21.

Internet Connectivity

When troubleshooting Internet access, always begin with the scale of the problem. If it is a single user, the problem is likely on that user's device or at least within the local segment to which the user is connected. If it involves many users and all other network functions are working as expected, the problem is likely with the Internet gateway (either the router or the service provider's network).

To troubleshoot Internet connectivity, consider the following points of failure:

- **Client configuration:** ensure that the IP configuration is accurate, including the DNS server and default gateway and any required Internet proxy configuration settings.
- **Infrastructure:** ensure that all switches and routers along the path to the Internet gateway are configured and operating as designed.

- **Internet gateway:** ensure that the connection to the service provider is still operational, and that the configuration is correct.
- **DNS:** ensure that the DNS server, if local, is configured to forward requests to a valid Internet server. Small- and medium-sized businesses often point to the Google public DNS servers at 8.8.8.8 and 8.8.4.4.
- **Captive portal:** ensure that the captive portal is responsive and configured properly. Additionally, clients often get confused over captive portals based on cached information. At times, clearing the cache (DNS and browser) may be required to reactivate the portal logon screen.

DHCP Issues

DHCP is used to configure the device IP settings on the network and, at times, to allow Aps to locate controllers. A very common problem for WLANs is DHCP pool depletion. This occurs because many wireless clients come and go from the network quickly. If a client connects for only two or three minutes, and the lease duration is set to multiple days (3-8 days is not uncommon), the IP address will be lost for that entire time. To resolve such issues, create more pools and reduce the lease duration to hours, instead of days. Look for DHCP negative acknowledgement or server log errors to determine if the IP pool is depleted.

Additional DHCP problems include location problems, the devices cannot locate the DHCP server, and configuration errors, the DHCP pool is not configured properly. For example, is it providing the DNS IP address and the default gateway IP address? A very useful and free tool, called DHCPTEST, is useful to have in your toolkit. It can test a DHCP server and report to you the details of the configuration it received from it.

Chapter Summary

In this chapter, you learned about troubleshooting methodologies, tools and common problems. With this information, you are ready to tackle some of the most common issues that occur in WLANs.

Points to Remember

Remember the following important points:

- The first step in any troubleshooting process is to clearly define the problem.
- A protocol analyzer is useful in the troubleshooting process as it allows you to view the actual network communications.
- A spectrum analyzer is very useful for locating sources of RF interference.
- Centralized management consoles and WLAN monitoring solutions allow you to view dashboards that are extremely helpful in resolving both localized and system-wide problems.
- System throughput problems may be resolved by installing more Aps, upgrading or replacing older PHY devices, and reducing the number of SSIDs on the Aps, among other things.
- CCI occurs between clients and Aps, clients and clients, and Aps and Aps.
- CCI does not happen within a BSS, but between STAs in different BSSs on the same channel.
- Microwave ovens are common sources of temporary interference in the 2.4 GHz band.
- Hidden node problems occur when two devices can receive RF transmissions from the AP, but they cannot receive them from each other.
- RTS/CTS can help reduce hidden node issues.
- DHCP pool depletion is a common issue on WLANs, particularly on guest WLANs.

Review Questions

1. What step comes after "Define the possible causes" in the CWNP troubleshooting methodology?
    a. Create a plan of action or escalate the problem
    b. Discover the scale of the problem
    c. Perform corrective actions
    d. Narrow to the most likely cause

2. What should you do immediately after you have positively verified the solution to a problem?
    a. Document the results
    b. Move on to the next ticket
    c. Delete the ticket
    d. Discover the scale of the problem

3. What are you not testing in any common throughput testing tool when you test throughput?
    a. UDP communications
    b. Data rate
    c. TCP communications
    d. QoS

4. What feature of Wireshark can be used to make particular frames stand out when viewing a capture file?
    a. I/O graph
    b. Coloring rules
    c. View > Full Screen
    d. None of these

5. Which one of these is not a good use of a protocol analyzer?
    a. Gathering details about unsupported networks
    b. Validating compliance with requirements
    c. Locating non-Wi-Fi interferers
    d. Identifying devices on the network

6. Which one of the following is not a good use of a spectrum analyzer?
    a. Determining channel utilization
    b. Viewing signal strength in important coverage areas
    c. Determining if QoS is enabled on the WLAN
    d. Selecting the least busy channel for a new BSA

7. What NETSH command shows information about WLAN drivers?
    a. NETSH WLAN SHOW DRIVERS
    b. NETSH WLAN SHOW NETWORKS
    c. NETSH WLAN SHOW PROFILES
    d. NETSH WLAN SHOW INTERFACES

8. What is a good solution for dealing with RF interference?
    a. Implement only 5 GHz channels
    b. Strategically plan the channels used to work around RF noise
    c. Implement only 2.4 GHz channels
    d. Use the new DMG PHY

9. What is a potential solution to hidden node problems?
    a. Move the client STAs
    b. Disable RTS/CTS
    c. Enable RSSI thresholds
    d. None of these

10. What can be used to test an Ethernet connection to verify PoE capabilities?
    a. Throughput tester
    b. Spectrum analyzer
    c. Protocol analyzer
    d. Line tester

Review Answers

1. **D is correct.** After defining the possible causes, based on the symptoms and your WLAN knowledge, you should narrow to the most likely cause.

2. **A is correct.** The last step in the troubleshooting process is to document the results.

3. **B is correct.** Throughput does not equal data rate in WLANs due to management overhead, interference, and the number of STAs contending for the medium.

4. **B is correct.** Wireshark coloring rules are very useful for viewing packet captures and bringing attention to particular frame types.

5. **C is correct.** Spectrum analyzers are best for locating non-Wi-Fi interference, not protocol analyzers.

6. **C is correct.** It is possible that a spectrum analyzer coupled with a Wi-Fi adapter may have the potential to reveal whether QoS is enabled or not, but more than likely you should use a protocol analyzer for this test.

7. **A is correct.** The obvious answer, in this case is the right answer. NETSH WLAN SHOW DRIVERS lists the driver files used by the WLAN NIC.

8. **B is correct.** To deal with known interference on certain frequencies in particular areas, it is best to strategically plan the channels used to work around the RF noise.

9. **A is correct.** One potential solution for dealing with hidden node problems is to move the client STAs.

10. **D is correct.** A line tester that supports PoE detection can be used to evaluate the PoE delivered on the cable.

Glossary: A CWNP Universal Glossary

40 MHz Intolerant: A bit, potentially set in the 802.11 frame, allowing STAs to indicate that 40 MHz channels should not be used in their BSS, or in surrounding networks. The bit is processed only in the 2.4 GHz band.

4-Way Handshake: The process used to generate encryption keys for unicast frames (Pairwise Transient Key (PTK)) and to transmit encryption keys for group (broadcast, multicast) (Group Temporal Key (GTK)) frames using material from the 802.1X/EAP authentication or the pre-shared key (PSK). The PTK and GTK are derived from the Pairwise Master Key (PMK) and Group Master Key (GMK) respectively.

802.11: A standard maintained by the IEEE for implementing and communicating with wireless local area networks (WLANs). Regularly amended, the standard continues to evolve to meet new demands. Several Physical Layer (PHY) methods are specified, and the Medium Access Control (MAC) sublayer is also specified.

802.11a: An 802.11 amendment that operates in the 5GHz band. It uses OFDM modulation and is called the OFDM PHY. It can support data rates of up to 54 Mbps.

802.11aa: An 802.11 amendment that added support for robust audio and video streaming through MAC enhancements. It specifies a new category of station called a Stream Classification Service (SCS) station. The SCS implementation is optional for a WMM QoS station.

802.11ac: An 802.11 amendment that operates in the 5GHz band. It uses MU-MIMO, beamforming, and 256 QAM technology, up to 8 spatial streams and OFDM modulation. Support is included for data rates up to 6933.3 Mbps.

802.11ae: An 802.11 amendment that provides prioritization of management frames. It defines a new Quality of Service Management Frame (QMF). When the QMF service is used, some management frames may be transmitted using an access category other than the one used for voice (AC_VO). When communicating with stations that do not support the QMF service, the station uses access category AC_VO to transmit management frames. When QMF is supported, the beacon frame includes a QMF Policy element.

802.11ah: An 802.11 draft that specifies operations in the sub-1 GHz range. Frequencies used vary by regulatory domain. The draft supports 1, 2, 4, 8 and 16 MHz channels with OFDM modulation.

802.11ax: An 802.11 draft that will support bi-directional MU-MIMO, higher modulation rates and sub-channelization. It is too early to know the final details of this amendment at the time of writing; however, it is planned to operate in the 2.4 GHz and 5 GHz band.

802.11b: An IEEE 802.11 amendment that operates in the 2.4 GHz ISM band. It uses HR/DSSS and earlier technology. It can support data rates of up to 11 Mbps.

802.11e: An 802.11 amendment, now incorporated into the most recent rollup, that provided quality of service extensions to the wireless link through probabilistic prioritization based on the contention window. The Wi-Fi Multimedia (WMM) certification is based on this amendment.

802.11g: An IEEE 802.11 amendment that operates in the 2.4 GHz ISM band. It uses ERP-OFDM and earlier technology. It can support data rates of up to 54 Mbps.

802.11i: An 802.11 amendment, now incorporated into the most recent rollup, which provided security enhancements to the standard and resolved weaknesses in the original WEP encryption solution. It provided for

TKIP/RC4 (now deprecated), and CCMP/AES cipher suites and encryption algorithms.

802.11n: An IEEE 802.11 amendment that operates in the 2.4 GHz ISM and 5 GHz UNII/ISM bands. It uses MIMO, HT-OFDM and earlier technology. It can support data rates of up to 600 Mbps.

802.11k: An IEEE 802.11 amendment that specifies and defines WLAN characteristics and mechanisms.

802.11r: An IEEE 802.11 amendment that enables roaming between access points.

802.11u: An IEEE 802.11 amendment that adds features for mobile communication devices, such as phones and tablets.

802.11w: An IEEE 802.11 amendment to increase security for the management frames.

802.11y: An IEEE 802.11 amendment that allows registered stations to operate at a higher power output in the 3650-3700 MHz band.

802.1X: 802.1X is an IEEE standard that uses the Extensible Authentication Protocol (EAP) framework to authenticate devices attempting to connect to the LAN or WLAN. The process involves the use of a supplicant to be authenticated, authenticator, and authentication server.

802.11 State Machine: The 802.11 state machine defines the condition of the connection of a client STA to another STA and can be in one of three states: Unauthenticated/Unassociated, Authenticated/Unassociated, or Authenticated/Associated.

802.3: A set of standards maintained by the IEEE for implementing and communicating with wired Ethernet networks and including Power over Ethernet (PoE) specifications.

AAA Framework: Authentication, Authorization, and Accounting is a framework for monitoring usage, enforcing policies, controlling access to computer resources, and providing the correct billing amount for services.

AAA Server Credential: The AAA server credential is the validation materials used for the server. When mutual authentication is required, a server certificate is typically used as the AAA server credential.

Absorption: This occurs when an obstacle absorbs some or all of a radio wave's energy.

Access Category (AC): An access category is a priority class. 802.11 specifies four different priority classes – voice (AC_VO), video (AC_VI), best effort (AC_BE), and background (AC_BK).

Access Layer Forwarding: Data forwarding that occurs at the access layer, also called *distributed data forwarding*. The data is distributed from the access layer directly to the destination without passing through a centralized controller.

Access Point: An access point (AP) is a device containing a radio that is used to create an access network, bridge network or mesh network. The AP contains the Distribution System Service.

Access Port: An AP used for mesh networks and that connects to the wired or wireless network at the edge of the mesh.

Acknowledgement Frame (ACK): A frame sent by the receiving 802.11 station confirming the received data.

Access Control List (ACL): ACLs are lists that inform a STA or user what permissions are available to access files and other resources. ACLs are also used in routers and switches to control packets allowed through to other networks.

Active Mode: A power-save mode in which the station never turns the radio off.

Active Scanning: A scanning (network location) method in which the client broadcasts probe requests and records the probe responses in order to determine the network with which it will establish an association.

Active Survey: A wireless survey conducted on location that involves measuring throughput rates, round trip time, and packet loss, by connecting devices to an AP and transmitting data during the survey.

Ad-Hoc Mode: The colloquial name for an Independent Basic Service Set (IBSS). STAs connect directly with each other and an AP is not used.

Adjacent Overlapping Channels: Adjacent overlapping channels are channels whose bands interfere with their neighboring channels on the primary carrier frequencies. Non-overlapping channels are channels whose bands do not interfere with neighboring channels on the primary carrier frequencies.

Adjacent Channel Interference (ACI): ACI occurs when channels near each other (in the frequency domain) interfere with one another due to either partial frequency overlap on primary carrier frequencies, or excessive output power.

Advanced Encryption Standard (AES): The encryption cipher used with CCMP and WPA2 providing improved security over WEP/RC4 or TKIP/RC4.

Association ID (AID): An AID is an identification assigned by a wireless STA (AP) to another STA (client) in order to transmit the correct data to that device in an Infrastructure Basic Service Set.

AirTime Fairness: This solution transmits more frames to client STAs with higher data rates than those with lower data rates so that the STAs get fair access to the air (medium) instead of having to wait for slower data rate STAs.

Aggregated MAC Protocol Data Units (A-MPDU): A-MPDU transmissions are created by transmitting multiple MPDUs as one PHY frame as opposed to A-MSDU transmissions, which are created by passing multiple MSDUs down to the PHY layer as a single MPDU.

Aggregated MAC Service Data Unit (A-MSDU): See *Aggregated MAC Protocol Data Unit.*

Amplification: The process of increasing a signal's power level.

Amplifier: A device intended to increase the power level of a signal.

Amplitude: The power level of a signal.

Antenna: A device that converts electric power into radio waves and radio waves into electric power.

Association: The condition wherein a client STA is linked with an AP for frame transmission through the AP to the network.

Announcement Traffic Indication Message (ATIM): A traffic indication map (sent in a management frame) in an Ad-Hoc (IBSS) network to notify other clients of pending data transfers for power saving purposes.

Attenuation: The loss of signal strength as an RF wave passes through a medium.

Attenuator: A device that intentionally reduces the strength of an RF signal.

Authentication: The process of user or device identity validation.

Authentication and Key Management (AKM): The protocols used to authenticate a client STA on a WLAN and generate an encryption key for use in frame encryption.

Authentication Server: The authentication server validates the client before allowing access to the network. In an 802.1X/EAP implementation for WLANs, the authentication server is often a RADIUS server.

Authenticator: The device that provides access to authentication services to allow connected devices to access network resources. In an 802.1X/EAP implementation for WLANs, the authenticator is typically the AP or controller.

Automatic Power Save Delivery (APSD): APSD is a power-saving method which uses both scheduled (S-APSD) and unscheduled (U-APSD) frame delivery methods. S-APSD sends frames to a power save STA from the AP at a planned time. U-APSD sends frames to a power save STA from the AP when the STA sends a frame to the AP. The frame from the STA is considered a trigger frame.

Autonomous AP: An AP that can perform security functions, RF management, and configuration, without the need for a centralized WLAN controller or any other control platform.

Azimuth Chart: A chart showing the radiation pattern of an antenna as viewed from the top of the antenna. Also called an H-Plane Chart or H-Chart.

Backoff timer: The timer used during CSMA/CA to wait for access to the medium, which is selected from the contention window.

Band Steering: A method used by vendors to encourage STAs to connect to the 5 GHz band, instead of the 2.4 GHz band, which is more congested. Typically implemented by ignoring probe requests for some period of time before allowing connection to the 2.4 GHz radio, by clients known to have a 5 GHz radio, based on previous connections to the AP or controller.

Bandwidth: The frequencies used for transmission of data. For example, a 20 MHz-wide channel has 20 MHz of bandwidth.

Basic Service Area (BSA): The coverage area provided by an AP wherein client STAs may connect to the AP to transmit data on the WLAN, or through the AP to the network.

Basic Service Set (BSS): An AP and its associated STAs. Identified by the BSSID.

Basic Service Set Identification (BSSID): The ID for the BSS. Often the MAC address of the AP STA. When multiple SSIDs are used, another MAC address-like BSSID is generated.

Beacon Frame: A frame transmitted periodically from an AP that indicates the presence of a BSS network and contains capabilities and requirements of the BSS. Also, colloquially called a beacon instead of the full phrase, beacon frame.

Beamforming: Directing radio waves to a specific area or device by manipulating the RF waveforms within the different radio chains.

Beamwidth: The width of the radiated signal lobe from the antenna in the intended direction of propagation. It is usually measured at the point where 3 dB of loss is experienced.

Bill of Materials (BOM): A list of the materials and licenses required to assemble a system, in the case of WLANs, including Aps, controllers, PoE injectors, licenses, etc.

Bit: A basic unit of information for computer systems. A bit can have a value of 1 or 0. Used in binary math.

Block Acknowledgement: An acknowledgement frame that groups together multiple ACKs, instead of transmitting each individual ACK when a block transmission has been received.

Bridge: A device used to connect two networks. Wireless bridges create the connection across the wireless medium.

BSS Transition: Roaming that occurs between two BSSs that are part of the same ESS.

Byte: A basic unit of information that typically consists of 8 bits. Also called an octet.

Capacity: The number of clients and applications a network or AP can handle.

Captive Portal: Authentication technique that re-routes a user to a special webpage to verify their credentials before allowing access to the network. Commonly used in hotel and guest networks.

Guest Networks: A segregated network that is designed for use by temporary visitors.

CardBus: A PCMCIA PC Card standard interface that supports 32 bits and operates at speeds of up to 33 MHz. It is primarily used in laptops.

Carrier Frequencies: The frequency of a carrier signal, or the frequencies used to modulate information.

Carrier Sense Multiple Access (CSMA): CSMA is a protocol that allows a node to detect the presence of traffic before sending data on a shared network. Used in CSMA/CA.

Carrier Sense Multiple Access with Collision Avoidance (CSMA/CA): CSMA/CA is the method in 802.11 networks in which a node only sends data if the shared network is idle to avoid collisions.

CCMP: Counter Cipher Mode with Block Chaining Message Authentication Code Protocol (CCMP) is an key management solution that provides for improved security over WEP.

CCMP/AES: CCMP used with AES, as it is in 802.11 networks, is a key management and encryption protocol that provides more security than WEP. It is based on the AES standard and uses a 128-bit key and 128-bit block size.

Centralized Forwarding: Every forwarding decision is made by a centralized forwarding engine, such as the WLAN controller.

Certificate Authority (CA): A server that validates the authenticity of a certificate used in authentication and encryption systems. The CA may issue certificates, or it may authorize other servers to do the same.

CompactFlash (CF): Originally produced in 1994 by SanDisk, CF is a flash memory mass storage device format that can support up to 256 GB. CF devices can also function as 802.11 WLAN adapters.

Channel: A specified range of frequencies used in the 802.11 standard, used by devices to communicate on the network. Channels are commonly 20, 40, 80 and 160 MHz in width in WLANs. Newer standards will support 1, 2, 4, 8 and 16 MHz channels in sub-1 GHz networks.

Channel Width: The range of frequencies a single channel encompasses.

Clear Channel Assessment (CCA): CCA is a feature defined in the IEEE 802.11 standard that allows a client to determine idle or busy state of the medium, based on energy levels of a frame, or raw energy levels as specified in each PHY.

Client Utilities: Software installed on devices that allow the device to connect to, authenticate with and participate in a WLAN.

Co-Channel Interference (CCI): Congestion cause by the normal operations of CSMA/CA when multiple BSSs exist on the same channel. Commonly called co-channel congestion (CCC) today.

Collision Avoidance (CA): A method in which devices attempt to avoid simultaneous data transmissions in order to prevent frame collisions. Used in CSMA/CA.

Coding: A process used to encode bits to be transmitted on the wireless medium, such that error recovery can be achieved. Part of forward error correction (FEC) and defined in the modulation and coding schemes (MCSs) from 802.11n forward.

Containment: A process used against a detected rogue AP to prevent any connected clients from accessing the network.

Contention Window: A number range defined in the 802.11 standard, and varying by QoS category, from which a number is selected at random for the backoff timer in the CSMA/CA process.

Control Frame: An 802.11 frame that is used to control the communications process on the wireless medium. Control frames include RTS frames, CTS frames, PS-Poll frames and ACK frames.

Controlled Port: In an 802.1X authentication system, the virtual port that allows all frames through to the network, but only after authentication is completed.

Controller-Based AP: An AP managed by a centralized controller device. Also called a lightweight AP or thin AP.

Coverage: 1) The colloquial term used for the BSA of an AP. 2) The requirement of available WLAN connectivity throughout a facility, campus or area. Often specified in minimum signal strength as dBm; for example, -67 dBm.

Clear-to-Send (CTS) Frame: A CTS frame sent from one STA to another to indicate that the other STA can transmit on the medium. The duration value in the CTS frame is used to silence all other STAs by setting their NAV timers.

Data Frame: An 802.11 frame specified for use in carrying data based on the general frame format. Also used for some signaling purposes as null data frames.

Data Rate: The rate at which data is sent across the wireless medium. Typically represented as megabits per second (Mbps) or gigabits per second (Gbps). The data rate should not be confused with throughput rate, which is a measurement of Layer 4 throughput or useful user data.

De-authentication Frame: A notification frame sent from an 802.11 STA to another STA in order to terminate a connection between them.

Decibel (dB): A logarithmic, relative unit used when measuring antenna gain, signal attenuation, and signal-to-noise ratios. Strictly defined as 1/10 of a bel.

Decibel to Dipole (dBd): A relative measurement of antenna gain compared to a dipole antenna. Calculated as 2.14 dB greater than dBi, as a dipole antenna already has 2.14 dBi gain.

Decibel to Isotropic (dBi): A relative measurement of antenna gain compared to a theoretical isotropic radiator. When necessary, calculated as 2.14 dB less than dBd.

Decibel to Milliwatt (dBm): An absolute measurement of the power of an RF signal based on the definition of 0 dBm = 1 milliwatt (mW).

Delay: The time it takes for a bit of data to travel from one node to another. Also called latency.

Delivery Traffic Indication Message (DTIM): A message sent from an AP to clients in the Beacon frame indicating that it has data to transmit to the clients specified by the AIDs.

Differentiated Services Code Point (DSCP): A Layer 3 QoS marking system. IP packets can include DSCP markings in the headers. Eight precedence levels, 0-7, are defined.

Diffraction: The bending of waves around a very large object in relation to the wave.

Direct-Sequence Spread Spectrum (DSSS): A modulation technique where data is coupled with coding that spreads the data across a wide frequency range. Provides 1 or 2 Mbps data rates in 802.11 networks.

Disassociation Frame: A frame sent from one STA to another in order to terminate the association.

Distributed Coordination Function (DCF): A protocol defined in 802.11 that uses carrier sensing, backoff timers, interframe spaces and frame duration values to diminish collisions on the wireless medium.

Distributed Forwarding: See *Access Layer Forwarding*. Also called, *distributed data forwarding.*

Distribution System (DS): The system that connects a set of BSSs and LANs such that an ESS is possible.

Distribution System Medium (DSM): The medium used to interconnect Aps through the DS such that they can communicate with each other for ESS operations using either wired or wireless for the DS connection.

Domain Name System (DNS): A protocol and service that provides host name resolution (looking up the IP address of a given host name) and recursive IP address lookups (finding the host name of a known IP address). Also, colloquially used to reference the server that provides DNS lookups.

Driver: Software that allows a computer to interact with a hardware device such as a WLAN adapter.

Duty Cycle: A measure of the time a radio is transmitting, or a channel is consumed by a transmitting device.

Dynamic Frequency Selection (DFS): A setting on radios that dynamically changes the channel selection based on detected interference from radar systems. Many 5 GHz channels require DFS operations.

Dynamic Rate Switching (DRS): The process of reducing a client's data rate as frame transmission failures occur, or signal strength decreases. DRS results in lower data rates but fewer transmissions required to successfully transmit a frame.

Encryption: The process of converting data into a form that unauthorized users cannot understand, by encoding the data with an algorithm and a key or keys.

Enhanced Distributed Channel Access (EDCA): An enhancement to DCF introduced in 802.11e that implements priority-based queuing for transmissions in 802.11 networks, based on access categories.

Elevation Chart: A chart showing the radiation pattern of an antenna, as viewed from the side antenna. Also called an E-Plane Chart or E-Chart.

Equivalent Isotropically Radiated Power (EIRP): The output power required of an isotropic radiator to equal the measured power output from an antenna in the intended direction of propagation.

Extended Rate Physical (ERP): A physical layer technology introduced in 802.11g that uses OFDM (from 802.11a) in the 2.4 GHz band and offers data rates up to 54 Mbps.

Extended Service Set (ESS): A group of one or more BSSs that are interconnected by a DS.

Extensible Authentication Protocol (EAP): An authentication framework that defines message formats for authentication exchanges used by 802.1X WLAN authentication solutions.

Fade Margin: An amount of signal strength, in dB, added to a link budget to ensure proper operations.

Fast Fourier Transform (FFT): A mathematical algorithm that takes in a waveform as represented in the time or space domain and shows it in the frequency domain. Used in spectrum analyzers to show real-time views in the frequency domain (Real-Time FFT).

Fragmentation: The process of fragmenting 802.11 frames based on the fragmentation threshold configured. Fragmented frames have a greater likelihood of successful delivery in the presence of sporadic interference.

Frame Aggregation: A feature in the IEEE 802.11n PHY and later PHYs that increases throughput by sending more than one frame in a single transmission. Aggregated MSDUs or aggregated MPDUs may be supported.

Frame: A well-defined, meaningful set of bits used to communicate management and control information on a network, or transfer payloads from higher layers. Frames are defined at the MAC and PHY layer.

Free Space Path Loss (FSPL): The natural loss of amplitude that occurs in an RF signal as it propagates through space and the wave front spreads.

Fresnel Zones: Ellipsoid-shaped zones around the visual LoS in a wireless link. The first Fresnel zone should be 60% clear and would preferably be 80% clear to allow for environmental changes.

Frequency: The speed at which a waveform cycles in a second.

Full Duplex: A communication system that allows an endpoint to send data to the network at the same time as it receives data from the network.

Gain: The increase in signal strength in a particular direction. Can be accomplished passively by directing energy into a smaller area, or actively by increasing the strength of the broadcasted signal before it is sent to the antenna.

Group Key Handshake: Used to transfer the GTK among STAs in an 802.11 network if the GTK requires updating. Initiated by the AP/controller in a BSS.

Group Master Key (GMK): Used to generate the GTK for encryption of broadcast and multicast frames and is unique to each BSS.

Group Temporal Key (GTK): Used to encryption broadcast and multicast frames and is unique to each BSS.

Guard Interval (GI): A period of time between symbols within a frame used to avoid intersymbol interference.

Half Duplex: A communication system that allows only sending or receiving data by an endpoint at any given time.

Hidden Node: The problem that arises when nodes cannot receive each other's frames, which can lead to packet collisions and retransmissions.

High Density: A phrase referencing a WLAN network type that is characterized by large numbers of devices requiring access.

Highly Directional Antenna: An antenna, such as a parabolic dish or grid antenna, that has a high gain in a specified direction and a low beamwidth measurement, as compared to semi-directional and omni-directional antennas.

High Rate Direct Sequence Spread Spectrum (HR/DSSS): An amendment-based PHY (802.11b) that increased the data rate in 2.4 GHz from the original 1 or 2 Mbps to 5.5 and 11 Mbps, while maintaining backward compatibility with 1 and 2 Mbps.

High Throughput (HT): An amendment-based PHY (802.11n) that increased the data rate up to 600 Mbps and added support for transmit beamforming and MIMO.

Hotspot: A term referencing a wireless network connection point that is typically open to the public or to paid subscribers.

Independent Basic Service Set (IBSS): A set of 802.11 devices operating in ad-hoc (peer-to-peer) mode without the use of an AP.

Institute of Electrical and Electronics Engineers (IEEE): A standardization organization that develops standard for multiple industries, including the networking industry, with standards such as 802.3, 802.11 and 802.16.

Intentional Radiator: Any device that is purposefully sending radio waves. Signal strength of the intentional radiator is measured at the point where energy enters the radiating antennas.

Interference: In WLANs, an RF signal or incidental RF energy that is radiated in the same frequencies as the WLAN and that has sufficient amplitude and duty cycle to prevent 802.11 frames from successful delivery.

Interframe Space (IFS): A time interval that must exist between frames. Varying lengths are used in 802.11 and a references as DIFS, SIFS, EIFS and AIFS in common use.

Internet Engineering Task Force (IETF): An open group of volunteers who develops Internetworking standards through request for comments (RFC) documents. Examples include RADIUS, EAP and DNS.

Isotropic Radiator: A theoretical antenna that spreads the radiation equally in every direction, as a sphere. None exist in reality, but the concept is used to measure relative antenna gain in dBi.

Jitter: The variance in delay between packets sent on a network. Excessive jitter can result in poor quality for real-time applications, such as voice and video.

Jumbo Frame: An Ethernet frame that contains more than 1500 bytes of payload and up to 9000 to 9216 bytes.

Latency: The time taken by data to move between places. Typically, synonymous with delay in computer networking.

Layer 1: The Physical layer (PHY) that is responsible for framing and transmitting bits on the medium. In 802.3 and 802.11 the entirety of Layer 1 is defined.

Layer 2: The Data-Link layer that deals with data frames moving within a local area network (LAN). In 802.3 and 802.11, the MAC sublayer of Layer 2 is defined.

Layer 3: The Network layer where packets of data are routed between sender and receiver. Most modern networks use Internet Protocol (IP) at Layer 3.

Layer 4: The Transport layer where segmentation occurs for upper layer data and TCP (connection-oriented) and UDP (connectionless) are the most commonly used protocols.

Lightning Arrestor: A device that can redirect ambient energy from a lightning strike, away from attached equipment.

Line of sight (LoS): When existing, the visual path between two ends. RF LoS is different from visual LoS. RF LoS does not require the same clear path for the remote receiver to hear the signal. When creating bridge links, visual LoS is often the starting point.

Link Budget: The measurement of gains and losses through an intentional radiator, antenna, and over a transmission medium.

Loss: The reduction in the amplitude of a signal.

MAC filtering: A common setting that only allows specific MAC addresses onto a network. Ineffective against knowledgeable attackers because the MAC address can be spoofed to impersonate authorized devices.

Management Frame: A frame type defined in the 802.11 standard that encompasses frames used to manage access to the network including beacon, probe request, probe response, authentication, association, reassociation, de-authentication and disassociation frames.

Master Session Key (MSK): A key derived between an EAP client and EAP server and exported by the EAP method. Used to derive the PMK, which is used to derive the PTK. The MSK is used in 802.1X/EAP authentication implementations. In personal authentication implementations, the PMK is derived from the pre-shared key.

Maximal Ratio Combining (MRC): A method of increasing the signal-to-noise ratio (SNR) by combining signals received on multiple radio chains (multiple antennas and radios).

Mesh: A network that uses interconnecting devices to form a redundant set of connections offering multiple paths through the network. 802.11s defined mesh for 802.11 networks.

Mesh BSS: A basic service set that forms a self-contained network of mesh stations.

Milliwatt (mW): A unit of electrical energy used in measuring output power of RF signals in WLANs. A mW is equal to 1/1000 of a watt (W).

Mobile User: A user that physically moves while connected to the network. The opposite of a stationary user.

Modulation: The process of changing a wave by changing its amplitude, frequency, and/or phase, such that the changes represent data bits.

Modulation and Coding Scheme (MCS): Term used to describe the combination of the radio modulation scheme and the coding scheme used when transmitting data, first introduced in 802.11n.

MPDU: A MAC protocol data unit (MPDU) is a portion of data to be delivered to a MAC layer peer on a network and it is data prepared for the PHY layer by the MAC sublayer. The MAC sublayer receives the MSDU from upper layers on transmission and creates the MPDU. It receives the MPDU from the lower layer on receiving instantiation and removes the MAC header and footer to create the MSDU for the upper layers.

MSDU: A MAC service data unit is a portion of transmitted data to be handled by the MAC sublayer that has yet to be encapsulated into a MAC Layer frame.

Maximum Transmission Unit (MTU): The largest amount of data that can be sent at a particular layer of the OSI model. Typically set at Layer 4 for TCP.

Multi-User MIMO (MU-MIMO): An enhancement to MIMO that allows the AP STA to transmit to multiple client STAs simultaneously.

Multipath: The phenomenon that occurs when multiple copies of the same signal reach a receiver based on RF behaviors in the environment.

Multiple Channel Architecture (MCA): A wireless network design using multiple channels strategically designed, so that the implemented BSSs have minimal interference with one another.

Multiple-Input/Multiple-Output (MIMO): A technology used to spread a stream of data bits across multiple radio chains using spatial multiplexing at the transmitter, and to recombine these streams at the receiver.

Narrowband Interference: Interference that covers a very narrow band of frequencies and typically not the full width of an 802.11 channel, when used in reference to WLAN interferers.

Near-Far: A problem that occurs when a high-powered device is closer to the AP in a BSS, and a low-powered device is farther from the AP. Most near-far problems are addressed with standard CSMA/CA operations in 802.11 networks.

Network Allocation Vector (NAV): The NAV is a virtual carrier sense mechanism used in CSMA/CA to avoid collisions and is a timer set based on the duration values in frames transmitted on the medium.

Network Segmentation: The process used to separate a larger network into smaller networks often utilizing Layer 3 routers or multi-layer switches.

Noise: RF energy in the environment that is not part of the intentional signal of your WLAN.

Noise Floor: The amount of noise that is consistently present in the environment, which is typically measured in dBm.

Network Time Protocol (NTP): A protocol used to synchronize clocks in devices using centralized time servers.

Octet: A group of eight ones and zeros. An 8-bit byte. Sometimes simply called a byte.

Orthogonal Frequency Division Multiplexing (OFDM): A modulation technique and a named physical layer in 802.11 that provides data rates up to 54 Mbps and operates in the 5 GHz band. The modulation is used in all bands, but the named PHY operates only in the 5 GHz band.

Omni-Directional Antenna: An antenna that propagates in all directions horizontally. Creates a coverage area similar to a donut shape (toroidal). Also known as a dipole antenna.

Dipole Antenna: An antenna that propagates in all directions horizontally. Creates a coverage area similar to a donut (toroidal) shape. Also known as an omni-directional antenna.

Open System Authentication: A simple frame exchange, providing no real authentication, used to move through the state machine in relation to the connection between two 802.11 STAs.

Opportunistic Key Caching (OKC): A roaming solution for WLANs wherein the keys derived from the 802.1X/EAP authentication are cached on the AP or controller, such that only the 4-way handshake is required at the time of roaming.

Open Systems Interconnection (OSI) Model: A theoretical model for communication systems that works by separating the communications process into seven, well-defined layers. The seven layers are Application, Presentation, Session, Transport, Network, Data Link and Physical.

Packet: Data as represented at the Network layer (Layer 4) for TCP communications.

Pairwise Transient Key (PTK): A key derived during the 4-way handshake and used for encryption only between two specific endpoints, such as an AP and a single client.

Passive Gain: An increase in strength of a signal by focusing the signal's energy, rather than increasing the actual energy available, such as with an amplifier.

Passive scanning: A scanning (network location) method, wherein a STA waits to receive beacon frames from an AP which contain information about the WLAN.

Passive survey: A survey conducted on location that gathers information about RF interference, signal strength and coverage areas by monitoring RF activity without active communications.

Passphrase Authentication: A type of access control that uses a phrase as the pass key. Also called personal in WPA and WPA2.

Phase: A measurement of the variance in arrival state between two copies of a wave form. Waves are said to be in phase or out of phase by some degree. The phase can be manipulated for modulation.

PHY: A shorthand notation for physical layer, which is the physical means of communication on a network to transmit bits.

Physical (PHY) Layer: The physical (PHY) layer refers to the physical means by which a message is communicated. Layer 1 of the OSI model.

PLCP: Physical Layer Convergence Protocol (PLCP) is the name of the service within the PHY that receives data from the upper layers and sends data to the upper layers. It is the interaction point with the MAC sublayer.

PMD: Physical Medium Dependent (PMD) is the service within the PHY responsible for sending and receiving bits on the RF medium.

PMK Caching: Stores the PMK so a device only has to perform the 4-way handshake when connecting to an AP to which it has already connected.

Pairwise Master Key (PMK): The key derived from the MSK, which is generated during 802.1X/EAP authentication. Used to derive the PTK. Used in unidirectional communications with a single peer.

PLCP Protocol Data Unit (PPDU): The prepared bits for transmission on the wired or wireless medium. Sometimes also called a PHY Layer frame.

PLCP Service Data Unit (PSDU): The name for the contents that are contained within the PPDU, the PLCP Protocol Data Unit. It is the same as the MPDU as perceived and received by the PHY.

PoE Injector: Any device that adds Power over Ethernet (PoE) to ethernet cables. Come in two variants, endpoint (such as switches) and midspan (such as inline injectors).

Point-to-Multipoint (PtMP): A connection between a single point and multiple other points for wireless bridging or WLAN access.

Point-to-Point (PtP): A connection between two points, often used to connect two networks via bridging.

Polarization: The technical term used to reference the orientation of antennas related to the electric field in the electromagnetic wave.

Power over Ethernet (PoE): A method of providing power to certain hardware devices that can be powered across the Ethernet cables. Specified in 802.3 as a standard. Various classes are defined based on power requirements.

Pre-authentication: Authenticating with an AP to which the STA is not intending to immediately connect so that roaming delays are reduced.

Pre-shared Key (PSK): Refers to any security protocol that uses a password or passphrase or string as the key from which encryption materials are derived.

Primary Channel: When implementing channels wider than 20 MHz in 802.11n and 802.11ac, the 20 MHz channel on which management and control frames are sent, and the channel used by STAs not supporting the wider channel.

Probe Request: A type of frame sent when a client device wants information about APs in the area or is seeking a specific SSID to which it desires to connect.

Probe Response: A type of frame sent in response to a probe request that contains information about the AP and the requirements of BSSs it provides.

Protected Management Frame (PMF): Frames used for managing a wireless network that are protected from spoofing using encryption. Protocol defined in the 802.11w amendment.

Protocol Analyzer: Hardware or software used to capture and analyze networking communications. WLAN protocol analyzers have the ability to capture 802.11 frames from the RF medium and decode them for display and analysis.

Protocol Decodes: The way information in captured packets or frames is interpreted for display and analysis.

Quality of Service (QoS): Traffic prioritization and other techniques used to improve the end-user experience. IEEE 802.11e includes QoS protocols for wireless networks based on access categories.

QoS BSS: A BSS supporting 802.11e QoS features.

Radio Chains: A reference to the radio and antenna used together to transmit in a given frequency range. Multi-stream devices have multiple radio chains, as one radio chain is required for each stream.

Radio Frequency (RF): The electromagnetic wave frequency range used in WLANs and many other wireless communication systems.

Radio Resource Management (RRM): Automatic management of various RF characteristics like channel selection and output power. Known by different terms among the many WLAN vendors but referencing the same basic capabilities.

Remote Authentication Dial-In User Service (RADIUS): A network protocol that handles AAA management, which allows for authentication, authorization and accounting (auditing). Used in 802.11 WLANs as the authentication server in an 802.1X/EAP implementation.

Real-Time Location Service (RTLS): A function provided by many WLAN infrastructure and overlay solutions allowing for device location based on triangulation and other algorithms.

Reassociation: The process used to associate with another AP in the same ESS. May also be used when a STA desires to reconnect to an AP to which it was formerly connected.

Received Channel Power Indicator (RCPI): Introduced in 802.11k, a power measurement calculated as INT ((dBm + 110) * 2). Expected accuracy is +/- 5 dB. Ranges from 0-220 are available, with 0 equaling or less than -110 dBm and 220 equaling or greater than 0 dBm. The value is calculated as an average of all received chains during the reception of the data portion of the transmission. All PHYs support RCPI and, though 802.11ac does not explicitly list its formulation, it references the 802.11n specification for calculation procedures.

Received Signal Strength Indicator (RSSI): A relative measure of signal strength for a wireless network. The method to measure RSSI is not standardized, though it is constrained to a limited number of values in the 802.11 standard. Many use the term RSSI to reference dBm, and the 802.11 standard uses terms like DataFrameRSSI and BeaconRSSI and defines them as the signal strength in dBm of the specified frames, so the common vernacular is understandable. However, according to the standard, "absolute accuracy of the RSSI reading is not specified" (802.11-2012, Clause 14.3.3.3).

Reflection: An RF behavior that occurs when a wave meets a reflective obstacle larger than the wavelength, similar to light waves in a mirror.

Refraction: An RF behavior that occurs as an RF wave passes through material, causing a bending of the wave and possible redirection of the wave front.

Regulatory Domain: A reference to geographic regions management by organizations like the FCC and ETSI that determine the allowed frequencies, output power levels and systems to be used in RF communications.

Remote AP: An AP designed to be implemented at a remote location and managed across a WAN link using special protocols.

Request to Send/Clear to Send (RTS/CTS): A frame exchange used to clear the channel before transmitting a frame in order to assist in the reduction of collisions on the medium. Also used as a backward compatible protection mechanism.

Resolution Bandwidth (RBW): The smallest frequency that can be extracted from a received signal by a spectrum analyzer, or the configuration of that frequency. Many spectrum analyzers allow for the adjustment of the RBW within the supported range of the analyzer.

Retry: That which occurs when a frame fails to be delivered successfully. A bit set in the frame to specify that it is a repeated attempt at delivery.

Return Loss: A measure of how much power is lost in delivery from a transmission line to an antenna.

RF Cables: A cable, typically coaxial, that allows for the transmission of electromagnetic waves between a transceiver and an antenna.

RF Calculator: A software application used to perform calculations related to RF signal strength values.

RF Connector: A component used to connect RF cables, antennas and transmitters. RF connectors come in many standardized forms and should match in type and resistance.

RF Coverage: Synonymous with coverage in WLAN vernacular. Reference to the BSA provided by an AP.

RF Link: An established connection between two radios.

RF Line of Sight (LoS): The existence of a path, possibly including reflections, refractions and pass-through of materials, between two RF transceivers.

RF Propagation: The process by which RF waves move throughout an area including reflection, refraction, scattering, diffraction, absorption and free space path loss.

RF Signal Splitter: An RF component that splits the RF signal with a single input and multiple outputs. Historically used with some antenna arrays, but less common today in WLAN implementations.

RF Site Survey: The process of physically measuring the RF signals within an area to determine resulting RF behavior and signal strength. Often performed as a validation procedure after implementation based on a predictive model.

Rivest Cipher 4 (RC4): An encryption cipher used in WEP and with TKIP. A stream cipher.

Roaming: That which occurs when a wireless STA moves from one AP to another, either because of end-user mobility or changes in the RF coverage.

Robust Security Network (RSN): A network that supports CCMP/AES or WPA2 and optionally TKIP/RC4 or WPA. To be an RSN, the network must support only RSN Associations (RSNAs), which are only those associations that use the 4-way handshake. WEP is not supported in an RSN.

Robust Security Network Association (RSNA): An association between a client STA and an AP that was established through authentication resulting in a 4-way handshake to derive unicast keys and transfer group keys. WEP is not supported in an RSNA.

Rogue Access Point: An access point that is connected to a network without permission from a network administrator or other official.

Rogue Containment: Procedures used to prevent clients from associating with a rogue AP, or to prevent the rogue AP from communicating with the wired network.

Rogue Detection: Procedures used to identify rogue devices. May include simple identification of unclassified APs or algorithmic processes that identify likely rogues.

Role-Based Access Control (RBAC): An authorization system that assigns permissions and rights based on user roles. Similar to group management of authorization policies.

RSN Information Element: A portion of the beacon frame that specifies the security used on the WLAN.

RTS Threshold: The minimum size of a frame required to use RTS/CTS exchanges before transmission of the frame.

S-APSD: See *Automatic Power Save Delivery*.

Scattering: An RF behavior that occurs when an RF wave encounters reflective obstacles that are smaller than the wavelength. The result is multiple reflections or scattering of the wave front.

Secondary Channel: When implementing channels wider than 20 MHz in 802.11n and 802.11ac, the second channel used to form a 40 MHz channel for data frame transmissions to and from supporting client STAs.

Semi-Directional Antenna: An antenna such as a Yagi or a patch that has a propagation pattern, which maximizes gain in a given direction rather than an omni-directional pattern, having a larger beamwidth than highly directional antennas.

Service Set Identifier (SSID): The BSS and ESS name used to identify WLAN. Conventionally made to be readable by humans. Maximum of 32 bytes long.

Signal Strength: A measure of the amount of RF energy being received by a radio. Often specified as the RSSI, but referenced in dBm, which is not the proper definition of RSSI from the 802.11 standard.

Single Channel Architecture (SCA): A WLAN architecture that places all APs on the same channel and uses a centralized controller to determine when each AP can transmit a frame. No control of client transmissions to the network is provided.

Single-Input/Single-Output (SISO): A radio transmitter that supports one radio chain and can send and receive only a single stream of bits.

Signal to Noise Ratio (SNR): A comparison between the received signal strength and the noise floor. Typically presented in dB. For example, given a noise floor of -95 dBm and a signal strength of -70 dBm, the SNR is 25 dB.

Space-Time Block Coding (STBC): The use of multiple streams of the same data across multiple radio chains to improve reliability of data transfer through redundancy.

Spatial Multiplexing (SM): Used with MIMO technology to send multiple spatial streams of data across the channel using multiple radio chains (radios coupled with antennas).

Spatial Multiplexing Power Save (SMPS): A power-saving feature from 802.11n that allows a station to use only one radio (or spatial stream).

Spatial Streams: The partitioning of a stream of data bits into multiple streams transmitted simultaneously by multiple radio chains in an AP or client STA.

Spectrum Analysis: The inspection of raw RF energy to determine activity in an area on monitored frequencies. Useful in troubleshooting and design planning.

Spectrum Analyzer: A hardware and software solution that allows the inspection of raw RF energy.

Station (STA): Any device that can use IEEE 802.11 protocol. Includes both APs and clients.

Supplicant: In 802.1X, the device attempting to be authenticated. Also, the term used for the client software on a device that is capable of connecting to a WLAN.

Sweep Cycle: The time it takes a spectrum analyzer to sweep across the frequencies monitored. Often a factor of the number of frequencies scanned and the RBW.

System Operating Margin (SOM): The actual positive difference in the required link budget for a bridge link to operate properly, and the received signal strength in the link.

Temporal Key Integrity Protocol (TKIP): The authentication and key management protocol supported by WPA systems and implemented as an interim solution between WEP and CCMP.

Transition Security Network (TSN): A network that allows WEP connections during the transition period over to more secure protocols and an eventual RSN. An RSN does not allow WEP connections.

Transmit Beamforming (TxBF): The use of multiple antennas to transmit a signal strategically with varying phases, so that the communication arrives at the receiver such that the signal strength is increased.

Transmit Power Control (TPC): A process implemented in WLAN devices allowing for the output power to be adjusted according to local regulations, or by an automated management system.

U-APSD: See *Automatic Power Save Delivery*.

Uncontrolled Port: In an 802.1X authentication system, the virtual port that allows only authentication frames/packets through to the network and, when authentication is successfully completed, provides the 802.1X service with the needed information to open the controlled port.

User Priority (UP): A value (from 0-7) assigned to prioritize traffic that corresponds to different access categories for WMM QoS.

Virtual Carrier Sense: The 802.11 standard currently defines the Network Allocation Vector (NAV) for use in virtual carrier sensing. The NAV is set based on the duration value in perceived frames within the channel.

Voltage Standing Wave Ratio (VSWR): The Voltage Standing Wave Ratio is the ratio between the voltage at the maximum and minimum points of a sanding wave.

Watt: A unit of power. Strictly defined as the energy consumption rate of one joule per second such that 1 W is equal to 1 joule per 1 second.

Wavelength: The distance between two repeating points on a wave. Wavelength is a factor of the frequency and the constant of the speed of light.

Wired Equivalent Privacy (WEP): A legacy method of security defined in the original IEEE 802.11 standard in 1997. Used the RC4 cipher like TKIP (WPA), but implemented it poorly. WEP is deprecated and should no longer be used.

Wi-Fi Alliance: An association that certifies WLAN equipment to interoperate based on selected portions of the 802.11 standard and other standards. Certifications include those based on each PHY as well as QoS and security.

Wi-Fi Multimedia (WMM): A QoS certification created and tested by the Wi-Fi Alliance using traffic prioritizing methods defined in the IEEE 802.11e.

Wi-Fi Multimedia Power Save (WMM-PS): A power-saving certification designed by the Wi-Fi Alliance and optimized for mobile devices and implementing methods designated in the IEEE 802.11e amendment.

Wireless Intrusion Prevention System (WIPS): A system used to detect and prevent unwanted intrusions in a WLAN by detecting and preventing rogue APs and other WLAN threats.

Wireless Local Area Network (WLAN): A local area network that connects devices using wireless signals based on the 802.11 protocol, rather than wires and the common 802.3 protocol.

WPA-Enterprise: A security protocol designed by the Wi-Fi Alliance. Requires an 802.1X authentication server. Uses the TKIP encryption protocol with the RC4 cipher. Implements a portion of 802.11i and the older, deprecated TKIP/RC4 solution.

WPA-Personal: A security protocol designed by the Wi-Fi Alliance. Does not require an authentication server. Uses the TKIP encryption protocol with the RC4 cipher. Also known as WPA-PSK (Pre-Shared Key).

WPA2-Enterprise: A security protocol designed by the Wi-Fi Alliance. Requires an 802.1X authentication server. Uses the CCMP key management protocol with the AES cipher. Also known as WPA2-802.1X. Implements the non-deprecated portion of 802.11i.

WPA2-Personal: A security protocol designed by the Wi-Fi Alliance. Does not require an authentication server. Uses the CCMP key management protocol with the AES cipher. Also known as WPA2-PSK (Pre-Shared Key).

Wi-Fi Protected Setup (WPS): A standard designed by the Wi-Fi Alliance to secure a network without requiring much user knowledge. Users connect either by entering a PIN associated with the device or by Push-Button, which allows users to connect when a real or virtual button is pushed.

Index

NOTES

NOTES

NOTES

NOTES

NOTES

NOTES

NOTES

NOTES

NOTES

NOTES

NOTES

NOTES

NOTES

CPSIA information can be obtained
at www.ICGtesting.com
Printed in the USA
BVHW022356180322
631711BV00001B/1

9 780999 304815